Coastal Hazards Related to Storm Surge

Special Issue Editor
Rick Luettich

MDPI • Basel • Beijing • Wuhan • Barcelona • Belgrade

MDPI

Special Issue Editor
Rick Luettich
University of North Carolina at Chapel Hill
USA

Editorial Office
MDPI AG
St. Alban-Anlage 66
Basel, Switzerland

This edition is a reprint of the Special Issue published online in the open access journal *JMSE* (ISSN 2077-1312) from 2015–2016 (available at: http://www.mdpi.com/journal/jmse/special issues/storm-surge).

For citation purposes, cite each article independently as indicated on the article page online and as indicated below:

Lastname, F.M.; Lastname, F.M. Article title. *Journal Name*. **Year**. *Article number*, page range.

First Edition 2018

ISBN 978-3-03842-711-7 (Pbk)
ISBN 978-3-03842-712-4 (PDF)

Table of Contents

About the Special Issue Editor

Rick Luettich, Professor of Marine Sciences and Environmental Sciences and Engineering Luettich has an undergraduate and masters degree in civil engineering from Georgia Tech and a doctor of science in civil engineering from MIT. He serves as the Director of UNC's Institute of Marine Science and Center for Natural Hazards Resilience, which he founded in 2008.

His research addresses modeling and observational studies of circulation and transport in coastal waters. He has pioneered the development and application of the ADCIRC computer modeling system that is widely used for coastal hazards prediction and event-based forecasting.

He is actively engaged in the coastal science and natural hazards resilience communities, serving as the lead PI for the Department of Homeland Securitys Coastal Resilience Center of Excellence and NOAAs Coastal and Ocean Modeling Testbed, as a member of three recent National Academies committees (chairing the 2013–2014 committee on Coastal Risk Reduction) and as the Vice-President of the Southeast Louisiana Flood Protection Authority-East.

Preface to "Coastal Hazards Related to Storm Surge"

Whether by necessity, convenience or preference, the worlds coastal regions are important places to live, work and recreate. Our coasts are home to population and economic centers, major port facilities, fishing fleets, oil and gas refineries, military complexes, recreational industries and have many other uses. Growth in many coastal regions is occurring at higher rates than inland; for example in the United States population growth along the southern and southeastern coasts is nearly twice that of the national average. However, coastal areas are also subject to some of the most powerful storms on earth, whose destructive potential is increasing due to relative sea level rise and climate change.

While a seemingly large number of weather-caused coastal disasters has occurred since the beginning of the 21st century, for any given coastal community these remain relatively low frequency (yet potentially high consequence) events. Thus it is difficult to develop public support and secure the often formidable resources required for courses of action such as abandoning current, or limiting future development, constructing barriers, imposing stronger building codes, and hardening infrastructure that may be necessary to lower risk in coastal areas. Rather, we typically wait to act until after a disaster has occurred and then often focus on putting things back as quickly and inexpensively as possible, as opposed to reducing risk. As a result, risk, in many coastal areas, has been accumulating over time and looms as a massive burden to be passed on to future generations.

Given the compelling reasons to inhabit the coast and the great challenges of doing so, it is imperative that we develop better capabilities for predicting the hazards that coastal regions face. One of the greatest of these hazards is the storm surge, waves and flooding associated with severe tropical and extratropical storms. For this reason, we have assembled a collection of papers representing state of the art research that extends our understanding of coastal hazards associated with storm surge. Fourteen papers cover topics ranging from predicting coupled surge and wave dynamics at multiple scales; erosion and scour; statistical considerations for hazard delineation; joint effects of climate change and storm surge; mitigation strategies and human response to storm surge threats. This work represents important advancements in our ability to predict, mitigate and respond to this threat to most of the world's coastal areas. Recognizing these advancements and translating them into policy and practice are essential if we are to effectively manage coastal risk and create more resilient coastal communities in which to live, work and enjoy.

Rick Luettich
Special Issue Editor

Journal of
Marine Science and Engineering

MDPI

Article

The Use of a Statistical Model of Storm Surge as a Bias Correction for Dynamical Surge Models and Its Applicability along the U.S. East Coast

Haydee Salmun [1,2,†,*] and Andrea Molod [3,4,†]

1   Department of Geography, Hunter College of the City University of New York, 695 Park Ave.,
    New York, NY 10065, USA
2   Earth and Environmental Science Doctoral Program, The Graduate Center of CUNY, 365 Fifth Ave.,
    New York, NY 10016, USA
3   Earth System Science Interdisciplinary Center, University of Maryland, College Park, MD 20742, USA;
    Andrea.Molod@nasa.gov
4   NASA Goddard Space Flight Center, Mail Code: 610.1, Greenbelt, MD 20771, USA
*   Author to whom correspondence should be addressed; hsalmun@hunter.cuny.edu; Tel.: +1-212-772-4159;
    Fax: +1-212-772-5268.
†   Authors contributed equally to this work.

Academic Editor: Rick Luettich
Received: 19 October 2014; Accepted: 29 January 2015; Published: 12 February 2015

Abstract: The present study extends the applicability of a statistical model for prediction of storm surge originally developed for The Battery, NY in two ways: I. the statistical model is used as a biascorrection for operationally produced dynamical surge forecasts, and II. the statistical model is applied to the region of the east coast of the U.S. susceptible to winter extratropical storms. The statistical prediction is based on a regression relation between the "storm maximum" storm surge and the storm composite significant wave height predicted ata nearby location. The use of the statistical surge prediction as an alternative bias correction for the National Oceanic and Atmospheric Administration (NOAA) operational storm surge forecasts is shownhere to be statistically equivalent to the existing bias correctiontechnique and potentially applicable for much longer forecast lead times as well as for storm surge climate prediction. Applying the statistical model to locations along the east coast shows that the regression relation can be "trained" with data from tide gauge measurements and near-shore buoys along the coast from North Carolina to Maine, and that it provides accurate estimates of storm surge.

Keywords: extratropical storms; storm surge; statistical methods; bias correction

1. Introduction

Storm surge is a potentially devastating rise in the coastal water levels caused by tropical cyclones such as hurricanes and typhoons and by extratropical storms. Approximately 50% of the world's population lives within 150 kilometers of the ocean, particularly in low-lying coastal areas susceptible to the consequences of storm surge. Although tropical cyclones, with their lower low-pressure centers and higher wind speeds, typically produce significantly higher surge than extratropical cyclones do (Hurricane Katrina in the U.S. in 2005, for example), extratropical cyclones are frequent occurrences during much of the year over large populated coastal regions such as the east coast of the United States, and therefore require similar research efforts and accurate prediction. These extratropical storms occur on average 10–12 times per year and the strongest and most destructive storms appear on average 2 to 3 times per year [1,2]. Storm surge can lead to large loss of human life, destruction of homes and civil infrastructure, and disruption of commerce. The Fifth Assessment Report of the Intergovernmental

Panel on Climate Change (IPCC) [3] estimates that over the next century global sea level is likely to rise from between 26 and 63 cm, and, depending on emissions, as much as 90 cm. This sea level rise will extend the zone of impact from storms, storm surge, and storm waves farther inland. Examples of extratropical storms and their associated damage along the east coast abound. The extratropical storm of 11–12 December 1992 [4] produced near-record flooding along the entire Atlantic Coast. Storm surge levels of over 1 meter persisted for a few days at Sandy Hook, NJ, with a maximum storm surge value of 2.13 m [4]. The storm surge for that storm recorded at The Battery, NY, was 1.75 m [5]. Flooding resulted in the closing of major highways and railroads, in millions of dollars of losses due to property damage, and in two storm-related deaths. For reference, a storm surge threshold value of 0.6 m (above Mean High Water Level) can cause minor flooding during a high tide at The Battery, NY, and is often used as a criterion to issue an advisory of coastal flood conditions for the New York Metropolitan region by the National Weather Service [6]. More recently, on 15–16 April 2007, an extratropical system resulted in storm surge at The Battery, NY, of 1 m [7]. In addition to flooding, power outages, property damage, evacuations and traffic disruption, there were several deaths reported for this storm.

The factors that contribute to the challenge of modeling and predicting storm surge include storm intensity and storm track, coastal bathymetry, pre-storm average water height, astronomical tides and ocean wave field and associated wave transport. The National Oceanic and Atmospheric Administration (NOAA) produces 96-h operational surge forecasts with the extratropical storm-surge model called ET-SURGE (ETSS) developed by Kim et al. [8]. A detailed description of the ETSS model output fields is available online at [9]. The operational storm surge forecast guidance consists of the ETSS output augmented by a bias correction. This error correction is computed as the five-day average of the difference between the observed and the predicted water levels, computed over the 5 days previous to the forecast start time and referred to as the "anomaly" [9]. This constant is added to the entire forecast produced by the dynamical model. NOAA now also provides surge forecast guidance from 96-h surge forecasts using the Extratropical Surge and Tide Operational Forecast System (ESTOFS) [10], which demonstrates error characteristics that are comparable to the bias-corrected ETSS [11], but has been operational for a short period of time and will not be discussed here. Despite recent advances in the development of dynamical models such as ETSS and ESTOFS and advances in computational capability, however, forecast errors remain quite large and improvements in performance are needed [6].

A time series of the NOAA bias correction (not shown) depicts significant temporal variability. This suggests that the existing method of bias correction is limited to near-future events and times (probably out to a few days), and could not be used, for example, to correct surge forecasts out to two weeks using the 14 day forecasts regularly released by many numerical weather prediction centers (see for example [12]), or to correct surge forecasts based on future climate projections. Both of these goals are desirable and feasible, and have motivated studies of hurricane-related storm surge [13,14]. Similar studies for extratropical storm surge have not been reported despite the indications from many Coupled Model Intercomparison Project—5 (CMIP5) climate forecasts that the frequency and intensity of extratropical storms may increase in a warmer climate [15].

There are other existing methods to correct errors in dynamical surge forecasts using observed water level data. For example the German Federal Marine and Hydrographic agency uses a Model Output Statistics (MOS) approach [16]. This approach produces a modified water level and surge time series based on a regression model. The regression includes over a dozen predictors, the most important of which are the dynamical model output itself and recent error in surge forecast. The evaluation of this technique shows substantial reduction in forecast error for lead times up 24 h. There was little to no benefit from the MOS approach after 33 h lead time. Based on these results we infer that, like NOAA' surge correction, this approach would also not be beneficial for long lead time surge forecasts.

Part I of the present study describes an improved bias correction method for use in operational storm surge forecasts that may not have the issue of degrading skill for long lead time forecasts. The new bias correction makes use of the statistical model developed by Salmun et al. [2] to predict the maximum value of storm surge during a storm event which is referred to as the "storm maximum storm surge" (SSMAX). The model was validated for surge forecasts by Salmun et al. [17]. SSMAX is an event based metric for each storm event. A storm event is defined by the period over which the sea level pressure (as measured at a nearby location) remains below two standard deviations from its seasonal mean. The authors presented a regression equation relating the value of SSMAX at The Battery, NY, to the average of the top one-third of significant wave heights during the storm event measured at the National Data Buoy Center (NDBC, [18]) station off Fire Island, NY (NDBC Sta. 44025). The regression relation was developed based on storm surge computed from water level available for The Battery, NY [5], for the extratropical storm season during the period 1991–2008. The statistical model based on the storm composite significant wave height was chosen for its simplicity since the model with this single predictor was shown to perform as well as models with up to seven different predictors. The regression relation they obtained was:

$$SSMAX_{44025} = 0.2055H_{44025} - 0.0851 \qquad (1)$$

with RMS error of 0.167 m. H_{44025} denotes the significant wave height at NDBC Sta. 44025. The capability for predictions of the maximum value of the storm surge due to a storm event by the statistical method was verified for The Battery, NY, based on a series of retrospective forecasts using predicted wave heights from NOAA's WAVEWATCH III™ (WWIII) wave model [19] for storm events determined from surface pressure from NOAA's North American Mesoscale (NAM-WRF) [20] weather forecast model [8]. Salmun et al. [17] showed that the prediction capability of the statistical model is comparable to that of NOAA's operational forecast of the storm maximum storm surge.

The relationship between the height of waves and the magnitude of the storm surge that is reflected in the regression relation can be conceptualized in terms of basic physical processes. The wave field in the ocean is the result of energy transfer from the wind field to the surface. The transfer depends strongly on wind velocity, wind duration and the distance affected by sustained winds or fetch. Significant wave heights are determined largely by the energy imparted to the water over the fetch, hence they are a good indication of the intensity of the winds (storm) that produced them. SSMAX represents a single value, the maximum, of storm surge that is experienced at a location during the entire period of a storm and as such SSMAX does not give information about timing of the potentially most damaging storm surge. Its usefulness derives from its simplicity, from being easy to compute and from its ability to capture local information about a storm from a near-shore single location. Although the prediction of an event-based storm maximum storm surge does not entirely satisfy the public's need for surge forecasting, an accurate prediction of SSMAX can provide valuable information, and can be used in conjunction with NOAA's dynamical forecast to provide better operational forecasts. In addition, because SSMAX makes use of statistical relationship between the state of the ocean and the surge there should be fewer restrictions in the use of SSMAX as a bias correction related to forecast lead time.

In Part II of the present work we examine the applicability of the statistical model to the other areas along the U.S. east coast. In this way we examine the extent to which the local geography and geomorphology affect the specifics of the relationship between the significant wave height measured at a local buoy and SSMAX. Following this introduction, the statistical storm surge model and its validation are briefly described. In Section 3 we describe, discuss and evaluate the use of SSMAX as a bias correction to current NOAA storm surge forecast. The extension of the statistical method for storm maximum storm surge estimated along the North Atlantic east coast affected by extratropical systems is presented in Sections 4 and 5 summarizes the major results of the study and Section 6 discusses the present findings and future extensions of the study.

J. Mar. Sci. Eng. **2015**, 3, 73–84

2. Description of the Statistical Method

A brief summary of the relevant details of the statistical method of Salmun et al. [2,17] and the choice of storms used for its evaluation is presented here. Salmun et al. [2] established a statistical relation between the maximum value of the surge (SSMAX) reached during any particular storm event and the storm composite near-shore significant wave height. Values of SSMAX were computed using water level data at The Battery, NY, obtained from NOAA's Center for Operational Oceanographic Products and Services (CO-OPS) [5], and nearby significant wave heights were determined from data measured at the National Data Buoy Center (NDBC, [18]) Sta. 44025 for extratropical storm events during the storm season (September–April) of the period 1991–2008. The storm composite significant wave height is defined as the average of the "top third" largest significant wave heights during the storm period. Extratropical storm events were determined using a threshold of two standard deviations below the climatological mean of the sea level pressure time series, also using data from the NDBC station. A single event that satisfied this sea level pressure criterion, lasted longer than four hours and was separated from all other continuous groups of measurements by at least 24 h defined a storm event. Extensive examination of surface weather maps was used to validate the list of storms.

Salmun et al. [17] tested the statistical method by performing a series of retrospective forecasts of the events identified by the earlier study that occurred during the period from February 2005 to December 2008. They used existing operational forecasts of surface pressure from NOAA's North American Mesoscale (NAM-WRF) [20] weather forecast model, operational forecasts of wave height from NOAA's WAVEWATCH III™ (WWIII) model [19] and the regression relation obtained by Salmun et al. [2]. The list of storms identified at NDBC station 44025 as described above for the test period was the starting list of candidate events for the testing process. Retrospective forecasts of sea level pressure, available at three-hour time intervals, were then used to verify the existence of candidate events in the forecast record. Any storm not forecasted was eliminated from the final list of test storms (which resulted in the elimination of one storm from the original record), and 41 storms comprised the testing sample. Starting and ending times for each storm event were determined from the forecasted sea level pressure. Storm composite significant wave heights at the NDBC station were computed based on retrospective forecasts from [19] for each storm event, and used as the predictor for SSMAX at The Battery, NY, USA.

The statistical predictions of SSMAX thus obtained were compared to values of "storm maximum" storm surge obtained from the NOAA ETSS model output, from the NOAA operational (bias corrected) forecasts, and from the observations at that location for each storm event. The results of this evaluation showed that the mean error of the statistical method for 12-, 24- and 48-h lead time forecasts is smaller than the mean error of the ETSS model forecasts with 95% confidence and that the predictions made with the statistical method are statistically indistinguishable from the NOAA bias-corrected operational forecast, which is the ETSS output with an anomaly correction.

3. Part I—The Statistical Approach as a Bias Correction

The statistical event-based forecast of SSMAX is used here to perform a bias correction for the duration of a forecasted storm event, and applied at each time during the storm event in a manner similar to the ETSS anomaly correction. This type of correction could be used during any storm event for which forecasts of sea level pressure and wave heights are available. We applied this approach to the ETSS forecasts of all storm events considered by Salmun et al. [17] in their study of storm surge at The Battery, NY. The use of the statistical model as a bias correction is evaluated here for the NOAA ETSS forecast, but could be used in a similar manner as part of a multi-model ensemble surge forecast, such as the one advocated by Di Liberto et al. [21].

The SSMAX predicted by the statistical model was used to correct ETSS model forecast by adding a constant value computed as the difference between the statistical prediction of SSMAX ($SSMAX_{stat}$)

and the ETSS prediction of SSMAX ($SSMAX_{model}$). For each time during a particular storm event, we denote the corrected time series by ET_{stat} and we write:

$$ET_{stat}(i,j) = ETSS_{model}(i,j) + (SSMAX_{stat}(j) - SSMAX_{model}(j)) \qquad (2)$$

where i refers to time during a particular event and j refers to the storm event number. The test cases used for evaluation here are the 41 storm events identified between 2005 and 2008 as described in Section 2. The simplest implementation of the bias correction is to apply the correction uniformly over the duration of the storm event as was done here. More complex weighting functions were attempted but found to underperform relative to the constant weighting and so are not justified. The corrected event time series, ET_{stat}, and the event time series corrected with NOAA anomaly, denoted by ET_{anom}, were compared to observations (OBS) at The Battery, NY, for each storm event. Figure 1 shows the comparison among the different surge predictions for two typical storm events. Figure 1a shows the time series of a storm for which ET_{stat} is higher than ET_{anom}, and closer to the observed surge. This occurred in approximately one third of the storm events, and is consistent with the determination made in Salmun et al. [17] that NOAA's ETSS generally underpredicts surge. Figure 1b shows the time series of a storm for which ET_{stat} is higher than ET_{anom} and further away from the observed surge. This also occurred in approximately one third of storm events. Many of these storms were those for which Salmun et al. [17] determined that the error in the SSMAX forecasts was attributable to errors in the predicted wave heights rather than to a failure of the regression.

Means and standard deviations of the differences between each of the estimates and observations were computed for 12 h, 24 h and 48 h lead-time forecasts for each storm event. Lead-time was measured as the time elapsed between the forecast initial condition and the onset of the storm event, and the average storm event duration was 22.5 h, with a standard deviation of 19.1 h. Results for the means are summarized in Table 1. The variance of ($ET_{stat} - OBS$) is the same as the variance of ($ETSS_{model} - OBS$) by definition, and almost identical to the variance of ($ET_{anom} - OBS$). For the 12 h and 24 h lead-time forecasts, the mean difference between ET_{stat} and OBS and the mean difference between ET_{anom} and OBS are statistically indistinguishable as determined by a Student's t-test, while at the 48 h lead-time the forecast using ET_{stat} yields statistically more accurate results.

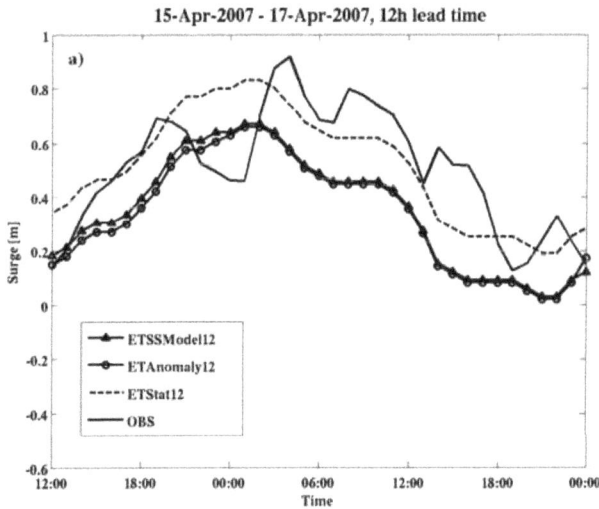

Figure 1. *Cont.*

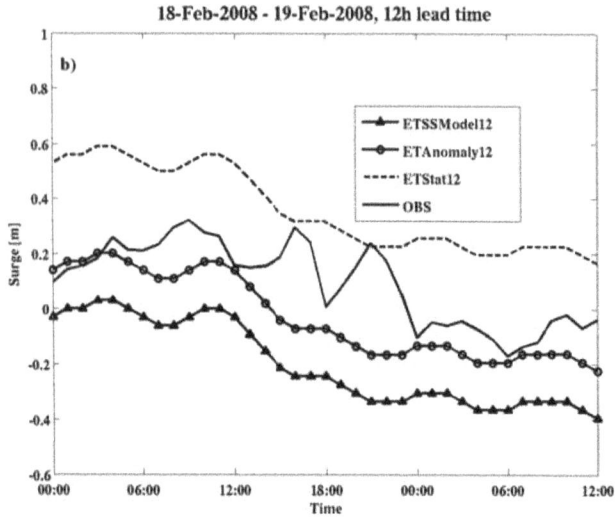

Figure 1. Hourly time series of storm surge from tide gauge observations and different model estimates. Solid black line is the gauge observations, dashed line is NOAA extratropical storm-surge model (ETSS) 12 h forecast with our statistical correction, the line with circles is NOAA ETSS 12 h forecast with NOAA's anomaly correction, and the line with triangles is NOAA ETSS 12 h forecast. (**a**) Time series from storm 26 of Salmun et al. [17] beginning on 15 April 2007 12Z; (**b**) Time series from storm 30, beginning on 18 February 2007 0Z.

Table 1. Means of the differences between National Oceanic and Atmospheric Administration (NOAA)'s storm surge predictions, $ETSS_{model}$, corrected with the statistical estimate, ET_{stat}, and with NOAA's anomaly, ET_{anom}, respectively, and observations at The Battery, NY, USA, and for 12 h, 24 h and 48 h lead-time forecasts.

Lead Time	$ET_{stat} - OBS$	$ET_{anom} - OBS$	$ETSS_{model} - OBS$
12 h	0.1976 m	0.1499 m	0.2459 m
24 h	0.1622 m	0.1274 m	0.2020 m
48 h	0.1682 m	0.1826 m	0.2854 m

4. Part II—Statistical Technique Applied along the East Coast of the U.S

Extratropical storms affect a large portion of the U.S. east coast and the validity of any statistical method must be evaluated throughout. The region affected by East Coast Winter Storms chosen here was similar to the region defined in Hirsh et al. [1] and extends from Portland, ME, to Charleston, SC. Data from thirteen NOAA tide gauge stations and nine NDBC [20] buoys were selected for analysis and are indicated on the map in Figure 2 and listed and described in Tables 2 and 3. The observed storm surge data were calculated using water level data from NOAA CO-OPS [5], and details of the computation can be found in Colle et al. [6]. Data from NDBC buoys were examined if the historical record ending in 2010 included at least five years of consecutive meteorological and oceanic data measurements. Data availability for the buoys ranges from 8 to 30 years, as seen in column 4 of Table 3. The buoys in this study are located on the continental shelf at 23.5 to 65.8 m water depth.

Table 2. List of tide gauge stations along the east coast of the U.S., their reference number, their location and a brief description of the specifics of their location.

Tide Gauge Station	Station #	Location	Description of Location
Portland, ME	8418150	43.66 N 70.25 W	South corner on the off shore end of the Maine State Pier
Boston, MA	8443970	42.35 N 71.05 W	Deep within the Boston Harbor near Logan International Airport
Nantucket Island, MA	8449130	41.28 N 70.10 W	Lee side of the island, inside a small harbor
Montauk, NY	8510560	41.05 N 71.96 W	Fort Pond Bay on the northern side of the Montauk peninsula at the entrance to the Long Island Sound
The Battery, NY	8518750	40.7 N 74.02 W	Deep within the Upper Bay of the New York/New Jersey Harbor
Sandy Hook, NJ	8531680	40.47 N 74.01 W	Just inside the Lower Bay (southern side of its entrance to the Lower Harbor), on the inside of the spit
Atlantic City, NJ	8534720	39.36 N 74.42 W	On a pier extending out from the barrier island
Cape May, NJ	8536110	38.97 N 74.96 W	Two tide gauges about 25 km apart, across the entrance to the Delaware Bay just behind the northern and southern points that form the entrance to the bay.
Lewes, DE	8557380	38.78 N 75.12 W	
Ocean City Inlet, MD	8570283	38.33 N 75.09 W	55 km south of the Delaware Bay just inside a small inlet between two barrier islands
Wrightsville Beach, NC	8658163	34.21 N 77.79 W	Two tide gauges on piers extending out from barrier islands
Springmaid Pier, SC	8661070	33.65 N 78.92 W	
Charleston, SC	8665530	32.78 N 79.93 W	On the Cooper River near the Port of Charleston

Table 3. List of the NDBC stations along the east coast of the U.S., their reference numbers, positions, water depths, length of the available data and number of storms identified at each station.

NDBC Station #	Location	Water Depth (m)	Length of Record (Year)	Number of Storms
44007	43.531 N 70.144 W	23.7	30	366
44013	42.346 N 70.651 W	61	26	366
44018	41.255 N 69.305 W	63.7	10	132
44017	40.692 N 72.048 W	45	10	104
44025	40.250 N 73.166 W	36.3	21	254
44009	38.464 N 74.702 W	28	28	351
41013	33.436 N 77.743 W	23.5	8	89
41004	32.501 N 79.099 W	38.4	18	241

A regression analysis similar to that described in Salmun et al. [2] was performed for thirteen selected pairings of tide gauges and nearby buoys. The only new aspect of the regression analysis reported here is the use of non-dimensional wave heights as predictors (rather than actual wave heights, as was done in Salmun et al. [2]), designed to help assess the relative importance of the different terms in the regression. The values of wave heights were non-dimensionalized by the mean significant wave height computed over all events considered at a single location.

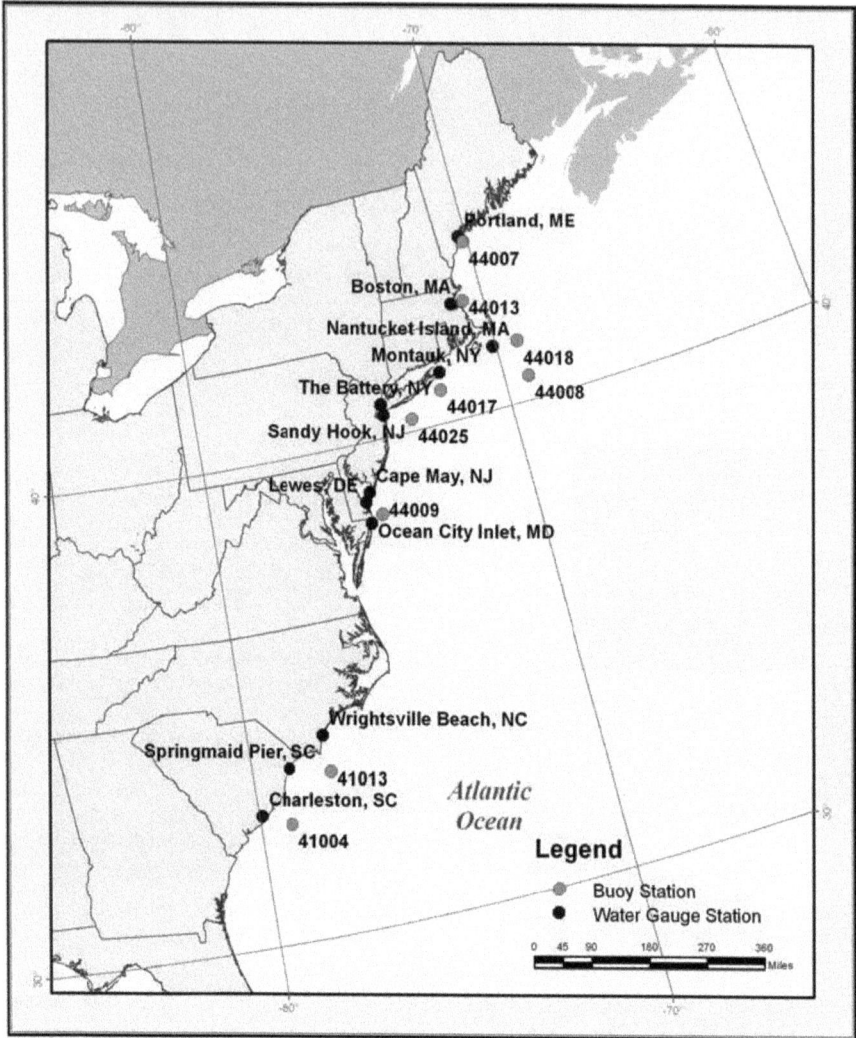

Figure 2. A reference map of the study area showing the locations of all tide gauges and National Data Buoy Center (NDBC) stations used in the analysis.

The choice of tide gauge—buoy pairs to use for the remainder of the regression analysis was informed by the "best performing" regression based on the magnitude of the root mean square error of the regression. Consideration for the selection of pairings was given to quality of data at the buoy, length of data record at the buoy, data record continuity, and proximity of the buoy to the location and setting (e.g., bay, barrier island) of the tide gauge. Based on these considerations, some tide gauges were initially paired with several different buoys, and the buoy that resulted in the smallest regression error was chosen for examination here. A list of the selected pairings appears in Table 4. The buoys selected are generally within 100 km of the tide gauge stations. For most pairings, the bearing (the angle between a line connecting buoy and tide gauge and a North-South line, measured from the North direction) from the buoy to the tide gauge is between 270° (buoy to the East of tide gauge) and

360° (buoy to the South of tide gauge). A notable exception is the bearing of 10.6° from NDBC Station 44017 to the tide gauge at Montauk, NY, USA.

Table 4. List of the tide gauge-NDBC station pairings selected in the study, the bearing from NDBC stations to tide gauges, distance separating them and the general orientation of the buoys with respect to the tide gauges, indicated in the fifth column as "dir to NDBC Sta". Also shown in the table are the parameters of the regression equation for each pairing.

Pair	Tide Gauge	NDBC Station	Bearing—from NDBC Sta. to WG (deg from N)	Distance to (Great Circle, km) &dir to NDBC Sta.	Regression Equation Slope	Intercept	RMS Error
1	Portland, ME	44007	328.54	16.95 SE	0.2236	0.1180	0.1255
2	Boston, MA	44013	270.14	32.87 E	0.2599	0.0975	0.1309
3	Nantucket Island, MA	44018	273.12	66.95 E	0.3620	0.0377	0.1272
4	Montauk, NY	44017	10.68	40.74 SSW	0.3558	0.0080	0.1816
5	The Battery, NY	44025	305.11	87.59 SE	0.5308	0.1338	0.2031
6	Sandy Hook, NJ	44025	289.24	75.26 SE	0.5122	0.0994	0.1988
7	Atlantic City, NJ	44025	227.57	145.6 NE	0.4477	0.1191	0.1952
8	Cape May, NJ	44009	338.39	61.3 SSE	0.3915	0.0552	0.1608
9	Lewes, DE	44009	314.41	50.97 SE	0.4506	0.1125	0.1856
10	Ocean City Inlet, MD	44009	247.08	36.93 ENE	0.2891	0.0588	0.1420
11	Wrightsville Beach, NC	41013	356.93	85.74 S	0.1016	0.0565	0.1767
12	Springmaid Pier, SC	41013	282.37	111.8 ESE	0.1798	0.0404	0.1911
13	Charleston, SC	41004	292.06	83.72 SE	0.1804	0.0193	0.1929

The results of the regression are shown in Table 4, alongside the list of pairings. Based on the large variation in length of data at the NDBC stations shown in Tables 2 and 3, there is also a large variation in the length of data records used to develop the regression equations shown here. To insure the stability of the regression coefficients, two of the regressions were repeated (pairs 1 and 2) using only a subset of years and found to be statistically equivalent to regressions based on the full time records. Root mean square regression errors, indicative of the uncertainly in the estimated surge, are all comparable to or lower than the value for The Battery, suggesting that the statistical model based on significant wave height can be adequately "trained" at locations throughout the region affected by winter extratropical storms.

We also investigated the validity of a single set of regression coefficients that could be used across different locations, either universally or regionally. The statistical regression with coefficients obtained for the pair The Battery—NDBC Sta. 44025 (pair 5 from Table 4) was used as the test regression for the "universal" case. For the "regional" regression canonical coefficients, we used coefficients obtained from the three pairings (pair 2) Boston—NDBC Sta. 44013, (pair 5) The Battery—NDBC Sta. 44025 and (pair 12) Springmaid Pier—NDBC Sta. 41013. Statistical regressions were evaluated using the significant wave height as the best predictor for storm surge and also using all other predictors as measured at the nearby buoy. In all cases the universal model was statistically inferior to the model determined locally. Regionally, the regression obtained from pairing 5 with all predictors was statistically equivalent to the local model estimate of storm maximum storm surge at three other geographical locations: Nantucket Island, MA Sandy Hook, NJ and Lewes, DE The other regional models were statistically inferior to the local models. On the basis of these sparse positive results, we concluded that regional models were also generally statistically inferior to the model determined locally.

5. Summary

In Part I of the present study the applicability of the statistical storm surge prediction of Salmun et al. [17] as a bias correction for NOAA's dynamical surge prediction was demonstrated. The new bias correction was evaluated using data at The Battery, NY, and was found to be statistically equivalent to the existing bias-corrected NOAA operational surge forecast at 12 and 24-h lead time and statistically improved at 48-h lead time.

In Part II the applicability of the statistical approach at a series of locations along the east coast of the U. S. that are susceptible to winter extratropical storms was investigated. The original methodology consisted of two essential elements, the first is the prediction of storm maximum storm surge using the storm composite significant wave height as a single predictor, and the second element is the training of the regression using tide gauge measurements and wave heights measured at a near-shore NDBC buoy. Although a set of universal or regional regression coefficients were statistically inferior to locally developed regression models, we showed here that the methodology using information from a local buoy to train a regression model is applicable to the other locations. The root mean square error from the local regressions were all found to be comparable to the root mean square error of the regression trained at The Battery, but the conclusive validation for the other locations based on retrospective forecasts still remains.

6. Discussion and Future Work

The new statistical bias correction presented in this article is a simple regression relation using the wave height information from a nearby offshore location as the single predictor. Despite its simplicity, it does reflect information about the basic underlying physical processes. The statistical method presented here does not depend on knowledge of either the forecast initial conditions or knowledge of previous errors. This lends support to the use of this method for longer forecast lead times. The only limiting factor is the accuracy of the wave field forecast. The use of this method for longer forecast lead times is further supported by the fact that the forecast error does not decrease with forecast lead time, as was shown in Section 3 (Table 1). In addition, because the method captures basic underlying physical relationships, it is plausible to investigate its use for prediction of storm surge in a changing climate.

The advent of ensemble mean surge forecasting, which would be expected to result in a lower overall forecast bias, could possibly diminish the positive impact of any simple bias correction, including the NOAA operational bias correction and the one advocated here based on the statistical model. The usefulness of the new bias correction method with ESTOFS and other more sophisticated models of surge forecast as well as with ensemble simulations will have to be evaluated in detail for each case separately.

With respect to the geographical extension of the statistical model, the reason for the statistical inferiority of the regional and universal regression models relative to the local model may be related to the complexity of the factors that determine the storm surge. As mentioned earlier, storm surge is determined by, among other factors, storm characteristics (surface winds and pressure fields) and coastal geomorphology and dynamics. Thus, the magnitude of the surge may not be inferred from knowledge about meteorological and water surface variables alone and it may be necessary to include explicit information about the geometry and bathymetry of the coast. The region covered in the present study, although characterized generally by wide continental shelves, presents many regional differences such as shelf widths that range from 250 km in the eastern Nova Scotia Shelf to about 30 km at Cape Hatteras. The bathymetry of Gulf of Maine, for example, is extremely irregular and complex in the regions near Portland and Boston, MA. The South Atlantic Bight, extending from Southern North Carolina to Florida, has a series of cuspate embayments within an overall concave shelf. The area in between, from South New England to the Mid-Atlantic Bight where tide gauges from Nantucket Island to Ocean City Inlet are located, includes the New York Bight where the coastline changes direction and the Hudson Canyon significantly changes the bathymetry of the region. The statistical model of Salmun et al. [2] was developed using the meteorological data provided by the local NDBC stations and the state of the ocean surface, as described by the surface wave field, at a single location. To extend the applicability of a locally determined regression model to other regions of the coast it may be necessary to investigate the inclusion of predictor variables that explicitly describe the coastal geomorphology.

J. Mar. Sci. Eng. **2015**, *3*, 73–84

A remaining drawback of the statistical method used in all the parts of the present study is that it is an event-based approach: storm events are identified (from data or forecasts) and the pertinent variables are analyzed over the storm duration. An advantage of event-based approach is that it is easier statistically to predict the maximum storm surge during a storm event than to predict a time series of storm surge. The disadvantages of the event-based approach relates to the limited information that it can provide. In order to provide accurate information about flooding it is very important to determine the timing of the surge relative to the tidal cycle. In addition, improved water level predictions in the absence of storms are also needed. Therefore, the extension of the statistical methodology to a non-event-based approach needs to be investigated.

Acknowledgments: Partial support for this work was provided by the CUNY Research Foundation through PSC-CUNY Award 63088-00 41. The manuscript was partially written at the GAIN 2012 Workshop funded by NSF Grant Number 0620087. We cheerfully acknowledge the technical assistance provided by Kamila Wisniewska and Casey King in performing calculations for this work and of Mark Dempsey who provided the graphics for Figure 1. We appreciate the participation of Tracy Tran, who performed many of the early exploratory calculations, and of Aline Gjelaj, who verified storm events from weather maps. In addition, we acknowledge the thoughtful reviews by anonymous reviewers that resulted in a substantially improved manuscript.

Author Contributions: The two authors made equal contributions to the research and to the manuscript.

Conflicts of Interest: The authors declare no conflict of interest.

References

1. Hirsh, M.E.; DeGaetano, A.T.; Colucci, S.J. An East Coast winter storm climatology. *J. Clim.* **2001**, *14*, 882–899. [CrossRef]
2. Salmun, H.; Molod, A.; Buonaiuto, F.; Wisniewska, K.; Clarke, K. East Coast Cool-weather Storms in the New York Metropolitan Region. *J. Appl. Meteo. Climatol.* **2009**, *48*, 2320–2330. [CrossRef]
3. Intergovernmental Panel for Climate Change. Climate Change 2013: The Physical Science Basis. Available online: http://www.ipcc.ch/report/ar5/wg1/ (accessed on 13 October 2014).
4. Nor'easter, 11–14 December 1992. Available online: http://www.hurricanes-blizzards-noreasterns.com/1992noreaster.html (accessed on 14 October 2014).
5. NOAA Tides & Currents; CO-OPS. Available online: http://tidesandcurrents.noaa.gov/ (accessed on 13 October 2014).
6. Colle, B.A.; Rojowsky, K.; Buonaiuto, F. New York City storm surges: Climatology and an analysis of the wind and cyclone evolution. *J. Appl. Meteo. Climatol.* **2010**, *49*, 85–100. [CrossRef]
7. NOAA National Weather Service Forecast Office, Philadelphia/Mount Holly. Available online: http://www.erh.noaa.gov/phi/storms/04162007.html (accessed on 14 October 2014).
8. Kim, S.-C.; Chen, J.; Shaffer, W.A. An operational Forecast Model for Extratropical Storm Surges along the U.S. East Coast. In *Proceedings of the Conference on Coastal Oceanic and Atmospheric Prediction*; Atlanta, Georgia, American Meteorological Society, 1996; pp. 281–286. Available online: http://www.nws.noaa.gov/mdl/pubs/Documents/Papers/Shaffer1996AnOperationalForecast.png (accessed on 13 October 2014).
9. NOAA/NWS MDL Extratropical Water Level Guidance. Available online: http://www.nws.noaa.gov/mdl/etsurge/ (accessed on 13 October 2014).
10. NOAA/NWS OPC ESTOFS Storm Surge Model Guidance (0–96 Hour Forecasts). Available online: http://www.opc.ncep.noaa.gov/estofs/estofs_surge_info.shtml (accessed on 13 October 2014).
11. Funakoshi, Y.; Feyen, J.C.; Aikman III, F.; van der Westhuysen, A.; Tolman, H. *The Extratropical. Surge and Tide Operational Forecast. System (ESTOFS) Atlantic Implementation and Skill Assessment*; NOAA Technical Report NOS CS 32; U.S. Department of National Commerce, National Ocean Service, Coast Survey Development Laboratory: Silver Spring, MD, USA, October 2013. Available online: http://www.nauticalcharts.noaa.gov/csdl/publications/TR_NOS-CS32FY14_01_Yuji_ESTOFS_SKILL_ASSESSMENT.png (accessed on 13 October 2014).
12. NOAA/NWS CPC 8 to 14 Days Outlooks. Available online: http://www.cpc.ncep.noaa.gov/products/predictions/814day/ (accessed on 13 October 2014).
13. Lin, N.; Emanuel, K.A.; Smith, J.A.; Vanmarcke, E. Risk assessment of hurricane storm surge for New York City. *J. Geophys. Res.* **2010**, *115*, D18121. [CrossRef]

14. Lin, N.; Emanuel, K.A.; Smith, J.A.; Vanmarcke, E. Physically based assessment of hurricane surge threat under climate change. *Nat. Clim. Chang.* **2012**, *2*, 462–467. [CrossRef]

15. Colle, B.A.; Zhang, Z.; Lombardo, K.; Chang, E.; Liu, P.; Zhang, M. Historical Evaluation and Future Prediction of Eastern North American and Western Atlantic Extratropical Cyclones in the CMIP5 Models during the Cool Season. *J. Clim.* **2013**, *26*, 6882–6903. [CrossRef]

16. Müller-Navarra, S.H.; Knüpffer, K. Improvement of water level forecasts for tidal harbours by means of model output statistics (MOS). Part I (Skewed surge forecasts). *Berichte des Bundesamtes für Seeschifffahrt und Hydrographie.* Nr. 47/2010. ISSN-Nr. 0946-6010. Available online: www.bsh.de/de/Produkte/Buecher/Berichte_/Bericht47/BSH_Bericht_47.png (accessed on 4 December 2014).

17. Salmun, H.; Molod, M.; Wisniewska, K.; Buonaiuto, F. Statistical prediction of the storm surge associated with cool-weather storms at The Battery, New York. *J. Appl. Meteo. Climatol.* **2011**, *50*, 273–282. [CrossRef]

18. NOAA National Data Buoy Center. Available online: http://www.ndbc.noaa.gov/ (accessed on 13 October 2014).

19. NOAA EMC Marine Modeling & Analysis Branch. Available online: http://polar.ncep.noaa.gov/Data downloaded from public domain: ftp://polar.ncep.noaa.gov/pub/history/waves; (accessed on 13 October 2014).

20. NOAA NCDC National Operational Model Archive & Distribution System; NOMADS Data Access. Available online: http://nomads.ncdc.noaa.gov/data.php#hires_weather_datasets (accessed on 13 October 2014).

21. Di Liberto, T.; Colle, B.A.; Georgas, N.; Blumberg, A.F.; Taylor, A. Verification of a multi-model storm surge ensemble around New York City and Long Island for the cool season. *Weather. Forecast.* **2011**, *26*, 922–939. [CrossRef]

Journal of
Marine Science and Engineering

MDPI

Article

Tolerable Time-Varying Overflow on Grass-Covered Slopes

Steven A. Hughes [1,*] **and Christopher I. Thornton** [2,†]

1   Senior Research Scientist, Engineering Research Center, Colorado State University, 1320 Campus Delivery, Fort Collins, CO 80523, USA

2   Director, Engineering Research Center, Colorado State University, 1320 Campus Delivery, Fort Collins, CO 80523, USA; thornton@engr.colostate.edu

*   Author to whom correspondence should be addressed; shughes2@engr.colostate.edu; Tel.: +1-970-491-8160; Fax: +1-970-491-8462.

†   These authors contributed equally to this work.

Academic Editor: Rick Luettich
Received: 5 December 2014; Accepted: 10 March 2015; Published: 19 March 2015

Abstract: Engineers require estimates of tolerable overtopping limits for grass-covered levees, dikes, and embankments that might experience steady overflow. Realistic tolerance estimates can be used for both resilient design and risk assessment. A simple framework is developed for estimating tolerable overtopping on grass-covered slopes caused by slowly-varying (in time) overtopping discharge (e.g., events like storm surges or river flood waves). The framework adapts the well-known Hewlett curves of tolerable limiting velocity as a function of overflow duration. It has been hypothesized that the form of the Hewlett curves suggests that the grass erosion process is governed by the flow work on the slope above a critical threshold velocity (referred to as excess work), and the tolerable erosional limit is reached when the cumulative excess work exceeds a given value determined from the time-dependent Hewlett curves. The cumulative excess work is expressed in terms of overflow discharge above a critical discharge that slowly varies in time, similar to a discharge hydrograph. The methodology is easily applied using forecast storm surge hydrographs at specific locations where wave action is minimal. For preliminary planning purposes, when storm surge hydrographs are unavailable, hypothetical equations for the water level and overflow discharge hydrographs are proposed in terms of the values at maximum overflow and the total duration of overflow. An example application is given to illustrate use of the methodology.

Keywords: time-varying overflow; grass levee resiliency; discharge hydrograph; tolerable overtopping

1. Introduction

Earthen levees and embankments are used extensively as protective barriers between water and land areas intended for residential, commercial, or agricultural purposes. Ideally, the crest elevations of these man-made structures would extend beyond the imaginable highest water levels under the most extreme flood conditions. However, this ideal design is seldom realized because of the uncertainty in predicting flood levels, and the expense of constructing structures with crest elevations beyond the flood levels associated with extremely rare events.

Numerous studies have been conducted that discuss and describe breaching of earthen dams and embankments by steady overflowing water, and an extensive summary of earthen embankment breaching knowledge was published in 2011 by the ASCE/EWRI Task Committee on Dam/Levee Breaching [1]. In addition, numerical models [2–7] have been developed to predict the progression of breaching, starting with erosion of the grass. These models incorporate varying degrees of the physical processes based on observation from full-scale tests, such as those described by Hanson and Temple [8], Hunt et al. [9], and Hanson et al. [10].

Resiliency against overtopping for earthen structure landward-side slopes is provided by grass or stronger alternatives, such as turf-enforcement mats, articulated concrete blocks, rip-rap, or roll-compacted concrete. Grass-covered slopes provide a surprising amount of erosion resistance to overflow provided the grass is well maintained and has developed a healthy, dense root system. If the grass is eventually eroded in spots by overflow, head-cut erosion may develop quickly; and this could eventually result in breach formation, particularly where the bare soil is highly erodible. The situation of imminent breaching should be avoided at all costs.

"Tolerable (or limiting) steady overflow" is an ill-defined phrase that could refer to any point between initial localized loss of grass on the landward-side slope up to initial loss of crest width due to head-cutting. Hewlett et al. [11] stated that *"The condition when soil is directly exposed to flowing water is classified as the onset of failure and is unacceptable."* Consequently, prudent designs require earthen structures to withstand some duration of steady overflow without experiencing significant grass erosion of the landward-side slope. This conservative definition for tolerable steady overflow is adopted for the developments given herein.

This paper develops a simple methodology (spreadsheet-level) for estimating tolerable cumulative overflow on grass-covered slopes caused by slowly-varying (in time) overtopping discharge. Examples of events with time-varying discharge include storm surges in coastal estuaries, passing of flood waves in rivers, or reservoir overfilling caused by rainwater run-off. The methodology adapts well-known curves for tolerable limiting velocity as a function of overflow duration, and expresses the limiting criterion in terms of excess work above the work associated with a critical threshold velocity. Site-specific applications are best performed using forecast storm surge or flood hydrographs. But for preliminary analyses, hydrographs of slowly-varying water level above the structure crest and corresponding discharge can be characterized by somewhat realistic hypothetical water-level hydrographs defined by the maximum water height above the crest and the total duration of overflow. An example application is given to illustrate use of the preliminary analysis methodology.

2. Tolerable Steady Overflow Limit for Grass Slopes

The development in this paper is based on a design nomogram for erosional stability of grass-covered slopes during steady overflow presented by Hewlett et al. [11]. This nomogram used full-scale steady overflow test data to propose stability curves for three different grass qualities and other slope protection surfaces (turf reinforcement mats and concrete block systems). Each curve presents the limiting steady overflow velocity as a function of overflow duration that results in an acceptable level of erosion without putting the grass cover at risk of failure (as defined earlier). Figure 1 shows a reproduction of the limiting velocity curves for grass as given in Hewlett et al. [11].

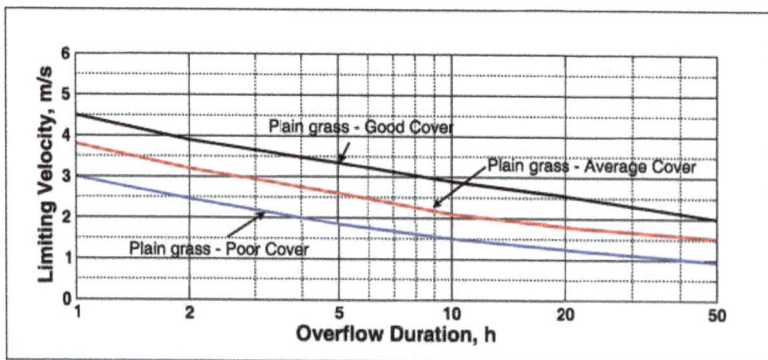

Figure 1. Limiting velocity *versus* overflow duration for grass (after Hewlett et al. [11]).

Hewlett et al. [11] had curves for good-, average-, and poor-cover plain grass as shown on Figure 1. As might be expected, the grass tolerates faster steady overflow velocities for only short durations, whereas slower velocities can be tolerated for longer durations. Hewlett et al. described the grass qualities as follows:

"Good grass cover is assumed to be a dense, tightly-knit turf established for at least two growing seasons. Poor grass cover consists of uneven tussocky grass growth with bare ground exposed or a significant proportion of non-grass weed species. Newly sown grass is likely to have poor cover for much of the first season".

(Hewlett et al. [11])

Presumably, average grass cover is something in between good and poor cover. The Hewlett et al. curves are based on earlier work by Whitehead et al. [12]. Figure 2, based on the figure in the Whitehead et al. report, includes the sparse data points on which the curves in the Whitehead et al. and the Hewlett et al. reports are based. The three design curves proposed by Whitehead et al. are shown as dashed lines, whereas the three grass curves presented by Hewlett et al. are shown by heavy solid lines. It is interesting to note that the Hewlett et al. curves for grass were adjusted downward on the low-duration side of the plots in comparison to the curves and data given by Whitehead et al. There was no explanation for this adjustment, but the result is that the Hewlett et al. curves become more conservative on the low-duration end when compared to the guidance given by Whitehead et al.

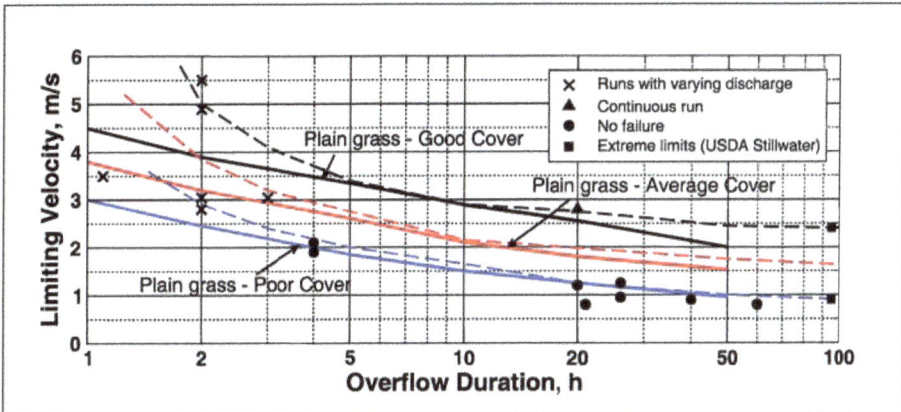

Figure 2. Erosion resistance of grass-lined spillways (after Whitehead et al. [12]).

For unchanging steady overflow the Hewlett et al. curves can be interpreted as duration limits for tolerable overflow velocities resulting from discharge flowing over a slope having some flow resistance due to grass. Every point on a curve represents an equivalent cumulative flow condition, i.e.,

$$[u_1, T_1] \text{ is erosionally equivalent to } [u_2, T_2] \tag{1}$$

where u is the limiting velocity and T is the flow duration at the point where the limiting velocity intersects the curve for a specific quality of grass. Erosional equivalence means that any combination of u,T that intersects a curve will result in the same erosional damage (defined as the tolerable limit for that grass quality) as any other combination of u, T that intersects the same curve at a different location. If a combination of u,T goes above the designated curve, grass failure will occur. Likewise, if a combination of u, T is below the curve, the grass will withstand the overflow. Note: the expression given in Equation (1) is not a mathematical equation; but instead it is a statement of erosional equivalence. As an example

15

for good grass, a limiting velocity of u_1 = 3 m/s lasting for a duration of T_1 = 8 h is considered equivalent to a limiting velocity of u_2 = 2 m/s lasting for a duration of about T_2 = 50 h as seen in Figure 1 or Figure 2 because both of these combinations fall on the good grass curve.

Dean et al. [13] used numerical values manually extracted from the Hewlett et al. grass-cover curves (Figure 1) to examine the underlying physical mechanism that might be responsible for the form of the tolerable velocity limit curves. They assumed that cumulative tolerable erosion, E, took the form:

$$E_m = K_m(u^m - u_c^m)t \text{ for } u > u_c \qquad (2)$$

where K_m is a proportionality constant, u_c is a critical threshold erosion velocity, and t is time. Three possibilities were investigated: (a) flow velocity above a certain threshold velocity given when the exponent m = 1; (b) shear stress [τ_o proportional to u^2] above a certain threshold shear stress when m = 2; and (c) stream power [$P_s = (\tau_o u)$ proportional to u^3] above a certain threshold stream power when m = 3. (Note that Hanson and Temple [8] assumed an erosion rate based on excess shear stress.)

Dean et al. [13] performed least-squares best-fits of Equation (2) to the Hewlett et al. curves for grass covers using the three values of exponent m. Appropriate best-fit values were determined for the threshold velocities (u_c) and erosion limits (E_m/K_m) so that the error of the curve-fit was minimized. The analysis showed that the assumption of excess stream power (m = 3) had the smallest standard error by a significant margin when the three equations were fitted to all three grass quality curves. This yielded the following equation for the tolerable grass design curves:

$$\left(u^3 - u_c^3\right)t = \frac{E_3}{K_3} = G_F \qquad \text{for } u > u_c \qquad (3)$$

with the best-fit values shown in Table 1.

Table 1. Results for the best-fit of Equation (3) to the Hewlett et al. [12] grass curves.

Grass Quality	Threshold Velocity u_c (m/s)	Erosion Limit $E_3/K_3 = G_F$ (m³/s²)	Std. Error in Velocity (m/s)
Good Cover	1.80	0.492×10^6	0.38
Average Cover	1.30	0.229×10^6	0.12
Poor Cover	0.76	0.103×10^6	0.04

Stream power is the time rate of flow work per unit surface area, and the best-fit analysis of Equation (2) led Dean et al. to conclude that cumulative work (W = $P_s \cdot t$) done on the slope by the flowing water, in excess of some critical value of work, was the physical mechanism responsible for the trends given in the Hewlett et al. [12] limiting velocity *versus* duration curves. Figure 3 shows the best-fit obtained by Dean et al. compared to the original Hewlett et al. curves for three types of grass. The heavy solid lines are the Hewlett et al. curves, and the symbols connected by lighter dashed lines are the best-fit values. Note the very good fits for average and poor grass (also indicated by the standard deviations in Table 1).

The most deviation between velocities predicted using the best fit of Equation (3) and the original Hewlett velocities occurred at the short durations, and this is seen in Figure 3 for good grass at short durations. Given that the Hewlett curves have lower velocities at short-durations when compared to the Whitehead et al. curves (see Figure 2), this over-prediction by the best-fit procedure for good grass at short durations should not be too much of a concern. Note there could be additional "almost" best-fits of the two-parameter equation that might result in different values for the threshold velocity and the erosion limit.

The formulation based on cumulative excess work might imply that the grass surface is fairly homogeneous in structure and strength. If this were the case, then we should expect erosion to be somewhat uniform, and grass damage would occur nearly simultaneously over the entire region subjected to the same flow conditions. Of course, this is not realistic, and erosion will occur sporadically at isolated "hot spots" or weaknesses in the grass cover. The implicit assumption of Dean et al. cumulative excess work model is that the model gives an indication when the isolated grass cover damage begins to become locally problematic as determined by analysis of the original full-scale test data that resulted in the Hewlett et al. curves.

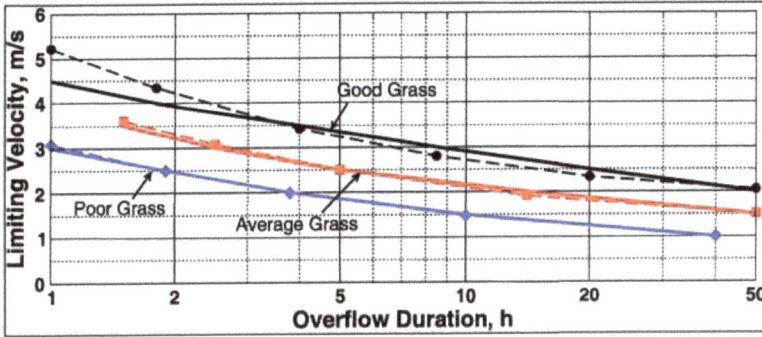

Figure 3. Dean et al. (2010) [13] best fit of Equation (3) to Hewlett et al. (1987) [11] curves.

3. Slowly-Varying Overflow on Grass Slopes

Many overflow situations are characterized by relatively slow variations in the overflow water depth that occur over durations of hours or days. For example, river levels can rise to a peak level and then fall again with passage of the flood wave downstream, and this causes a corresponding variation of the overflow velocity on the grass-covered slope. Likewise, storm surges propagating into estuarine systems can cause overflow variations at protective levees lasting for many hours. Rainwater run-off can cause reservoirs to overflow with similar overtopping hydrographs. Figure 4 illustrates a water level above the structure crest that slowly varies in time.

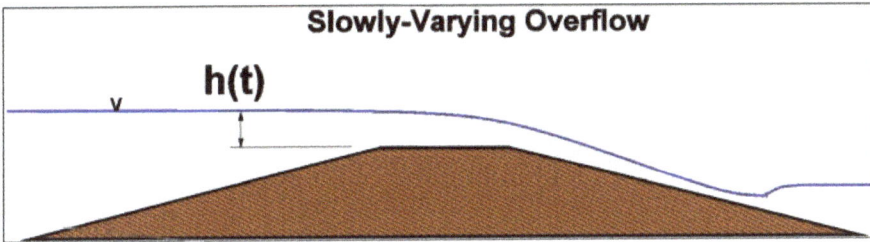

Figure 4. Slowly-varying overflow of a levee.

The following developments discount any influence of wave action that might accompany the coastal storm surge. Therefore, wave overtopping combined with storm surge overflow occurring at sea dikes situated at the coastline would not be a valid application of the methodology developed below because storm surge is typically accompanied by large waves. However, as storm surges propagate into estuaries, embankment or levee overtopping can occur where wave action may be limited by narrow fetches or the wave-generating wind may have abated by the time storm surge affects upper portions of the estuary. These situations would be a more appropriate application of the methodology. Future work will include the cases of tolerable time-varying overtopping due to waves only and to combined wave and surge overtopping by using erosional equivalence concepts with individual overtopping waves.

The tolerable erosion limit criterion given by Equation (3) is based on a constant limiting velocity over a given duration; but if cumulative excess flow work is responsible for eventual instability of grass slopes, there is no reason that velocity cannot vary slowly in time as the water level above the crest changes. Therefore, the criterion for tolerable cumulative excess work below the erosional limit for grass (G_F) can be given as the integral between $t = 0$ and $t = T$ as

$$\int_0^T \left([u(t)]^3 - u_c^3 \right) dt \quad \leq \quad G_F \qquad \text{for } u > u_c \tag{4}$$

where T is the duration of the overflow event. Dean et al. [13] noted that the limiting velocity, u, can be expressed in terms of overflow discharge under the conservative assumption that the flow velocity on the grass slope had reached terminal velocity, and the force balance described by the steady-flow resistance equation was piece-wise applicable to slowly-varying overflow. They used the Darcy-Weisbach steady-flow resistance equation at terminal velocity that is essentially the same as the Chezy equation with the Chezy coefficient represented by a friction factor, i.e.,

$$u^3 = \left(\frac{8g}{f_D} \sin\theta \right) q = \left(\frac{2g}{f_F} \sin\theta \right) q \tag{5}$$

where q is volumetric discharge per unit length of levee or embankment, θ is the landward-side levee or embankment slope angle, g is gravitational acceleration, f_D is the Darcy-Weisbach friction factor, and f_F is the Fanning friction factor. The two friction factors are related by the expression $f_D = 4 f_F$. The Fanning friction factor is more commonly used by chemical engineers, and it routinely appears in the European wave overtopping literature. Care must be taken not to confuse the two friction factors because often the identifying subscripts are omitted.

Substituting Equation (5) into Equation (4) and rearranging yields the integral cumulative discharge criterion for slowly-varying steady overflow:

$$V_{ET}(t) = \int_0^T [q(t) - q_c] \, dt \quad \leq \quad G_F \left(\frac{f_F}{2g \sin\theta} \right) \qquad \text{for } q > q_c \tag{6}$$

where the critical discharge is given by:

$$q_c = \left(\frac{f_F}{2g \sin\theta} \right) u_c^3 \tag{7}$$

and the parameter $V_{ET}(t)$ is used to designate the integration in Equation (6). A discrete version of the integral Equation (6) can be written as:

$$V_{ET}(t) = \sum_{n=1}^N (q_n - q_c) \, \Delta t_n \quad \leq \quad G_F \left(\frac{f_F}{2g \sin\theta} \right) \qquad \text{for } q_n > q_c \tag{8}$$

In Equation (8), the subscript n refers to a discrete value of overflow discharge, N is the total number of discrete discharges, and Δt_n is the uniform duration associated with each discrete discharge. The total duration of overflow is $T = N \cdot \Delta t_n$. It is interesting to note that both sides of the inequality in Equation (8) have units of m^3/m, so the summation of terms on the left side of the inequality in Equation (8) represents all the overflow water volume (per unit levee or embankment length) as a function of time greater than a critical discharge volume that is given by $q_c \cdot \Delta t_n$. The right-hand side of the inequality in Equation (8) represents the tolerable upper limit of cumulative overflow water above the critical discharge volume. Therefore, the parameter V_{ET} means "cumulative (or total) excess water volume" which is directly proportional to the cumulative excess work (CEW).

Application of Equation (8) to estimate tolerable cumulative overflow for grass slopes requires a discrete time-history of slowly-varying overflow discharge (q_n), estimates of the critical velocity

(u_c) and the grass erosional limit (G_F), the levee or embankment slope angle (θ), and a value for the Fanning friction factor (f_F). Values of critical velocity (u_c) and grass erosional limit (G_F) associated with the Hewlett et al. [11] curves are given in Table 1. However, alternate values for more robust levee grasses can be used if such values have been determined from full-scale overflow tests.

One difficulty is specifying a suitable value for the Fanning friction factor (f_F). Friction factors appropriate for grass are implicitly included in the Hewlett et al. velocity curves, but they are not known outright. Furthermore, the friction factor can be shown to be a function of flow depth; thus, the friction factor will vary in magnitude during slowly-varying overflow. Hughes [14] analyzed likely variations of the Fanning friction factor as reported in the literature, and he concluded that the range f_F = 0.01–0.02 was a reasonable representative average for overtopped grass-covered levee slopes for overflow water levels ranging between 0.15 m and 0.6 m above the levee crest. This range is expected to encompass many overflow conditions. For practical applications a constant value of f_F = 0.015 seems appropriate for typical grass-covered slopes, but some situations may require different values for the friction factor.

4. Idealized Representation of Slowly-Varying Overwash Discharge

In the absence of measured or predicted overflow water depth hydrographs (i.e., storm surge or flood hydrographs), an approximation can be made that mimics the time-varying overflow process. An idealized representation of slowly-varying overflow discharge should start with zero discharge when the water level reaches the levee or embankment crest elevation, increase monotonically over time to a maximum discharge when the water elevation is at its greatest elevation above the crest, and then decrease monotonically back to zero discharge. Whereas, a simple straight-line approximation for the increase and decrease of overflow water level could be used as a first approximation, it is better to have a more realistic continuous function without any singularities or discontinuities.

Several mathematical formulas are available that could be adapted for approximating time-varying overflow discharge. Cialone et al. [15] developed a synthetic symmetrical storm surge hydrograph in which the rising and falling portions of the hydrograph were represented by an exponential decay. Their synthetic hydrograph was given as the following:

$$S(t) = S_p\left(1 - e^{-|D/(t-t_0)|}\right) \tag{9}$$

where $S(t)$ is the time-dependent surge, S_p is the peak surge, D = half the storm surge duration, t = time, and t_0 is the time of peak surge.

Measurements indicated that the falling limbs of storm surges are steeper than the rising limbs, and this led Zevenbergen et al. [16] to represent the falling limb of the storm surge by adding a dimensionally nonhomogeneous term to the equation of Cialone et al. This resulted in the synthetic hydrograph being described by two equations:

$$\begin{aligned} S(t) &= S_p\left(1 - e^{-|D/(t-t_0)|}\right) & \text{for } t < t_0 \\ S(t) &= S_p\left(1 - e^{-|D/(t-t_0)|}\right) - 0.14\,(t-t_0)\,e^{-0.18(t-t_0)} & \text{for } t > t_0 \end{aligned} \tag{10}$$

More recently, Xu and Huang [17] proposed equations that successfully represented Gulf of Mexico storm surges when the empirical coefficients were determined from data. There equations were given as:

$$\begin{aligned} S(t) &= S_p\left(1 - a \cdot e^{-|b/(t-t_0)|}\right) & \text{for } t < t_0 \\ S(t) &= S_p\left(1 - c \cdot e^{-|d/(t-t_0)|}\right) & \text{for } t > t_0 \end{aligned} \tag{11}$$

where a, b, c, and d are dimensionless empirical coefficients. Fitting of Equation (11) to measured Gulf of Mexico category 4 and 5 hurricane storm surge hydrographs resulted in values for the four coefficients shown in Table 2.

Table 2. Empirical coefficients for Xu and Huang [17] synthetic storm surge hydrograph.

Hurricane	a	b	c	d
Category 3	1	3	0.9	4
Category 4	1	4	1.2	5

Of course, storm surges may not be good representations of the time-varying water elevations associated with flooding events that cause steady overflow (without wave action).

In this paper we propose an alternative hypothetical hydrograph to represent time-varying overflow of a levee or embankment. For convenience, we selected the well-known symmetric formula for the normal (or Gaussian) probability density function. However, in this application we are not using the formula for estimating probability; instead we are just applying a mathematical form that provides properties that are similar to an actual overflow hydrograph.

The equation for a normal probability distribution is:

$$p(x) = \frac{1}{\sigma\sqrt{2\pi}} \exp\left[-\frac{(x-\mu)^2}{2\sigma^2}\right]$$

(12)

where p is the probability associated with x, μ is the mean value of x, and σ is the standard derivation of the probability density function. The probability density function is symmetric about the mean, the maximum of the density function occurs at the mean, and the area beneath the curve given by Equation (12) is equal to unity.

Adopting Equation (12) for use in describing the time-varying overwash water level is accomplished by the following substitutions:

1. Let x equal time, t_s, in seconds. This means that standard deviation, σ, will also have units of seconds.
2. Let probability $p(x)$ equal instantaneous overwash water level, $h(t_s)$ as a function of time, as illustrated in Figure 4.
3. One feature of Equation (12) is that 99.6% of the area under the curve is contained by three standard deviations on each side of the peak that occurs at $x = \mu$. Therefore, let $6\sigma \approx T_s$ = total duration of overflow in seconds, and substitute $\sigma = T_s/6$. Also, let the mean value $\mu = T_s/2$. This results in overflow beginning very close to $t_s = 0$, and ending very close to $t_s = T_s$. Maximum overflow water depth and discharge occur at time $t_s = T_s/2$.
4. Add a scaling factor, C_A, to Equation (12) that represents the total area under the curve.

Making these substitutions into Equation (12) gives the following equation for the time-varying distribution of overwash elevation.

$$h(t_s) = \frac{6 C_A}{\sqrt{2\pi}\,T_s} \exp\left[\frac{-18\left(t_s - \frac{T_s}{2}\right)^2}{T_s^2}\right]$$

(13)

The peak overflow water depth (h_p) occurs at $t_s = T_s/2$, and from Equation (13) the scaling factor equals:

$$C_A = \frac{\sqrt{2\pi}\,T_s\,h_p}{6}$$

(14)

Substituting Equation (14) into Equation (13) gives the idealized hydrograph of overflow water depth in terms of the overflow duration and the peak water depth above the structure crest, i.e.,

$$h(t_s) = h_p \exp\left[\frac{-18\left(t_s - \frac{T_s}{2}\right)^2}{T_s^2}\right] \tag{15}$$

The corresponding slowly-varying overflow discharge hydrograph can be approximated by substituting Equation (15) into the broad-crested weir formula (e.g., Henderson [18])

$$q(t_s) = \left(\frac{2}{3}\right)^{3/2} \sqrt{g}\, [h(t_s)]^{3/2} \tag{16}$$

where q is overflow discharge per unit crest width, and g is gravitational acceleration. This substitution gives the expression

$$q(t_s) = \left(\frac{2}{3}\right)^{3/2} \sqrt{g}\, (h_p)^{3/2} \exp\left[\frac{-27\left(t_s - \frac{T_s}{2}\right)^2}{T_s^2}\right] \tag{17}$$

Equation (17) can also be expressed in terms of the peak discharge (q_p) as:

$$q(t_s) = q_p \exp\left[\frac{-27\left(t_s - \frac{T_s}{2}\right)^2}{T_s^2}\right] \tag{18}$$

where:

$$q_p = \left(\frac{2}{3}\right)^{3/2} \sqrt{g}\, (h_p)^{3/2} \tag{19}$$

Finally, the cumulative overflow water volume per unit crest width as a function of time is found by integrating Equation (18), i.e.,

$$V_W(t_s) = \int_{-\infty}^{t_s} q(t_s)\, dt = q_p \int_{-\infty}^{t_s} \exp\left[\frac{-27\left(t_s - \frac{T_s}{2}\right)^2}{T_s^2}\right] dt \tag{20}$$

or

$$V_W(t_s) = \sqrt{\frac{\pi}{3}}\, \frac{q_p\, t_s}{6} \left\{1 + \mathrm{erf}\left[\frac{3\sqrt{3}}{T_s}\left(t_s - \frac{T_s}{2}\right)\right]\right\} \tag{21}$$

where erf is the mathematical error function. The total cumulative overflow water volume per unit embankment width is given to a close degree of approximation when $t_s = T_s$ in Equation (21), i.e.,

$$V_W(T_s) = (V_W)_{Total} = \sqrt{\frac{\pi}{3}}\, \frac{q_p\, T_s}{3} \tag{22}$$

In summary, idealized formulations for slowly-varying water depth above the embankment crest (Equation 15) and overflow discharge (Equation 18) have been proposed in terms of time, the peak water depth expected during the flood event (h_p), and the duration of overflow (T_s). The water depth and discharge hydrographs are symmetric about the peak values, and the equations are dimensionally homogeneous so they can be used with any consistent set of units (i.e., all numerical coefficients are dimensionless).

Figure 5 illustrates three idealized overflow hydrographs having a maximum overflow depth of 0.25 m and overflow durations of 1, 3, and 5 h, respectively. The upper plot shows the variation in time of overflow water depth; the middle plot shows discharge per unit width; and the lower plot gives the corresponding cumulative overflow water volume per unit width as a function of time. Similar theoretical equations could be developed in the same manner, and equations with more variables, such as those proposed by Xu and Huang [17] could be fit to available measured water level hydrographs. However, in the absence of measured hydrographs and more elegant mathematical forms, the suggested hypothetical hydrographs provide a simple and reasonably realistic method for approximating time-varying overflow as part of a preliminary risk analysis.

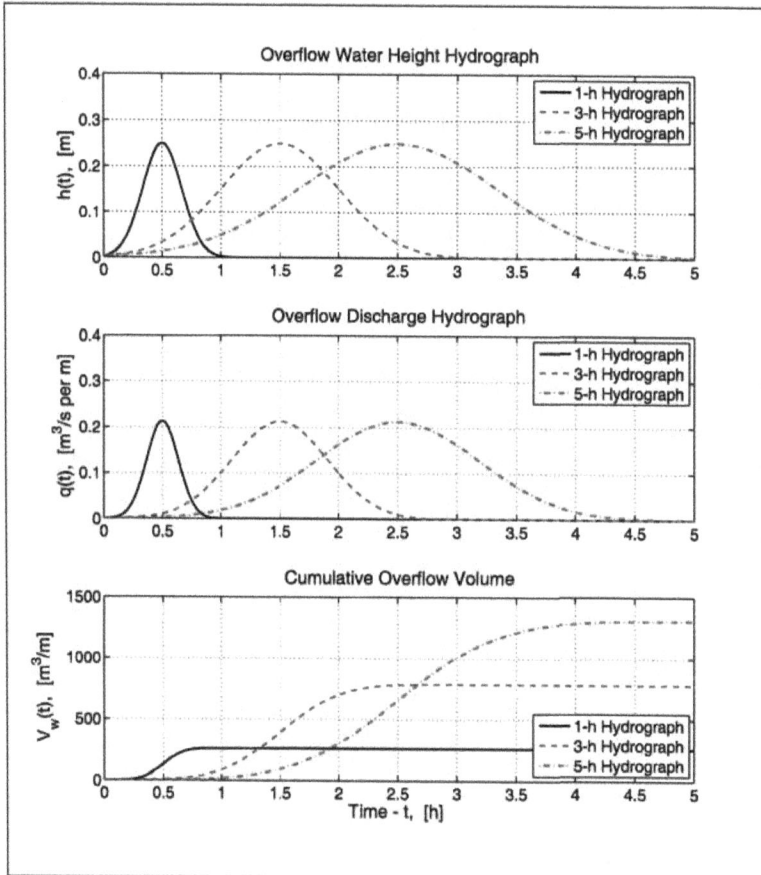

Figure 5. Idealized overflow hydrographs and cumulative overflow water volume.

5. Example Estimate of Tolerable Slowly-Varying Overflow

The following is an example that uses the proposed methodology given in this paper to estimate tolerable overflow as a function of duration during an event when the water level exceeds the embankment crest. Assume:

(1) The variation of water level above the embankment crest occurs according to Equation (15) over a duration of $T_s = 5$ h $= 18,000$ s.
(2) The peak water level above the crest is $h_p = 0.25$ m.

(3) The embankment grass is considered "good," so the parameters for critical threshold velocity and erosional limit in Table 1 for "good grass" are appropriate if no other full-scale test results are available for the particular grass/soil combination.

(4) The embankment landward-side slope is 1V-on-3H ($\theta = 18.4°$).

(5) The grass cover Fanning friction factor is approximately $f_F = 0.015$.

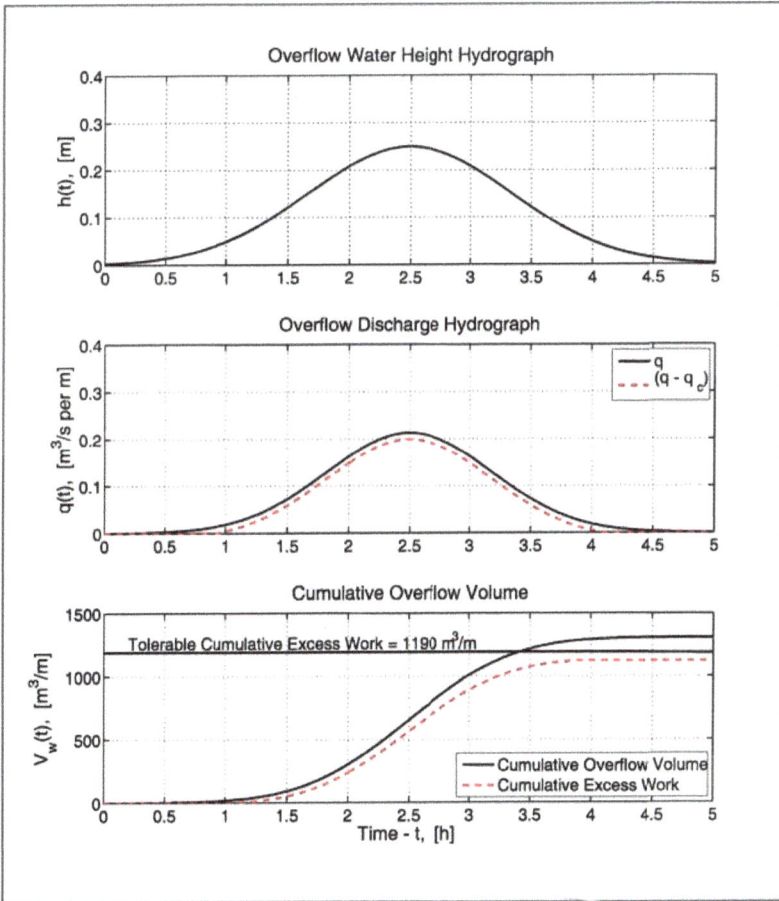

Figure 6. Example of cumulative excess work estimate for "good grass" ($u_c = 1.8$ m/s).

The solution is found by using Equation (18) in the discrete version of the cumulative excess work equation given by Equation (8). First, the peak discharge is determined using Equation (19), i.e.,

$$q_p = \left(\frac{2}{3}\right)^{3/2} \sqrt{g} \, (h_p)^{3/2} = \left(\frac{2}{3}\right)^{3/2} \sqrt{9.814 \, \text{m/s}^2} \, (0.25 \, \text{m})^{3/2} = 0.213 \, \text{m}^3/\text{s per m} \tag{23}$$

and the time-varying discharge (Equation 18) becomes:

$$q_n(t_s) = (0.213 \, \text{m}^3/\text{s per m}) \, \exp\left[\frac{-27 \left(t_s - \frac{18{,}000 \, \text{s}}{2}\right)^2}{(18{,}000 \, \text{s})^2}\right] \tag{24}$$

23

Next, find the critical discharge associated with the critical threshold velocity (u_c = 1.8 m/s from Table 1) using Equation (7).

$$q_c = \left(\frac{f_F}{2g\sin\theta}\right)u_c^3 = \left(\frac{0.015}{2\left(9.814\,\text{m/s}^2\right)\sin(18.4^o)}\right)(1.8\,\text{m/s})^3 = 0.014\,\text{m}^3/\text{s per m} \quad (25)$$

Finally, the cumulative excess work is found from Equation (8) as the summation:

$$V_{ET}(t) = \sum_{n=1}^{N}(q_n - q_c)\,\Delta t_n \quad \le \quad G_F\left(\frac{f_F}{2g\sin\theta}\right) \quad \le \quad 492,000\,\text{m}^3/\text{s}^2\left(\frac{0.015}{2\left(9.814\,\text{m/s}^2\right)\sin(18.4^o)}\right)$$

or

$$V_{ET}(t) = \sum_{n=1}^{N}(q_n - q_c)\,\Delta t_n \quad \le \quad 1,190\,\text{m}^3/\text{m} \qquad \text{for } q_n > q_c$$

$$(26)$$

where G_F = 492,000 m^3/s^2 was used from Table 1. Equation (26) can be solved in a spreadsheet using Equation (24) for q_n, Equation (25) for q_c, and an appropriate incremental time (say Δt_n = 0.01 h = 36 s).

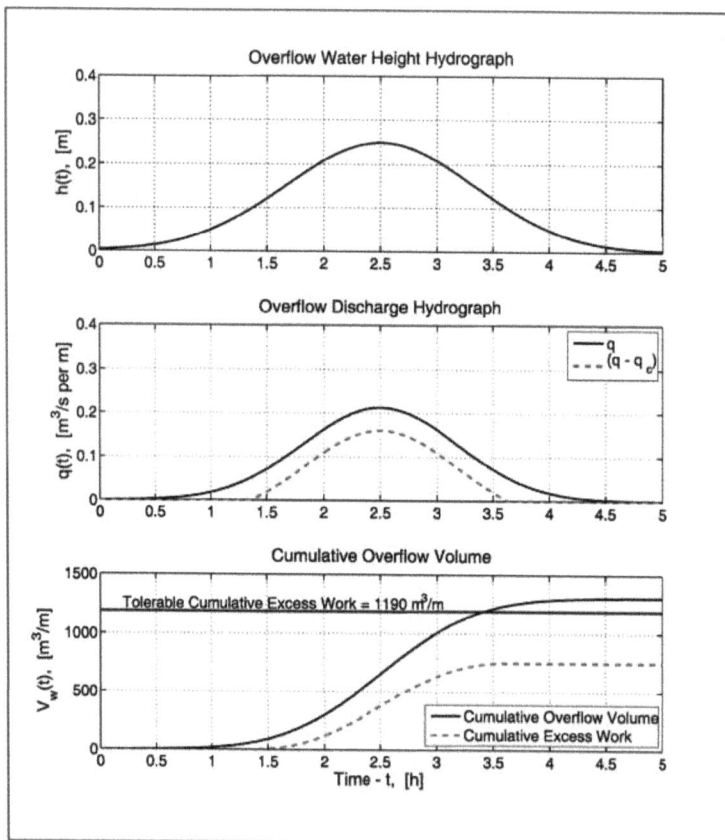

Figure 7. Example of cumulative excess work estimate for grass with u_c = 2.8 m/s.

Figure 6 presents the solution for this example plotted as a function of time (hours). The dashed line in the lower plot of Figure 6 shows the cumulative excess work remains just below the erosional limit for "good grass" suggested by Dean et al. [13]. For the peak water depth above the crest of h_p

= 0.25 m, shorter overflow duration would increase the safety margin and larger durations would exceed the tolerable erosional limit.

Note in this example with a critical velocity of only 1.8 m/s (Figure 6), there is not much difference between the cumulative excess work volume (dashed line in Figure 6) and the total cumulative overflow volume (solid line). Figure 7 shows the cumulative excess work estimated for a more robust embankment grass cover having a critical velocity of u_c = 2.8 m/s. All other parameters for this example are the same as used for the results shown in Figure 6.

Through trial and error it is possible to determine "*equivalence pairs*" of peak water depth and overflow duration (h_p, T_s) that just reach the tolerable limit for overflow. However, these equivalence pairs would only apply to grass of the same quality, same critical threshold velocity, same landward-side slope, and same friction factor subjected to an overflow discharge hydrograph similar to that given by Equation (18). The plotted lines in Figure 8 show equivalence pairs that result if the cumulative excess work is equal to the tolerable cumulative excess work of 1190 m^3/m. Iterative calculations were performed in which the duration was set to an integer hour, and the peak overflow depth was changed until the target cumulative excess work value was reach. All other parameters remained as given in this example. The two curves represent different values of the critical threshold velocity. Diagrams, such as those shown in Figure 8, can be used in risk analyses to assess the likelihood of grass failure associated with different projected flooding events. The estimated joint probabilities of peak surge and overtopping duration, as obtained from Monte Carlo storm surge simulations, could be used to assess whether or not the tolerable overflow limit has been surpassed. Bear in mind that grass failure is just the first step in dike or embankment failure, but a conservative criterion for tolerable overtopping would be no grass failure.

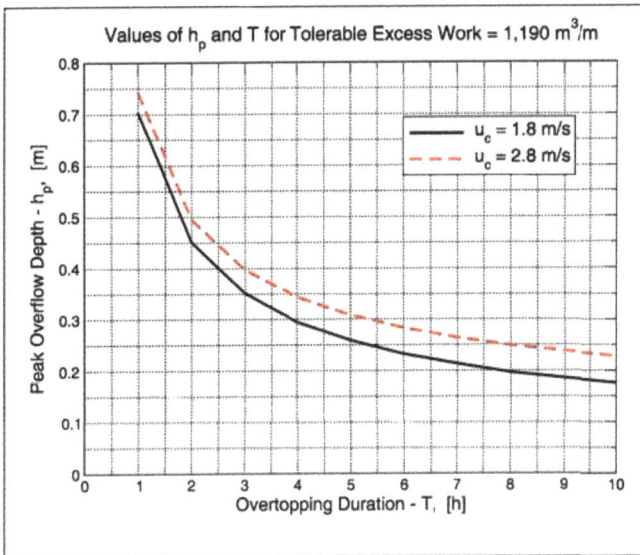

Figure 8. Equivalence pairs of peak overflow depth and overtopping duration that produce tolerable cumulative excess work of V_{ET} = 1190 m^3/m.

6. Summary and Conclusions

Engineers require estimates of tolerable overtopping limits for grass-covered levees, dikes, and embankments that might experience overflow due to storm surges or flooding. Realistic estimates can be used in both resilient design and in risk assessment. Hewlett et al. [12] presented design curves for grass-covered slopes in the form of steady overflow limiting velocity *versus* overflow duration. The curves were based on earlier full-scale overflow tests. The design curves were provided for three descriptions of grass cover (good, average, and poor), and these curves are used frequently for evaluation of steady overflow conditions on grass slopes. Dean et al. [13] analyzed the Hewlett et al. curves to determine the physical mechanism most likely to explain the curve shape. They concluded that flow work above a threshold velocity cubed (referred to as cumulative excess work) provided the best explanation. In this paper the concept of cumulative excess work for overflow on grass slopes is expressed in terms of slowly-varying overflow discharge (Equation 8) typical of realistic situations where the water level hydrograph above the structure crest increases to a maximum and then recedes similar to storm surges in estuaries and the passing of a flood wave in a river.

Whereas the cumulative excess work concept could be applied for any specified discharge hydrograph, this information may not be available for preliminary analyses. Hypothetical, but somewhat realistic, forms for the water level hydrograph above the embankment crest (Equation (15)) and the corresponding overflow discharge (Equations (17) or (18)) are given in terms of the values at maximum overflow (h_p or q_p) during the event and the total duration of overflow (T_s). An example application is given to illustrate use of the equations.

In the absence of additional grass and soil performance information, the values of critical threshold velocity, u_c, and the erosional limit, G_F, based on the Hewlett et al. [12] curves (given in Table 1) can be used. However, well-maintained grass with dense root systems can withstand greater cumulative overflow than the limits suggested by the Hewlett curves. Design of critical grass-protected structures should consider conducting full-scale overflow tests to establish reliable values for critical threshold velocity and tolerable cumulative excess work. Such testing could result in significant cost savings and decreased risk uncertainty.

Acknowledgments: The authors wish to thank Robert Ettema, Steven Apt, and the anonymous reviewers for their thoughtful comments and suggestions.

Author Contributions: Steven A. Hughes and Christopher I. Thornton conceived the analyses presented in this paper; and both authors contributed to the development. S. Hughes prepared the figures and wrote the paper.

Conflicts of Interest: The authors declare no conflict of interest.

References

1. ASCE/EWRI. Earthen embankment breaching, ASCE/EWRI Task Committee on Dam/Levee Breaching. *J. Hydraul. Eng.* **2011**, 137, 1549–1564.
2. Wahl, T.L. *Prediction of Embankment Dam Breach Parameters: Literature Review and Needs Assessment*; DSO-98-004, Dam Safety Research Report; U.S. Department of the Interior, Bureau of Reclamation, Dam Safety Office: Washington, DC, USA, 1998.
3. Temple, D.M.; Hanson, G.J.; Neilsen, M.L.; Cook, K.R. Simplified breach analysis model for homogeneous embankments: Part 1, background and model components. In Proceedings of the 25th Annual USSD (United States Society on Dams) Conference, Salt Lake City, UT, USA, 6–10 June 2005.
4. Hanson, G.J.; Cook, K.R.; Hunt, S.L. Physical modeling of overtopping erosion and breach formation of cohesive embankments. *Trans. Am. Soc. Agric. Eng.* **2005**, 48, 1783–1794.
5. Temple, D.M.; Hanson, G.J.; Neilsen, M.L. WINDAM—Analysis of overtopped earth embankment dams. In Proceedings of the ASABE Annual International Meeting, American Society of Agricultural and Biological Engineers, Portland, OR, USA, 9–12 July 2006.
6. Wu, W. Simplified physically based model of earthen embankment breaching. *J. Hydraul. Eng.* **2013**, 139, 837–851.
7. Sabbagh-Yazdi, S.-R.; Jamshidi, M. Depth-averaged hydrodynamic model for gradual breaching of embankment dams attributable to overtopping considering suspended sediment transport. *J. Hydraul. Eng.* **2013**, 139, 580–592.
8. Hanson, G.J.; Temple, D.M. Performance of bare-earth and vegetated steep channels under long-duration flows. *Trans. Am. Soc. Agric. Eng.* **2002**, 45, 693–701.

9. Hunt, S.L.; Hanson, G.J.; Cook, K.R.; Kadavy, K.C. Breach widening observations from earthen embankment tests. *Trans. Am. Soc. Agric. Eng.* **2005**, *48*, 1115–1120.
10. Hanson, G.J.; Temple, D.M.; Morris, M.W.; Hassan, M.A.; Cook, K.R. Simplified breach analysis model for homogeneous embankments: Part II, parameter inputs and variable scale model comparisons. In Proceedings of the 25th Annual USSD (United States Society on Dams) Conference, Salt Lake City, UT, USA, 6–10 June 2005.
11. Hewlett, H.W.M.; Boorman, L.A.; Bramley, M.E. *Design of Reinforced Grass Waterways*; CIRIA Report 116; Construction and Industry Research and Information Association: London, UK, 1987.
12. Whitehead, E.; Schiele, M.; Bull, W. *A Guide to the Use of Grass in Hydraulic Engineering Practice*; CIRIA Technical Note 71; Construction and Industry Research and Information Association: London, UK, 1976.
13. Dean, R.G.; Rosati, J.D.; Walton, T.L.; Edge, B.L. Erosional equivalences of levees: Steady and intermittent wave overtopping. *Ocean Eng.* **2010**, *37*, 104–113.
14. Hughes, S.A. *Adaptation of the Levee Erosional Equivalence Method for the Hurricane Storm Damage Risk Reduction System (HSDRRS)*; ERDC/CHL TR-11-3; U.S. Army Engineer Research and Development Center: Vicksburg, MS, USA, 2011.
15. Cialone, M.A.; Butler, L.; Amein, M. *DYNLET1 Application to FHWA Projects*; CERC-93-6; U.S. Army Engineering Research and Development Center: Vicksburg, MS, USA, 1993.
16. Zevenbergen, L.W.; Edge, B.L.; Lagasse, P.F.; Richardson, E.V. *Development of Hydraulic Computer Models to Analyze Tidal and Coastal Stream Hydraulic Conditions at Highway Structures, Phase III Report*; Research Project No. 591; South Carolina Department of Transportation: Columbia, SC, USA, 2002.
17. Xu, S.; Huang, W. An improved empirical equation for storm surge hydrographs in the Gulf of Mexico, U.S.A. *Ocean Eng.* **2014**, *75*, 174–179.
18. Henderson, F.M. *Open Channel Flow*; MacMillian Publishing Co.: New York, NY, USA, 1966.

Journal of
Marine Science and Engineering

MDPI

Article

Directional Storm Surge in Enclosed Seas: The Red Sea, the Adriatic, and Venice

Carl Drews

Atmospheric Chemistry Observations & Modeling, National Center for Atmospheric Research, P.O. Box 3000, Boulder, CO 80307, USA; drews@ucar.edu; Tel.: +1-303-497-1429; Fax: +1-303-497-1400

Academic Editor: Rick Luettich
Received: 30 March 2015; Accepted: 25 May 2015; Published: 29 May 2015

Abstract: Storm surge is dependent on wind direction, with maximum surge heights occurring when strong winds blow onshore. It is less obvious what happens when a port city is situated at the end of a long narrow gulf, like Venice at the northwestern end of the Adriatic Sea. Does the narrow marine approach to the port city limit the dangerous wind direction to a span of only a few degrees? This modeling study shows that the response in surge height to wind direction is a sinusoidal curve for port cities at the end of a long inlet, as well as for cities exposed along a straight coastline. Surge height depends on the cosine of the angle between the wind direction and the major axis of the narrow gulf. There is no special protection from storm surge afforded by a narrow ocean-going approach to a port city.

Keywords: storm surge; Red Sea; Adriatic Sea; Venice; COAWST; ROMS; wind direction; wind setdown

1. Introduction

Wind-driven storm surge can cause great damage and loss of life in coastal regions. In August 2005, Hurricane Katrina struck the New Orleans, Louisiana, area and directly caused 1500 fatalities [1] (p. 11). In November 2013 Super Typhoon Haiyan made landfall near Tacloban City in the Philippines, generating a 6 m surge that drowned thousands of people [2]. Numerical models such as ADCIRC and Coupled Ocean Atmosphere Wave Sediment Transport (COAWST) seek to simulate the effects of tropical cyclones mathematically, and thereby provide timely forecasts for a threatened population. A real-time forecasting system can give advance warning to emergency managers in time to take action and perhaps order an evacuation. Static analyses of a port city's vulnerability can assist civil authorities in their long-term planning.

COAWST is a modeling framework that integrates the atmospheric model WRF (Weather Research and Forecasting) with the ocean model ROMS (Regional Ocean Modeling System), the wave model SWAN (Simulating WAves Nearshore), and the Community Sediment Transport Model [3]. COAWST provides a wetting and drying algorithm that simulates inundation and retreat of the ocean from the original shoreline [4]. Prior studies have demonstrated that COAWST can simulate storm surge accurately in its two-dimensional mode, which runs faster than 3-D [5,6]. For this study COAWST/ROMS was driven by idealized winds, without coupling to WRF or SWAN.

Drews and Galarneau (2015) presented a new technique for analyzing the surge potential of a coastal city when the wind is blowing from eight different compass points [6]. The present study expands this directional analysis and applies it to oblong enclosed bodies of water. The objective is to determine how a port city at the end of a long inlet is affected by storm surge. Is the port protected from winds outside a narrow range of directions? Or does a long narrow sea channel the surge toward the downwind end? The study results are applicable to several large coastal cities, such as New York City, NY, USA; Bangkok, Thailand; and St. Petersburg, Russia.

The Red Sea is a long narrow body of water between northeast Africa and the Arabian Peninsula. The northern end splits around the Sinai Peninsula into the shallow Gulf of Suez to the west, and the much deeper Gulf of Aqaba to the east (Figure 1). The Red Sea proper (south of Sinai) is also deep. The Suez Canal terminates at the port of Suez, providing an important marine connection between the Mediterranean Sea and the Indian Ocean. Deep-draft vessels should recognize the effects of wind on water levels in the Gulf of Suez. Table 1 lists the dimensions of bodies of water included in this study.

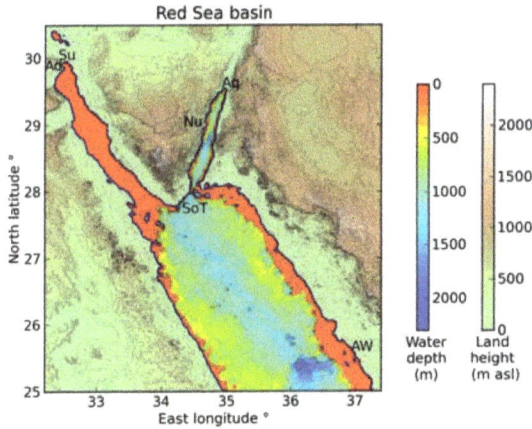

Figure 1. Red Sea domain. The northern section of the Red Sea splits into the shallow Gulf of Suez to the west (**left**); and the deeper Gulf of Aqaba to the east (**right**). The contour interval is 100 m. **Su** = Suez; **Ad** = Adabiya; **SoT** = Straits of Tiran; **Nu** = Nuweiba; **Aq** = Aqaba; **AW** = Al-Wajh. Sill depths are: Nuweiba 780 m, Enterprise Passage 250 m through the Straits of Tiran.

Table 1. Dimensions of seas and gulfs. The average depths are calculated by averaging all SRTM30 grid points below sea level. The Red Sea proper is south of the Sinai Peninsula and the Straits of Tiran. The modeled Red Sea domain extends south only to 25° North latitude.

Name	Length (km)	Width (km)	Average Depth (m)
Gulf of Suez	300	30	41
Gulf of Aqaba	180	18	546
Red Sea proper	1940	190	750
Adriatic Sea	790	130	218

Storm surge height increases with the fetch distance of open water along which the wind blows. For example, at Aqaba/Eilat the fetch distance is fairly short (~10 km) unless the wind direction is exactly aligned with the Gulf. The Gulf of Aqaba stretches 174 km from the Straits of Tiran to Aqaba. Figure 2 shows the fetch distance at Aqaba plotted according to compass direction. One might expect the surge height at Aqaba to display a similarly narrow spike when the wind blows directly out of Tiran.

The Adriatic Sea lies between Italy to the southwest and Croatia to the northeast. The UNESCO World Heritage City of Venice is situated in a lagoon at the northwestern end of the Adriatic, about 53 km north of the Po River delta (Figure 3). The Adriatic Sea becomes shallow toward Venice, and southeast winds cause flooding (*Acqua alta* in Italian) every few years. If sea level rises, more frequent high water could damage the city. What range of wind directions poses a danger to Venice?

29

Fetch Distance, Northern Red Sea

Straight-line distance to land

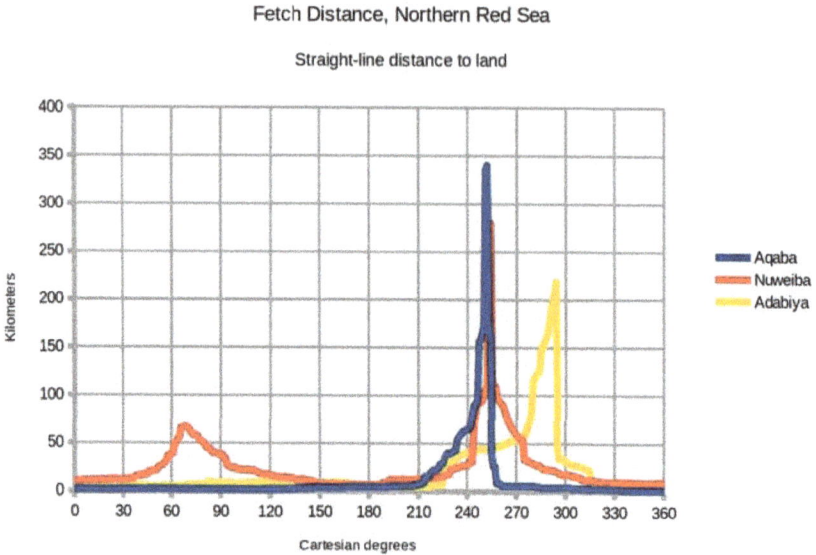

Figure 2. Fetch distance by compass direction at Aqaba; Nuweiba; and Adabiya (10 km southwest of Suez). 0° and 360° = east; 90° = north; 180° = west; 270° = south. The fetch is defined as the straight-line distance across water before hitting land.

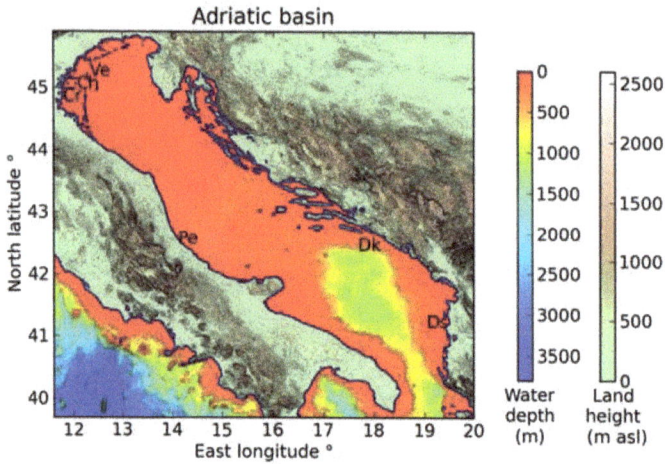

Figure 3. Adriatic Sea domain. The contour interval is 100 m. **Ve** = Venice; **Ch** = Chioggia; **Cr** = Crespino; **Pe** = Pescara; **Dk** = Dubrovnik; **Ds** = Durres. The SRTM30 terrain data shows the Po coastal plain as below sea level.

Figure 4 shows fetch distances along the coast of the Adriatic Sea, plotted according to compass direction. The fetch is the straight-line distance from the city across open water before hitting land, at the given angle. There are gaps in the fetch curve for Venice because small nearby islands block the fetch across the open sea.

Drews and Galarneau (2015) used angle increments of 45° for their directional analysis. The present study used increments of 10° to achieve more accurate resolution. I applied wind stress to

J. Mar. Sci. Eng. **2015**, *3*, 356–367

the model domain in a single direction, allowed the water's free surface to react to that wind stress over 24 h, and measured the resulting water level at several selected points along the coast. The wind direction was then rotated and the 24 h repeated, throughout the full 360° range of compass points. Winds blowing offshore generated wind setdown, a lowering of the water level, which is the opposite of storm surge.

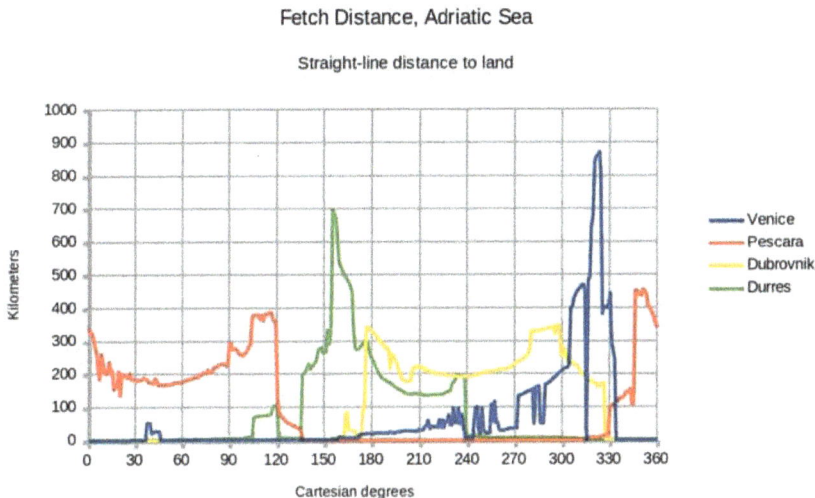

Figure 4. Fetch distance by compass direction for ports on the Adriatic Sea.

2. Results

Figure 5 shows the water level in the northern Red Sea when winds are blowing out of the southeast (from 300° Cartesian). This wind direction generates the maximum surge at Suez, since the incoming wind is aligned with the Gulf of Suez and the Red Sea proper along the longest possible fetch distance. The surge height is 1.68 m at Suez and 1.54 m at Adabiya. Table 2 shows the site names and locations for recording the simulated water level.

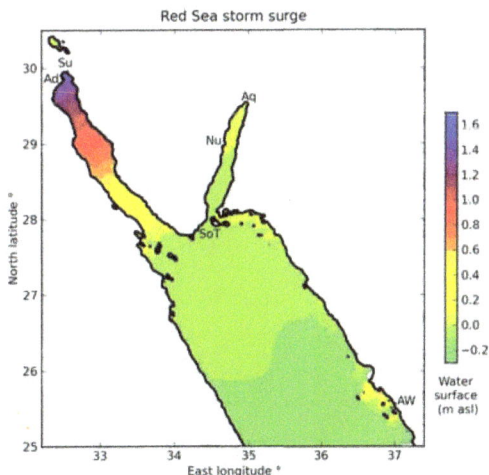

Figure 5. Red Sea surge height with wind blowing from southeast (300° Cartesian).

Table 2. Site names and locations for the Red Sea and Adriatic Sea. Latitude and longitude are in decimal degrees north and east. The maximum simulated surge heights are in meters, and the wind directions are given in Cartesian degrees.

Name	Latitude	Longitude	Country	Max Surge	Direction
Aqaba	29.5397	34.9764	Jordan	0.055	230
Nuweiba	28.9838	34.7576	Egypt	0.026	220
Tiran	27.9858	34.4444	Egypt/Saudi Arabia	0.034	100
Al-Wajh	25.5467	36.9883	Saudi Arabia	1.43	205
Adabiya	29.8885	32.5486	Egypt	1.54	300
Suez	29.9449	32.5486	Egypt	1.68	300
Venice	45.4240	12.3505	Italy	2.02	320
Chioggia	45.2304	12.2729	Italy	1.94	330
Crespino	44.9834	11.9167	Italy	2.72	350
Pescara	42.4751	14.2251	Italy	0.57	70
Dubrovnik	42.6405	18.0942	Croatia	0.36	170
Durres	41.2620	19.4988	Albania	0.74	190

Simulation experiment RS11 extends the Red Sea domain southward from 25° to 20° North latitude. This change increases the storm surge height to 1.77 m at Suez and 1.62 m at Adabiya. Adding 555 km of fetch to the Red Sea increases the storm surge at Suez by less than 10 cm because the Red Sea proper is very deep and does not generate much surge. The original southern boundary of 25° North latitude is satisfactory for the present modeling study.

Figure 6 shows the water level when winds are from the southwest (from 230° Cartesian). This direction generates the maximum possible surge at Eilat/Aqaba, since the wind is aligned with the Gulf of Aqaba. The surge height is 0.055 m at Aqaba, 0.025 m at Nuweiba, and −0.014 m at the Straits of Tiran. Notice the extreme surge height (1.3 m) at Al-Wajh, a shallow shelf along the Saudi Arabian coast. Al-Wajh offers an exposed point on a straight shoreline to compare with the enclosed ports of Suez and Aqaba.

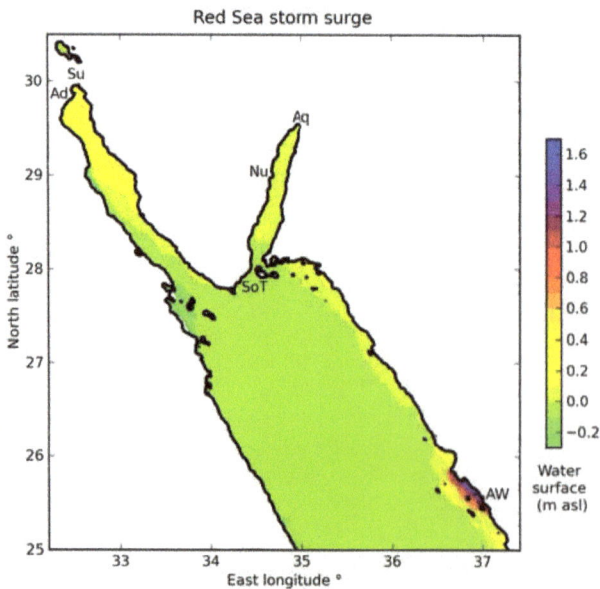

Figure 6. Red Sea surge height with wind from southwest (230° Cartesian). Note the extreme storm surge at Al-Wajh.

The directional analysis plots the surge height and wind setdown as a function of the wind angle. Figure 7 displays the results of experiment RS8 for Suez, Adabiya, and Al-Wajh. The lowest water levels reached are 1.72 m below sea level for Suez, 1.57 m below sea level for Adabiya, and 2.58 m below sea level for Al-Wajh. Figure 8 displays the results of RS8 for Aqaba, Nuweiba, and the Straits of Tiran; the vertical scale is different because the changes in water level are so much smaller in the Gulf of Aqaba. The lowest water levels reached are 0.050 m below sea level for Aqaba, 0.020 m below sea level for Nuweiba, and 0.032 m below sea level for Tiran. Table 3 lists the simulation experiments performed for the Red Sea and the Adriatic Sea.

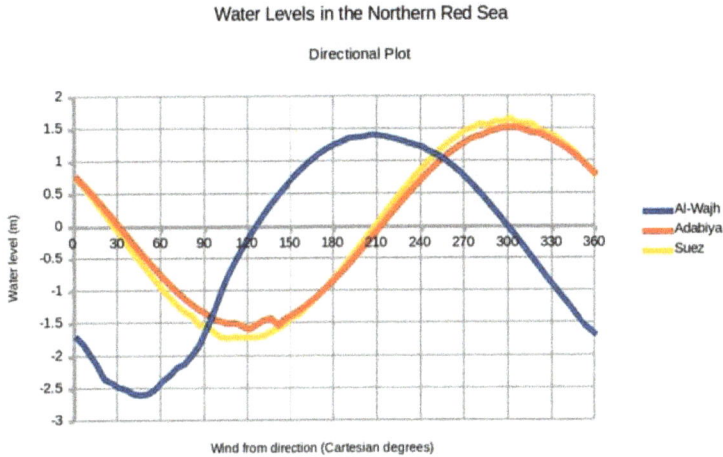

Figure 7. Directional plot for Suez, Adabiya, and Al-Wajh. Cartesian degrees are: 0° and 360° = east; 90° = north; 180° = west; 270° = south.

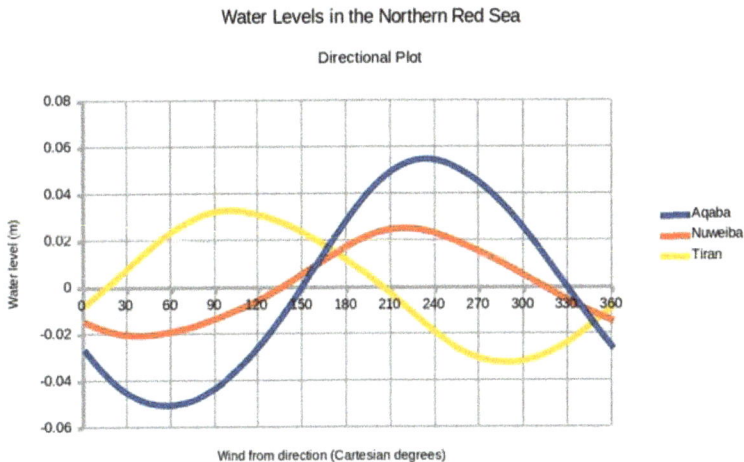

Figure 8. Directional plot for Aqaba, Nuweiba, and Tiran.

The Suez Canal Authority is responsible for the safe navigation of ships that transit the canal. The extreme tidal range is 0.65 m in the north at Port Said, and 1.9 m in the south at Suez (peak-to-peak) [7]. Strong and sustained winds blowing from the southeast or northwest would approximately add to these tides. Using 1.7 m as the maximum change in water level for storm surge and wind setdown

(Figure 7), the water level at Suez could drop to 2.65 m below mean sea level when setdown coincides with an extreme low tide. Suezmax deep-draft vessels should be made aware of such conditions. Both surge and setdown would cause unusually strong currents within the canal when they coincide with extreme tides.

Table 3. Simulation experiments performed.

Number	Domain	Purpose
RS8	Red Sea	Directional analysis
RS9	Red Sea	Shallow Red Sea proper
RS10	Red Sea	Shallow Gulf of Aqaba
RS11	Red Sea	Domain extended southward
A2	Adriatic	Directional analysis

Figure 9 shows water levels in the Adriatic Sea when wind blows from the southeast. Storm surge at Venice reaches its maximum height of 2.02 m when wind is blowing from the southeast (from 320° Cartesian). The maximum wind setdown of 2.28 m below sea occurs at Venice when wind is blowing from the northwest (from 140° Cartesian). Wind setdown is slightly greater in magnitude than storm surge because shallow waters are more affected by wind stress.

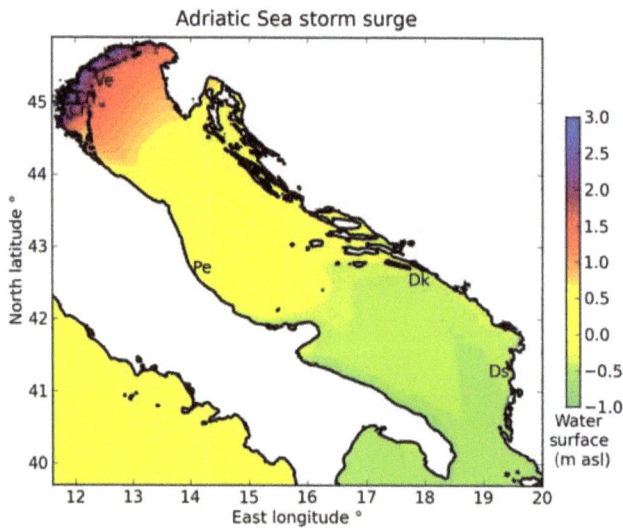

Figure 9. Adriatic surge height with wind from southeast (320° Cartesian).

The highest water level recorded in Venice occurred on November 4, 1966; the Adriatic Sea rose to 1.94 m above mean sea level [8]. The maximum surge height of 2.02 m calculated by the directional analysis exceeded this observed maximum by 4%, demonstrating that these model results are realistic simulations of extreme surge events at Venice.

Figures 10 and 11 display the results of experiment A2 for ports on the Adriatic Sea; Figure 10 shows the directional analysis for Venice, Chioggia, and Crespino. Crespino is a village on the Po River 52 km inland from the river's mouth. The ocean model and SRTM30 data calculate extensive inland flooding of the Po coastal plain when wind blows from the southeast. The curve for Crespino is flattened on the bottom because water takes longer to drain out the Po river channel than to drain from the Venetian lagoon.

The maximum surge at Venice, and therefore the maximum vulnerability of the city to flooding, occurs when the wind blows from the southeast at 320° Cartesian. Surge height remains over 90% of the peak value for wind directions between 300° and 340°, or within ±20° of the "optimum" wind direction (Figure 10). Venice is vulnerable to storm surge driven by winds blowing from directions substantially different than the theoretical optimum of 320°.

Water Levels in the Adriatic Sea

Directional Plot

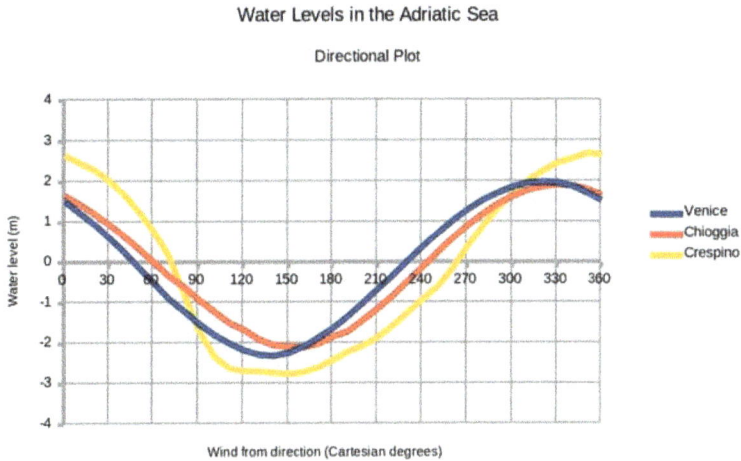

Figure 10. Directional plot for Venice, Chioggia, and Crespino.

Figure 11 shows the results of the directional analysis for Pescara (Italy), Dubrovnik (Croatia), and Durres (Albania). Pescara was chosen as an exposed port city along a straight coastline; it serves as a comparison for the sheltered location of Venice. Water levels range from −0.58 to 0.57 m at Pescara, from −0.37 to 0.36 m at Dubrovnik, and from −0.78 to 0.74 m at Durres.

Water Levels in the Adriatic Sea

Directional Plot

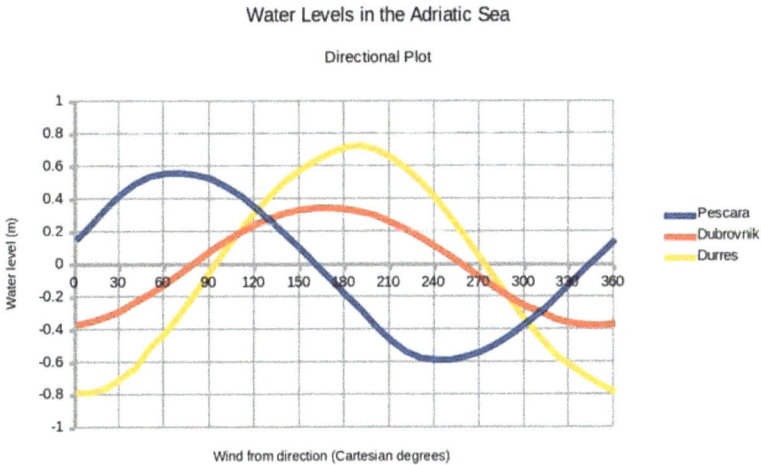

Figure 11. Directional plot for Pescara, Dubrovnik, and Durres.

Plots of the directional analysis reveal an interesting result: There is no significant narrowing of the peak for port cities at the end of a long inlet. The plotted curves of water level as a function of wind angle are all sinusoidal. There is no qualitative difference between protected coastal sites and the exposed sites. Storm surge at Suez is affected by wind direction in the same way as storm surge at Pescara.

Why are surge heights so much smaller for the Gulf of Aqaba than for the Gulf of Suez? Part of the reason is that the Red Sea proper is aligned with the Gulf of Suez, and therefore water levels in the Red Sea tend to enhance the surge at Suez and diminish the surge at Aqaba. But the dominant reason is that the bathymetry of Aqaba is too deep to generate significant storm surge or wind setdown. Experiment RS10 tests this hypothesis by reducing the depth of the Gulf of Aqaba by 95%, making it comparable to the Gulf of Suez. The average depth becomes 48 m. When the bathymetry northeast of Tiran is modified in this manner, the water levels vary from −0.73 m to 0.73 m at Aqaba, from −0.42 to 0.42 at Nuweiba, and from −0.025 to 0.031 at the Straits of Tiran. Shallow water surges more. Figure 12 shows the values of surge height and wind setdown that result from a range of shallow depths for the Gulf of Aqaba. Wind setdown is comparable in magnitude to storm surge.

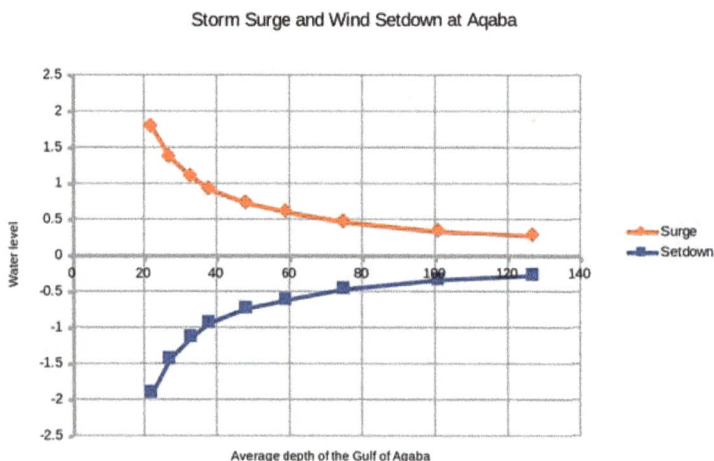

Figure 12. Storm surge and wind setdown at Aqaba when the Gulf of Aqaba is made shallow.

To further illustrate the effect of water depth on storm surge, experiment RS9 reduced the depth of the Red Sea proper (south of Sinai) and measured the resulting water levels at Adabiya for wind angles of 120° and 300° (Cartesian). Storm surge at Adabiya rose to 1.95 m above sea level, and wind setdown dropped to 2.01 m below sea level. Thus the water displacement for a shallow Red Sea proper increased by 27% and 28%, respectively.

3. Materials and Methods

COAWST is configured to run in its two-dimensional mode, without coupling to WRF or SWAN. The air-to-sea drag coefficient is described by Oey et al. [9] To compensate for 2-D mode and the lack of waves, the calculated wind stress is multiplied by the same factor of 1.63 used in previous studies for Lake Erie [5] and the Atlantic coast of the United States [6]. The quadratic bottom drag coefficient (RDRG2) for ROMS is 1.0×10^{-3}. The ocean model does not include tides. The critical depth for wetting and drying is 0.5 m, and the time step is 2.0 s. Any open domain boundaries that are not land surface are no-slip walls.

The Shuttle Radar Topography Mission supplied bathymetry and topography at a grid resolution of 30 arc-seconds (SRTM30) [10]. Grid cells for the Red Sea domain are about 900 m wide, and 800 m wide for the Adriatic Sea.

The wind speed ramps up from 0 to 28 m/s over a period of 24 h. This ramp-up time is necessary to avoid seiche oscillations within the enclosed seas. 28 m/s approximates a medium-strength tropical storm on the Saffir-Simpson scale of hurricane winds. Sea level measurements are recorded at the end of 24 h. The wind field for each test case is uniform in speed and direction across the entire domain. The wind direction rotates around the compass in increments of 10°, producing 36 test cases for each full experiment in the directional analysis.

Table 3 lists all the model experiments. The Gulf of Aqaba and the Red Sea proper were reduced in depth for experiments RS10 and RS9, respectively. Bathymetry deeper than the 25 m isobath was reduced by 95% according to the following transformation:

$$newDepth = ((originalDepth - 25.0) \times 0.05) + 25.0 \tag{1}$$

Underwater values in Equation (1) are positive. This transformation makes the Gulf of Aqaba and the Red Sea proper comparable in depth to the Gulf of Suez, while retaining some of the bottom roughness. Experiment RS10 replaces the fraction 0.05 in Equation (1) with a range of values from 0.00 to 0.20 (see Figure 12).

4. Discussion and Conclusions

The directional analysis proposed by Drews and Galarneau [6] has been implemented here in angle increments of 10° in order to determine the potential of a coastal city for storm surge with respect to wind direction. For the cases studied here, the maximum and minimum water levels were generated by angles differing by 180°, and the response across all compass directions closely approximated a sinusoidal curve. Ports at the end of a long narrow inlet exhibit this response because the wind stress on the inlet acts in proportion to the cosine between the wind direction and the long axis of the inlet.

Storm surge maps are commonly created by simulating a series of hurricane tracks making landfall, then combining the maximum surge heights into a single map [6]. The directional analysis provides another way of looking at the risk posed by tropical cyclones and other windstorms. The directional plot reveals at a glance what is the most dangerous wind direction for storm surge, and the opposite angle that will interfere with ship navigation when water levels drop.

This paper examined the influence on storm surge of ocean depth, fetch distance, and wind direction. (The wind speed was left uniform across all experiments.) Shallow waters are more susceptible to storm surge than deep ocean basins. In general, a longer fetch will produce higher storm surge, but the exact relationship is complex and dependent on the local geography. Changes in wind direction produced a simple curve resembling the cosine function for the cases studied here.

Acknowledgments: The National Center for Atmospheric Research provided computational support. NCAR is sponsored by the National Science Foundation. Any opinions, findings and conclusions or recommendations expressed in the publication are those of the author and do not necessarily reflect the views of the National Science Foundation. I thank Yu-heng Tseng and Thomas Galarneau for their manuscript review and comments.

Author Contributions: Carl Drews conducted the research and wrote the paper.

Conflicts of Interest: The author has written and published a book titled "Between Migdol and the Sea" (Carl Drews, 2014) about the Hebrew Exodus from Egypt.

References

1. Knabb, R.; Rhome, J.; Brown, D. Tropical Cyclone Report Hurricane Katrina 23–30 August 2005. Available online: http://www.disastersrus.org/katrina/TCR-AL122005_Katrina.png (accessed on 27 March 2014).
2. Teves, O.; Bodeen, C. Typhoon Haiyan Storm Surges Caught Philippines by Surprise. Available online: http://www.weather.com/news/weather-hurricanes/typhoon-haiyan-storm-surges-20131111 (accessed on 11 November 2013).
3. Warner, J.C.; Armstrong, B.; He, R.; Zambon, J.B. Development of a coupled ocean-atmosphere-wave-sediment transport (coawst) modeling system. *Ocean Model.* **2010**, *35*, 230–244. [CrossRef]
4. Drews, C.; Han, W. Dynamics of wind setdown at Suez and the Eastern Nile Delta. *PLoS ONE* **2010**, *5*, e12481. [CrossRef] [PubMed]

5.   Drews, C. Using wind setdown and storm surge on lake erie to calibrate the air-sea drag coefficient. *PLoS ONE* **2013**, *8*, e72510. [CrossRef] [PubMed]
6.   Drews, C.; Galarneau, T.J.J. Directional analysis of the storm surge from hurricane sandy 2012, with applications to charleston, new orleans, and the philippines. *PLoS ONE* **2015**, *10*, e0122113. [CrossRef] [PubMed]
7.   Authority, S.C. About Suez Canal. Available online: http://www.suezcanal.gov.eg/sc.aspx?show=17 (accessed on 27 March 2015).
8.   di Venezia, C. 4 Novembre 1966. Available online: http://www.comune.venezia.it/flex/cm/pages/ServeBLOB.php/L/IT/IDPagina/2053 (accessed on 25 March 2015).
9.   Oey, L.Y.; Ezer, T.; Wang, D.P.; Fan, S.J.; Yin, X.Q. Loop current warming by hurricane wilma. *Geophys. Res. Lett.* **2006**, *33*, L08613. [CrossRef]
10.   Survey, U.S.G. Shuttle Radar Topography Mission. Available online: http://dds.cr.usgs.gov/srtm/version2_1/ (accessed on 28 July 2014).

Journal of
Marine Science and Engineering

MDPI

Article

Exploring Water Level Sensitivity for Metropolitan New York during Sandy (2012) Using Ensemble Storm Surge Simulations

Brian A. Colle [1,*], Malcolm J. Bowman [1], Keith J. Roberts [1], M. Hamish Bowman [2], Charles N. Flagg [1], Jian Kuang [1], Yonghui Weng [3], Erin B. Munsell [3] and Fuqing Zhang [3]

[1] School of Marine and Atmospheric Sciences, State University of New York, Stony Brook, New York, NY 11794-5000, USA; malcolm.bowman@stonybrook.edu (M.J.B.); keithrbt0@gmail.com (K.J.R.); charles.flagg@stonybrook.edu (C.N.F.); kuangjian2011@gmail.com (J.K.)

[2] Department of Geology, Otago University, Dunedin 9054, New Zealand; hamish.bowman@gmail.com

[3] Department of Meteorology, The Pennsylvania State University, University Park, PA 16802, USA; yhweng@psu.edu (Y.W.); munsell.erin@gmail.com (E.B.M.); fzhang@psu.edu (F.Z.)

* Author to whom correspondence should be addressed; brian.colle@stonybrook.edu; Tel.: +1-631-632-3174.

Academic Editor: Rick Luettich

Received: 20 May 2015; Accepted: 8 June 2015; Published: 19 June 2015

Abstract: This paper describes storm surge simulations made for Sandy (2012) for the Metropolitan New York (NYC) area using the Advanced Circulation (ADCIRC) model forced by the Weather Research and Forecasting (WRF) model. The atmospheric forecast uncertainty was quantified using 11-members from an atmospheric Ensemble Kalman Filter (EnKF) system. A control WRF member re-initialized every 24 h demonstrated the capability of the WRF-ADCIRC models to realistically simulate the 2.83 m surge and 4.40 m storm tide (surge + astronomical tide) above mean lower low water (MLLW) for NYC. Starting about four days before landfall, an ensemble of model runs based on the 11 "best" meteorological predictions illustrate how modest changes in the track (20–100 km) and winds (3–5 m s^{-1}) of Sandy approaching the New Jersey coast and NYC can lead to relatively large (0.50–1.50 m) storm surge variations. The ensemble also illustrates the extreme importance of the timing of landfall relative to local high tide. The observed coastal flooding was not the worst case for this particular event. Had Sandy made landfall at differing times, locations and stages of the tide, peak water levels could have been up to 0.5 m higher than experienced.

Keywords: storm surge; hurricane Sandy; ADCIRC; predictability; WRF; EnKF; ensemble; Battery; New York City; coastal flooding

1. Introduction

1.1. Background

Hurricane Sandy, which struck Metropolitan New York (NYC), New Jersey (NJ), and Long Island (LI) on 29 October 2012, resulted in 72 deaths and over $50 billion dollars (US$ 2012) in property damage [1]. Sandy began as a disturbance off the African coast on 11 October 2012 [1], and developed into a major hurricane at 06:00 UTC, 25 October, just south of Cuba.

An upper-level trough moving eastward along the US-Canadian border helped steer Sandy northward, while a blocking ridge downstream over the northern Atlantic ultimately forced Sandy westward towards the New Jersey coast [2]. After some weakening over Cuba due to topography, Sandy fell below hurricane strength. As Sandy moved northward along the U.S. east coast, the interaction of the upper-level trough and the warm Gulf Stream sea surface temperatures (SSTs) resulted in a re-intensification to a category 1. Additionally, the radial extent of tropical storm force winds around Sandy increased to 2300 km at 12:00 UTC, 28 October, as cold air wrapped around the vortex and formed a warm secluded frontal structure [2].

The National Weather Service (NWS) declassified Sandy as a hurricane just before landfall (21:00 UTC, 29 October) [3], but its unprecedented large size, slow progression, and track resulted in record-high storm surges along coastal New York and New Jersey [1]. The storm tide (surge + tide) for Sandy was around 4.40 m above mean lower low water (1.0 m; MLLW; 1983–2001) at the southern tip of Manhattan Island, which was the largest storm-tide in recorded history (150 year) at this location.

The sustained winds (36 m s^{-1}) were category 1 hurricane-equivalent (minimum pressure 945 hPa) when Sandy made landfall over southern New Jersey [1]. Tropical storm force winds extended over 500 km to the north and east of the system (see NHC Best Track 2014 Figure 1), and the significant wave height peaked at around 10 m at National Data buoy 44025 60 km south of Long Island [4]. The official National Hurricane Center (NHC) track errors for Hurricane Sandy were about 50.0% lower than the long-term mean track error for 48-h to 96-h track forecasts, which demonstrates that Sandy's operational forecast was relatively accurate [1].

Experimental real-time forecasts during Sandy, including the Pennsylvania State University (PSU) Weather Research and Forecasting (WRF)—Ensemble Kalman Filter (EnKF) system that assimilated airborne Doppler radar observations [5,6] were analyzed. At a forecast lead-time four to five days prior to landfall, nearly 18.0% of the PSU WRF-EnKF ensemble members did not predict United States landfall. Munsell and Zhang [6] showed that the uncertainty in the initial environmental steering-level flow over the tropics and Sandy's subsequent interaction with the approaching mid-level trough resulted in much of the spread of the Sandy track forecasts.

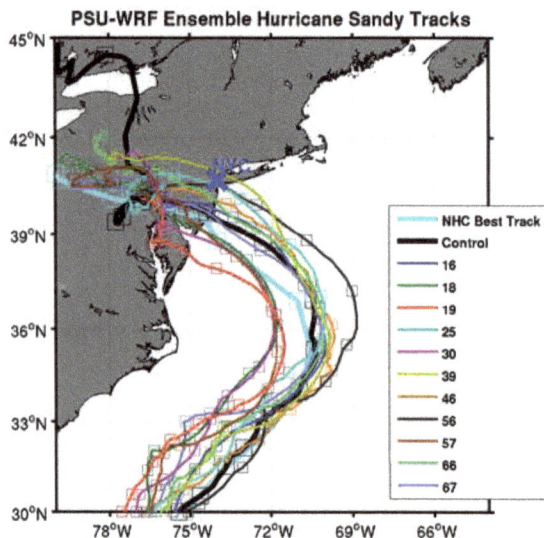

Figure 1. Track of hurricane Sandy for the best track (bold cyan), control WRF member (bold black), and 11 ensemble WRF members (see legend) starting at 00:00 UTC, 26 October 2012 (symbols plotted every 12 h). The location of The Battery tide gauge for the verification is shown at the NYC location (blue cross).

Even for those tracks that predicted landfall within about 150 km of observed, the operational surge models in the region, including the NOAA extra tropical surge model [7,8], Stevens Institute NYHOPS (SIT-NYHOPS, Hoboken, NJ, USA; [9], and the Stony Brook Surge Model [10,11], under-predicted the surge by 0.50 to 1.0 m starting at a lead time of 48 h (not shown).

The SIT-NYHOPS under-prediction was at least partially the result of inferior wind forecasts from the operational models [12]. Forbes et al. [13] showed that the NWS Sea, Lake, and Overland

Surges from Hurricanes (SLOSH) model [14] could realistically simulate the Sandy water levels using an ensemble of tracks and intensities from the National Hurricane Center. Both Forbes et al. [13] and Georgas et al. [12] illustrated that Sandy's observed flooding was not a worst-case scenario for this event given the variations in Sandy's track and phase of the tide landfall that might have occurred.

1.2. Motivation

Sandy's atmospheric predictability has been extensively discussed in [6], but little information exists on how small changes in the track and intensity of Sandy might have changed regional storm surges. Forbes et al. (2014) [13] showed a large ensemble spread of surge predictions using SLOSH for Sandy around NYC, but did not discuss the origin of that water level spread. Wind and pressure forcing for SLOSH are geometrically circular, thus unrepresentative of the large asymmetries observed during Sandy. Georgas et al. (2014) [12] illustrated the tidal uncertainty aspect of Sandy's water level prediction, but did not run SIT-NYHOPS days before landfall with an ensemble of forecasts.

Our study's goal is not to fully quantify the storm surge probabilities of Sandy since only a subset of WRF forecasts from the PSU EnKF system were available for analysis. Rather, using a more limited ensemble of ~11 members, this paper addresses:

- How well can a storm surge ocean model predict Sandy's surge if the track, winds and sea level pressure (SLP) from Sandy are reasonably well predicted?
- How does a shift in the track, timing (*vis-à-vis* tides), and wind intensity for Sandy translate into predicted storm-tide changes for the New York coastline?
- Given the ensemble envelope of storm surge predictions/timing uncertainties, how much worse might the flooding have been?

2. Data and Methods

Munsell and Zhang [6] presented the details of the WRF-EnKF Sandy runs, extending the analysis of Weng and Zhang [5] and Zhang et al. [15]. The EnKF system used an ensemble of WRF forecasts to estimate flow-dependent background error covariance for the data assimilation cycling. It assimilated NOAA hurricane hunter aircraft Doppler velocity observations to create a 60-member ensemble starting at 00:00 UTC, 26 October 2012, using the same model physics of Munsell and Zhang [6]. Three domains were used (27-, 9-, and 3-km resolution), with the inner two nested domains centered around and following the storm (dimensions of 2700×2700 km and 900×900 km, respectively, with 44 vertical levels and a model top at 10 hPa).

NOAA's operational global forecast system (GFS) [16] was used for initial and boundary conditions, with the WRF ensemble initialized 6 to 12-h before the Doppler observations were ingested. EnKF analyses were used every 12-h to generate successive WRF forecasts. The WRF runs every 24-h (starting at 00:00 UTC, 25 October 2012) were used to construct a "control" member, in which the first 24-h from each run was used to force the ADCIRC [17].

From the 60-member ensemble starting at 00:00 UTC, 26 October, a set of 11 members was chosen that were subjectively judged as the "best" performing members. Here "best" means that the track made landfall anywhere from coastal NY southward to the southern Maryland coastline (Figure 1). We chose these relatively accurate members to quantify how relatively small changes in the track can change NYC surge predictions, but not to determine the overall probability of the event with a 4-day lead-time.

The grid for ADCIRC extends from the eastern seaboard from Central Florida to Nova Scotia, Canada and eastward into the central North Atlantic Ocean in a semi-circular arc (Figure 2; 184,534 node points). This grid greatly improved near-shore coastal bathymetric resolution to the grid used previously by DiLiberto et al. [11] and extended much further along the eastern seaboard in both directions. A decision was made not to extend the grid further as then the ocean model's grid would have been considerably larger that the overlying WRF grid and various tests on smaller grids showed that the final range chosen was sufficient.

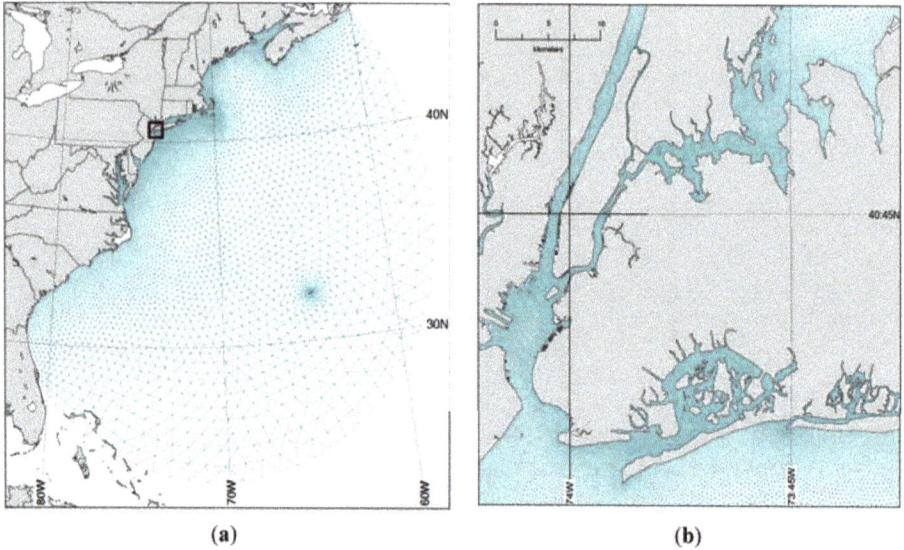

Figure 2. (a) Domain of the ADCIRC model, extending from mid State of Florida (FL) to the Atlantic coast of Nova Scotia, Canada (NS), indicated by the black nodes. (b) Zoom in for the small black box in (a) for the region around NYC and western Long Island. The sizes of elements range from ~7 m in inner bays and estuaries to ~70 km along the outer ocean boundary. The grid has 184,534 nodes.

ADCIRC is usually run in a 2-D vertically integrated mode for fastest operation for surge prediction studies [18,19]. It also has a 3-D mode, which allows for vertical shear in the horizontal tidal and wind-driven currents. (Note that this is not a baroclinic mode *per se* since water density is not specified and the layers are spaced equidistant between the surface and the bottom at each node). The operational version for the NYC region, described by Colle et al. [20] and DiLiberto et al. [11], is normally run in a 2-D (barotropic) configuration with no wave coupling [10]. However, for this analysis of Sandy storm surges, ADCIRC was run with 1, 3, 5, and 7-vertical layer configuration coupled with the Simulating Waves Nearshore (SWAN) wave generation model [21]. Orton et al. [22] showed that stratification can be important for this region, since ignoring it can lead to some surge underprediction, but this process was not included in our ADCIRC runs. SWAN ingests WRF 10-m wind predictions to create radiation wave stresses, which act as additional terms in the momentum equations that characterize coastal setup due to breaking waves along and near the shoreline. SWAN allows for shoaling, refraction and diffraction (including frequency shifting due to currents). Both ADCIRC and SWAN were run on the same non-structured triangular mesh grid with 184,534 node points. The ADCIRC time step was 4 s, while SWAN model ran stably with a 5-min time step.

ADCIRC was spun up two days before 00:00 UTC, 26 October, using the M2, K1, O1, N2, S2, M4, M6, S4, and MS4 tidal constituents [23] along the open boundary (Figure 2). Starting at 0000 UTC 25 October, 10-m winds from an earlier WRF-EnKF control forecast were taken every 15 min from the 9- and 3-km WRF domains to derive ADCIRC wind stresses using the Garratt [24] drag formulation without a cap since there is considerable uncertainty in regard to what the actual cap should be (e.g., Weisberg et al. [25]).

Between 00:00 UTC, 25 October, and 00:00 UTC, 26 October, the surface wind and pressures within ADCIRC were linearly increased during this final one-day spin up period. The 9-km WRF domain was used for the surface wind stresses and surface pressures for those portions of the ADCIRC domain lying outside the 3-km WRF domain. The observed storm-tide and storm surge levels at The Battery (NYC in Figure 1) were taken from NOAA [26].

3. Results

3.1. Control Run Surge Simulations

Figure 1 shows the National Hurricane Center (NHC) best track (bold cyan), the control member (bold black), and the 11 WRF ensemble members starting at 00:00 UTC, 26 October 2012. The control (CTL) run, which was restarted every 24 h, as described above, realistically predicted Sandy's track (the modeled storm made landfall +/−24 h to the observed landfall time and +/−300 km to the observed landfall location). The CTL provides one of the more realistic simulations as compared to the other ensemble members given its 24-h restarts, which minimize the growth of initial condition error. Starting at 12:00 UTC, 29 October, Sandy's track was directed northwestward towards the southern NJ coast.

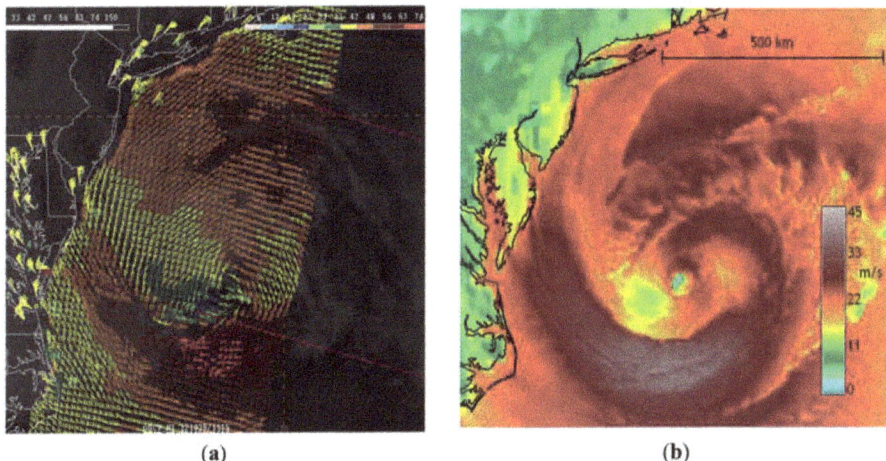

Figure 3. (a) Scatterometer surface winds (colored in kts and full barb = 10 kts) at 12:00 UTC, 29 October 2012; (b) Same as (a) except for the control WRF run (hour 12 forecast).

Figure 3 shows a 12-h forecast from the CTL run valid at 12:00 UTC, 29 October, as well as surface winds derived from Indian Space Research Organization's (IRSO) Oceansat-2 scatterometer [27]. The areal extent of Sandy (with tropical storm force winds >18 m s^{-1}) extended >500 km NE from the center. The strongest winds were to the south of center within a developing bent-back front (Galarneau et al [2]).

The Battery surge gradually increased from 0.30 to 1.0 m between 00:00 UTC, 28 October, to 12:00 UTC, 29 October (Figure 4a). The CTL underestimated this gradual increase in surge on the 28 October by 0.20–0.30 m, which was a consequence of previous WRF forecasted inaccuracies. The observed peak surge of 2.83 m occurred at 02:00 UTC, 30 October, whereas the simulated peak surge also occurred around that time at ~3.0 m, but with some members predicting a peak surge about 12-h too soon.

The peak storm tide (surge + astronomical tide) of around 4.30 m (MLLW) was predicted within 0.20 m by the CTL run (Figure 3b). Sandy's peak surge occurred close to a local high tide and a bi-weekly spring tide. If Sandy had made landfall about 6 h earlier, around low tide, the storm tide would have been around 3.0 m (MLLW), flooding would have been much less severe and comparable to the December 1992 nor-easter [20].

The ADCIRC was verified at several other stations around the region (not shown). The storm tide was well simulated (within 10% of the observations) at Bergen Point near the Battery, as well as to the south along the southern New Jersey coast (e.g., Atlantic City), while the ADCIRC peak water levels were 0.5 to 1.0 m too low over western and eastern Long Island Sound stations. This underprediction was mainly the result of the CTL storm arriving 1–2 h too early when the tide was rapidly increasing. This 1–2 h timing error was less critical for the Battery, since it was around a high tide.

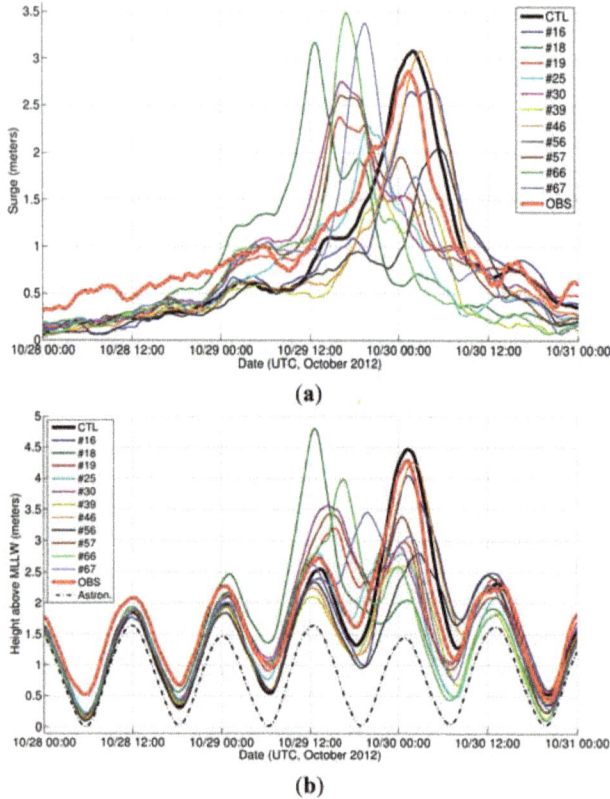

Figure 4. (**a**) Storm surge (in meters) at The Battery, NYC for the observed (bold red), control WRF-ADCIRC member (bold black), and 11 ensemble WRF-ADCIRC members (see legend) from 00:00 UTC, 28 October 2012, to 00:00 UTC, 31 October; (**b**) Same as (**a**) except for the total water level (surge + tide) in meters above mean lower low water (MLLW; 1.0 m; 1983–2001).

3.2. ADCIRC 2D and 3D Simulations

Water level simulations were sensitive to the number of ADCIRC vertical levels used. When ADCIRC was run in a barotropic 2-D (one vertical level) mode (e.g., DiLiberto et al. [11]), the surge was consistently under-forecasted by around 0.50 m and was delayed by a few hours (Figure 5). Weisberg and Zheng [28] showed that the 2D mode may overestimate the bottom stress and thus underestimate the surge.

3.3. Flooding Map for Sandy

Choosing three levels significantly improved the forecast but, when 5-layers were used, there was a modest change in the simulation of peak surge. Further, increasing the vertical resolution to seven levels did not significantly change the peak surge simulation (not shown). Additionally, the use of SWAN did not significantly alter the peak simulation of surge (not shown), which may be a consequence of the sheltered location of The Battery inside the upper bay of NY Harbor.

The ADCIRC surge forecast derived from the Sandy's control run was used to create a first-order coastal flooding map of Sandy for Manhattan Island, western Brooklyn, and Queens along the East

River (Figure 6). One challenge is that heights of seawalls surrounding the 870 km perimeter of New York Harbor are often not well documented. Rather than running ADCIRC in a wetting-drying mode, an estimate of local inundation using the control run was created by propagating sheets of water inland from the nearest ADCIRC coastal node (so-called "bathtubbing") using a high horizontal resolution (~1 m) digital elevation map (DEM) of land elevation created using LIDAR data sets now available in the public domain [29,30]. This approach is justified when the source river water carries significantly more flow that the flooding sheets themselves; the effect of local flooding does not change the underlying river dynamics.

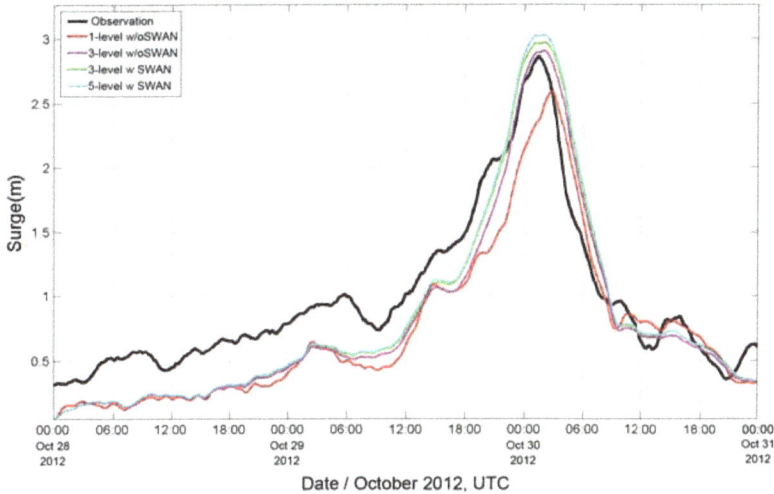

Figure 5. Simulations of surge using different configurations of 2D and 3D ADCIRC (3 and 5-vertical layers) and with or without SWAN in predicting.

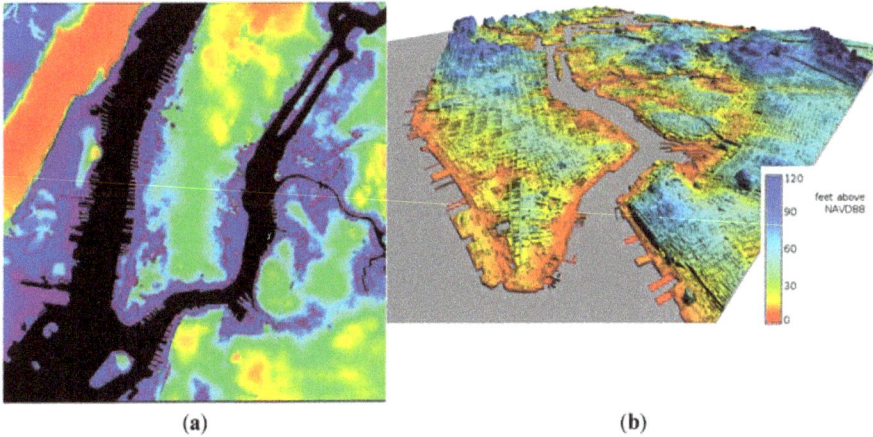

(a)   (b)

Figure 6. (a) 5-level ADCIRC w/SWAN-predicted Sandy inundation of central NYC and northern NJ during Sandy (colored blue), obtained with EnKF Control run, 2D ADCIRC, SWAN- and LIDAR-accurate topography; (b) Isometric projection of LIDAR-based 5-level ADCIRC w/SWAN-simulated flooding of the boroughs of Manhattan, Queens and Brooklyn (red and orange colors).

It is still a major challenge to create accurate street-by-street flooding maps since local flooding is very sensitive not only to elevation contours but also to local shifts in wind speed and direction at time and space scales atmospheric models cannot currently simulate. Additionally, street-level flooding cannot often be verified because of a lack of sensors and available operators during extreme weather events. Nevertheless, attempting to model inundation and disseminate this information at the street-by-street resolution remains an important challenge for emergency management and evacuation planning purposes and a topic of future scientific study.

3.4. Variations in Storm Tides around the Ensemble Forecast

The CTL simulations were compared with an 11-member ensemble started at 00:00 UTC, 26 October. There was a wide range of surge simulations derived from the ensemble (Figure 3a), ranging from 3.2–3.5 m in members 18, 66, 67 (nearly 0.1–0.4 m greater than observed and the CTL), to 1.5–2.0 m for 39 and 56. Member 46 generated the most accurate surge/storm tide forecast (Figure 3b), but its peak was delayed by 1–2 h. The timing of the surge with respect to the phase of the astronomical tide is critical to accurate peak storm-tide and surge simulations. Although members 66 and 67 had larger surges, they occurred close to local low tides, thus resulting in modest storm-tide levels.

To illustrate the importance of the timing of the arrival of the storm center on land, the predicted tidal phase was shifted in 30 min intervals from 13 h before landfall of CTL to 13 h after landfall and the tidal elevation added to the predicted surge at each time shift. This variation in the phase therefore represents all possible tidal phases over +/− one M_2 tidal period (T = 12.42-h). This addition of phase-shifted tides represents a convenient way of estimating the uncertainty in storm-tide (surge + tide) simulations using variations associated with each selected Sandy track (see Figure 1).

Thus, each of the 11 selected members produced 51 storm-tide simulations over two tidal cycles. These simulations were used to create a probability density function (PDF) of storm-tides (Figure 7). The PDF had a mean of 3.77 m (rel. to MLLW) and standard deviation of 0.59 m. Sandy's peak water level in the CTL (~4.25 m) lies in the right tail of the PDF since it occurred near a local high tide, but around 20.0% of the storm-tide simulations (red-filled bins in Figure 7) are still greater than the CTL.

The right tail of the PDF represents surge simulations by members 66 and 67. Overall, Figure 7 illustrates that Sandy's track did not lead to the worst possible coastal flooding for this particular event; the storm-tide could have been 0.50–0.75 m larger even using our small ensemble.

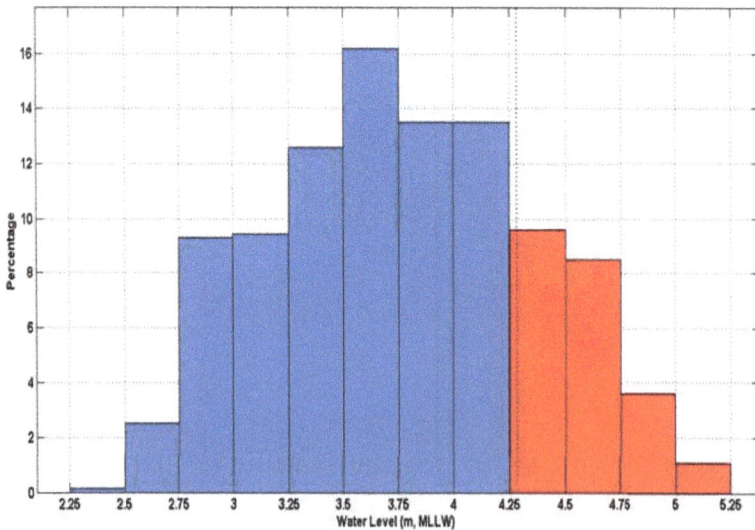

Figure 7. Probability distribution of peak water levels (storm tide rel. to MLLW) at The Battery created by shifting the phase of the tides in steps of 30 min between one tidal cycle before to one cycle following the time of landfall for each of the 11 wind ensemble field (see text for details).

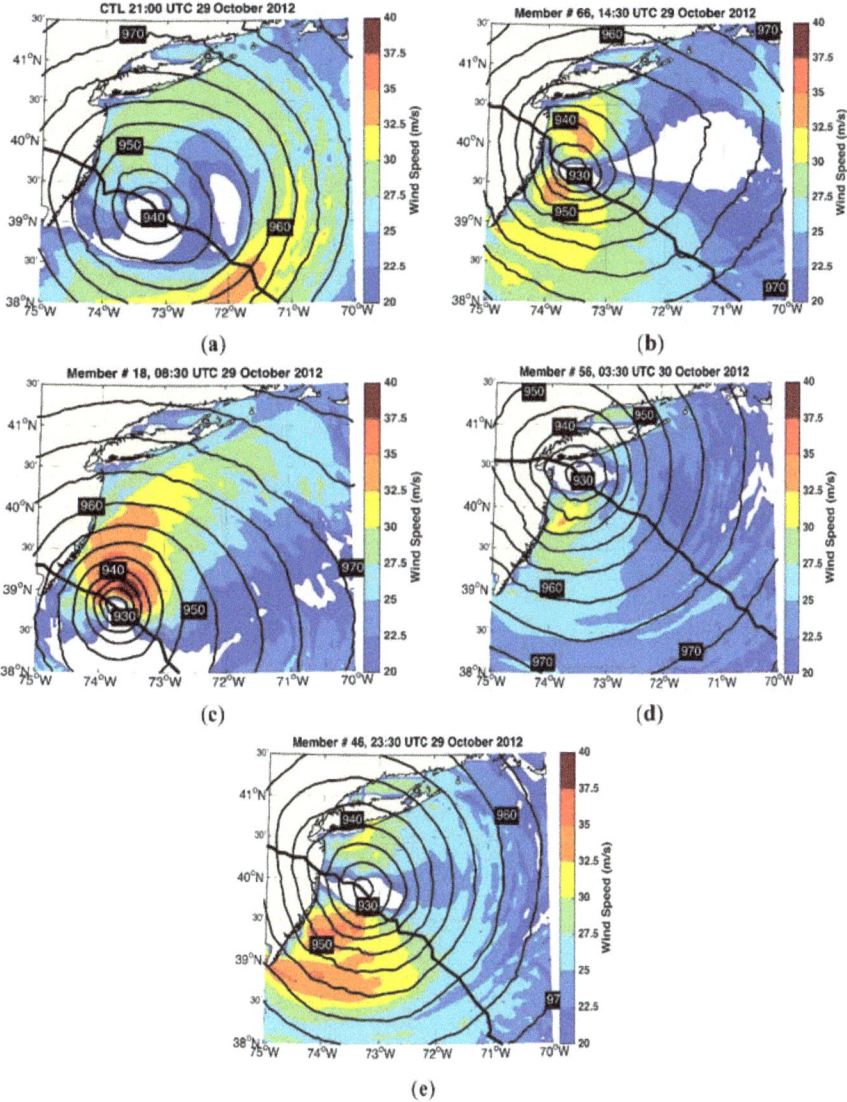

Figure 8. Mean sea-level pressure (every 10 hPa) and surface winds (color coded in m s^{-1}) for the (**a**) CTL; (**b**) 66; (**c**) 18; (**d**) 56; and (**e**) 46 members about an hour before landfall. The black line shows the track of the storm center.

3.5. Wind and Sea Level Pressure Forecasts

In order to better explain the complexity of simulated surge and water levels, the distributions of 10-m winds and mean sea level pressure (MSLP) predicted locally at 30 min shortly before landfall were analysed for four-selected ensemble members (CTL, 66, 18, 56, 46), chosen because they led to the largest and lowest predicted surges at The Battery.

Ensemble member 66 winds were 3–5 m s^{-1} stronger than the CTL north of the cyclone center (*cf.*, Figure 8a,b) since member 66 central MSLP was around 10 hPa lower (more intense), and the member

66's track was about 25 km further north. This resulted in about ~10.0% greater wind stress during the last few hours before landfall in member 66 than in the CTL south of Long Island (not shown).

Member 18 winds were stronger by 3–10 m s^{-1} than the CTL north of the cyclone's center (Figure 8c), but the track is 60 km to the south of the CTL. The winds in member 56 were similar to the CTL (Figure 8d), but its track took it right over NYC; therefore, the Battery was located away from the optimal surge generation region.

The winds north of the cyclone's center for member 46 were only 1–2 m s^{-1} greater than the CTL, and its track was located ~25 km farther to the north than the CTL, in a manner similar to member 66 (Figure 8e). Therefore, the surge for member 46 was relatively similar to the CTL's. Our analysis shows that relatively modest variations in wind speed (around 3–5 m s^{-1}) and track variation (around 20–50 km), even with the inherent uncertainty in a four-day forecast, can produce significant changes (+/−0.5 m) to the storm-tide.

3.6. The Second Surge Originating from Long Island Sound

It is not generally understood that during superstorm Sandy, New York City was subjected to two interacting surges; the first propagated into the Upper Bay of New York Harbor through the main connection to the Atlantic Ocean, the Verrazano Narrows. The second surge originating in Long Island Sound (LIS), propagated along the NE-SW oriented axis of the Sound and then through the East River. In fact the surge in western LIS was the larger of the two surges, peaking at 3.86 m at 23:00 UTC, 29 October 2012, at the Kings Point NOAA tide station [31]. The highest storm tide experienced at Kings Point (western Long Island Sound) at 02:12 UTC, Oct 30 2012, was 4.36 m above MLLW. The two Sandy surges met in the general region of the lower East River where much flooding and resulting damage occurred.

However, because of the three-hour phase shift in the tidal phase across the ends of the East River tidal strait between the Battery and Kings Point [32] the *storm tide* in the western Sound was small as the peak surge occurred between low to mid-tide. This phase shift across the East River is a result of the fact that the tide propagating up the Hudson River is close to a progressive wave, while the tides in Long Island Sound are closer to a resonant standing wave [33,34]. The surge simulations presented in this paper take all these dynamical considerations into account since the ADCIRC model accurately includes the tides in Long Island and Block Island Sounds [35], as well as in New York Harbor at high resolution.

4. Conclusions

The Advanced Circulation (ADCIRC) ocean model forced using surface winds and mean sea-level pressure (MSLP) from a nested 3-km Weather Research and Forecasting (WRF) model following hurricane Sandy (2012) was used to explore variations in Sandy storm surge simulations at The Battery, located at the southern tip of Manhattan Island, New York City. The 10-m wind and MSLP forecast uncertainties were quantified using an 11-member Ensemble Kalman Filter (EnKF) system, initialized (at 00:00 UTC, 26 October 2012) four days before landfall. A control WRF member re-initialized every 24 h reproduced the observed storm surge at the Battery to within 10–20 cm, so the model is capable of producing an accurate forecast for this event if the atmospheric forcing is well predicted.

An 11-member storm surge ensemble starting approximately four days before landfall illustrated how relatively modest differences in the track (around 20–50 km) and intensity (around 3–5 m s^{-1}) of Sandy's track, especially for a four-day forecast, can lead to relatively large storm surge variations of around 0.50 m. The ensemble approach also illustrates that the landfall timing relative to high tide is critical since Sandy's surge, had it occurred during low tide (+/−6-h), would have resulted in significantly less flooding and damage.

The ensemble approach also illustrates that storm-tide levels observed during Sandy were likely not worst-case for this event since several members predicted surges at least 0.5 m greater. More specifically, about 20.0% of the scenarios using the ensemble approach produced storm-tides greater

J. Mar. Sci. Eng. **2015**, *3*, 428–443

than observed. Overall, our study emphasizes the need for ensemble storm surge simulations for land-falling storm events.

Other tide stations in the area are being investigated and results will be published elsewhere.

Acknowledgments: This publication is a resulting product from New York Sea Grant project number R/CCP-18 funded under award NA10OAR4170064 from the National Sea Grant College Program of the U.S. Department of Commerce's National Oceanic and Atmospheric Administration, to the Research Foundation of State University of New York on behalf of New York Sea Grant. This work was also partially supported by the New York State Resiliency Institute for Storms and Emergencies. The statements, findings, conclusions, views, and recommendations are those of the author(s) and do not necessarily reflect the views of any of those organizations. The PSU team is partially supported by NOAA under the Hurricane Forecast Improvement Project (HFIP) and by NASA under Grant NNX12AJ79G. Computing is performed at NOAA and the Texas Advanced Computing Center (TACC). This research was supported in part by the Mary Jean and Frank P. Smeal Foundation.

Author Contributions: Conceived and designed the analysis: BAC MJB. Performed the EnKF runs: YW EBM FZ. Ran the ADCIRC model: JK CNF. Analysis and plotting of ADCIRC data: JK MHB KJR. Wrote and edited the paper: BAC MJB.

Conflicts of Interest: The authors declare no conflict of interest.

References

1. Blake, E.S.; Kimberlain, T.B.; Berg, R.J.; Cangialosi, J.P.; Beven, J.L. *Tropical Cyclone Report: Hurricane Sandy (AL182012), 22–29 October 2012*; National Hurricane Center: Miami, FL, USA, 2013; pp. 1–157.
2. Galarneau, T.J., Jr.; Davis, C.A.; Shapiro, M.A. Intensification of Hurricane Sandy (2012) through extra tropical warm core seclusion. *Mon. Weather Rev.* **2013**, *141*, 4296–4321. [CrossRef]
3. Background information page on hurricane Sandy. Available online: http://en.wikipedia.org/wiki/Meteorological_history_of_Hurricane_Sandy (accessed on 10 January 2014).
4. NOAA buoy 44025 data page. Available online: http://www.ndbc.noaa.gov/station_page.php?station=44025 (accessed on 10 June 2014).
5. Weng, Y.; Zhang, F. Assimilating airborne Doppler radar observations with an ensemble Kalman Filter for convection permitting hurricane initialization and prediction: Katrina (2005). *Mon. Weather Rev.* **2012**, *140*, 841–859. [CrossRef]
6. Munsell, E.B.; Zhang, F. Prediction and uncertainty of Hurricane Sandy (2012) explored through a real-time cloud-permitting ensemble analysis and forecast system assimilating airborne Doppler observations. *J. Adv. Model. Earth Sci.* **2014**, *6*, 1–20. [CrossRef]
7. Blier, W.; Keefe, S.; Shaffer, W.; Kim, S. Storm Surge in the Region of Western Alaska. *Mon. Weather Rev.* **1997**, *125*, 3094–3208. [CrossRef]
8. Burroughs, L.B.; Shaffer, W.A. *East Coast Extratropical Storm Surge and Beach Erosion Guidance*; NWS Technical Procedures Bulletin No. 436; U.S. Department of Commerce, National Oceanic and Atmospheric Administration: Silver Spring, MD, USA, 1997.
9. Georgas, N.; Blumberg, A. Establishing confidence in marine forecast systems: The design and skill assessment of the New York Harbor Observation and Prediction System, Version 3 (NYHOPS v3). In *Proceedings of the 11th International Conference on Estuarine and Coastal Modeling*, Seattle, WA, USA, 4–6 November 2009; pp. 660–685.
10. Stony Brook Storm Surge Model home page. Available online: http://stormy.msrc.sunysb.edu (accessed on 1 January 2013).
11. DiLiberto, T.; Colle, B.A.; Georgas, N.; Blumberg, A.; Taylor, A. Verification of a multiple model storm surge ensemble for the New York Metropolitan Region. *Weather Forecast.* **2011**, *26*, 922–939. [CrossRef]
12. Georgas, N.; Orton, P.; Blumberg, A.; Cohen, L.; Zarrilli, D.; Yin, L. The Impact of Tidal Phase on Hurricane Sandy's Flooding Around New York City and Long Island Sound. *J. Extrem. Events* **2014**, *01*. [CrossRef]
13. Forbes, C.; Rhome, J.; Mattocks, C.; Taylor, A. Predicting the storm surge threat of Hurricane Sandy with the National Weather Service SLOSH model. *J. Mar. Sci. Eng.* **2014**, *2*, 437–476. [CrossRef]
14. SLOSH home page. Available online: http://www.nhc.noaa.gov/surge/slosh.php (accessed on 12 January 2014).

15. Zhang, F.; Weng, Y.; Gamache, J.F.; Marks, F.D. Performance of convection-permitting hurricane initialization and prediction during 2008–2010 with ensemble data assimilation of inner-core airborne Doppler radar observations. *Geophys. Res. Lett.* **2011**, *38*, L15810. [CrossRef]
16. GFS model home page. Available online: http://www.emc.ncep.noaa.gov/index.php?branch=GFS (accessed on 2 January 2015).
17. ADCIRC model home page. Available online: http://adcirc.org (accessed on 6 June 2013).
18. Bowman, M.J.; Colle, B.A.; Bowman, M.H.E. Storm surge modeling for the New York City region. In *Climate Adaptation and Flood Risks in Coastal Cities*; Aerts, J., Botzen, W., Bowman, M., Ward, P., Diercke, P., Eds.; Earthscan Climate: London, UK, 2011.
19. Dietrich, J.C.; Tanaka, S.; Westerink, J.J.; Dawson, C.N.; Luettich, R.A.; Zijlema, M.; Holthuijsen, L.H.; Smith, J.M.; Westerink, L.G.; Westerink, H.J. Performance of the unstructured-mesh, SWAN+ADCIRC model in computing hurricane waves and surge. *J. Sci. Comput.* **2012**, *52*, 468–497. [CrossRef]
20. Colle, B.A.; Buonaiuto, F.; Bowman, M.J.; Wilson, R.E.; Flood, R.; Hunter, R.; Mintz, A.; Hill, D. Simulations of past cyclone events to explore New York City's vulnerability to coastal flooding and storm surge model capabilities. *Bull. Am. Meteorol. Soc.* **2008**, *89*, 829–841. [CrossRef]
21. SWAN model home page. Available online: http://www.swan.tudelft.nl/ (accessed on 10 January 2014).
22. Orton, P.; Georgas, N.; Blumberg, A.; Pullen, J. Detailed Modeling of Recent Severe Storm Tides in Estuaries of the New York City Region. *J. Geophys. Res.* **2012**, *117*, C09030. [CrossRef]
23. Westerink, J.J.; Luettich, R.A., Jr.; Scheffner, N.W. *ADCIRC: An Advanced Three-Dimensional Circulation Model for Shelves Coasts and Estuaries, Report 3: Development of a Tidal Constituent Data Base for the Western North Atlantic and Gulf of Mexico*; Dredging Research Program Technical Report DRP-92-6; U.S. Army Engineers Waterways Experiment Station: Vicksburg, MS, USA, 1993; p. 154.
24. Garratt, J.R. Review of drag coefficients over oceans and continents. *Mon. Weather Rev.* **1977**, *105*, 915–929. [CrossRef]
25. Weisberg, R.H.; Zheng, L.; Zijlema, M. Gulf of Mexico hurricane wave simulations using SWAN: Bulk formula-based drag coefficient sensitivity for Hurricane Ike. *J. Geophys. Res.* **2013**, *118*, 3916–3938. [CrossRef]
26. Battery, NY water level data at Tides and Currents home page. Available online: http://tidesandcurrents.noaa.gov/waterlevels.html?id=8518750 (accessed on 15 February 2015).
27. Indian Space Research Organization home page. Available online: http://www.isro.gov.in/applications/earth-observation. (accessed on 10 January 2014).
28. Weisberg, R.H.; Zheng, L. Hurricane storm surge simulations comparing three-dimensional with two-dimensional formulations based on an Ivan-like storm over the Tampa Bay, Florida region. *J. Geophys. Res.* **2008**, *113*, C12001. [CrossRef]
29. City of New York LIDAR data. Available online: https://data.cityofnewyork.us/City-Government/1-foot-Digital-Elevation-Model-DEM-/dpc8-z3jc (accessed on 15 February 2014).
30. New York State Departmental of Environmental Conservation LIDAR data. Available online: http://coast.noaa.gov/htdata/lidar1_z/geoid12a/data/1408/2012_NYDES_Lidar_metadata.html (accessed on 12 March 2014).
31. Kings Point, NY water level data at Tides and Currents home page. Available online: http://tidesandcurrents.noaa.gov/stationhome.html?id=8516945 (accessed on 22 February 2014).
32. Bowman, M.J. The tides of the East River, New York. *J. Geophys. Res.* **1976**, *81*, 1610–1616. [CrossRef]
33. Redfield, A.C. *Introduction to the Tides of New England*; Hutchinson Ross: London, UK, 1983; p. 108.
34. Swanson, R.L. Some Aspects of Currents in Long Island Sound. Ph.D. Thesis, Oregon State University, Corvallis, OR, USA, 1970. Available online: http://ir.library.oreganstate.edu/xmlui/handle/1957/29500 (accessed on 15 March 2014).
35. Bowman, M.J.; Esaias, W.E. Fronts, stratification and mixing in Long Island and Block Island Sounds. *J. Geophys. Res.* **1981**, *85*, 2728–2742. [CrossRef]

Journal of
Marine Science and Engineering

MDPI

Article

Climate Change, Coastal Vulnerability and the Need for Adaptation Alternatives: Planning and Design Examples from Egypt and the USA

S Jeffress Williams [1,*] and Nabil Ismail [2]

[1] U.S. Geological Survey, Woods Hole, Massachusetts and University of Hawaii, Honolulu, HI 96822, USA
[2] Coastal Engineering, Maritime Academy, Alexandria, Egypt and Director of Costamarine Technologies, Davis, CA 95616, USA; nbismail@usa.net
* Author to whom correspondence should be addressed; jwilliams@usgs.gov; Tel.: +1-508-457-2383.

Academic Editor: Rick Luettich
Received: 31 March 2015; Accepted: 3 July 2015; Published: 15 July 2015

Abstract: Planning and design of coastal protection for high-risk events with low to moderate or uncertain probabilities are a challenging balance of short- and long-term cost *vs.* protection of lives and infrastructure. The pervasive, complex, and accelerating impacts of climate change on coastal areas, including sea-level rise, storm surge and tidal flooding, require full integration of the latest science into strategic plans and engineering designs. While the impacts of changes occurring are global, local effects are highly variable and often greatly exacerbated by geophysical (land subsidence, faulting), oceanographic (ocean circulation, wind patterns) and anthropogenic factors. Reducing carbon emissions is needed to mitigate global warming, but adaptation can accommodate at least near future change impacts. Adaptation should include alternatives that best match region-specific risk, time frame, environmental conditions, and the desired protection. Optimal alternatives are ones that provide protection, accommodate or mimic natural coastal processes, and include landforms such as barrier islands and wetlands. Plans are often for 50 years, but longer-term planning is recommended since risk from climate change will persist for centuries. This paper presents an assessment of impacts of accelerating climate change on the adequacy of coastal protection strategies and explores design measures needed for an optimum degree of protection and risk reduction. Three coastal areas facing similar challenges are discussed: Abu-Qir Bay, Nile River delta plain, Egypt; Lake Borgne, New Orleans, Louisiana delta plain; and the New York City region.

Keywords: climate change; sea-level rise; Egypt; New York City; New Orleans; Louisiana; coastal vulnerability; deltas; coastal protection; coastal management; adaptation

1. Introduction

For over one hundred years, coastal protection structures were designed and constructed based mostly on assumptions that parameters were fairly constant and predictable and conditions driving forces of the past would persist pretty much unchanged into the future. Results of climate science studies over the past three decades, especially the past decade, show that climate change is unequivocal, due largely to carbon dioxide increase and other greenhouse gas (GHG) concentrations in the atmosphere. Observations show the effects are global, but vary greatly on regional scales. The impacts of climate change most germane to coastal protection are: global mean sea-level rise (GSLR), increase in storm surge and flooding levels, increase in high tide nuisance flooding, change in wave characteristics due to more intense storm events, and increase in extreme weather events. Natural disasters are becoming more frequent and more severe on highly vulnerable deltas and lowland coastal areas. This is evident for barrier islands and river delta coasts, which are experiencing high rates of erosion, flooding, and marine transgression. Projected future GSLR poses further significant threat, especially to coastal landforms, which are already undergoing deterioration due to anthropogenic impacts. The impacts of climate change on coasts vary temporally and regionally but all coastal regions are facing similar issues.

Adaptation measures responding to climate change impacts should be implemented in a phased and progressive manner over time and all vulnerable countries can benefit by international cooperation and sharing of experiences, expertise, and resources.

2. Objective

This paper presents an assessment of the impacts of accelerating climate change (GSLR, storm intensity, tidal nuisance flooding) on the adequacy of coastal protection strategies and explores additional measures needed for an optimum degree of protection and risk reduction. Three coastal regions with similar circumstances are considered: the Nile Delta Coast at Abu-Qir Bay, Egypt; the Louisiana coast including the Lake Borgne Surge Barrier; and the New York City region. Current plans are discussed and recommendations are made for employing adaptation to cope with projected climate change impacts and reduce risk to humans living in low coastal regions.

3. Climate Change—Global Sea-Level Rise

Climate science is expanding in information and its understanding rapidly. Applying 50 to 100 year conditions into the future and beyond will be different than the historic past or the present. This requires a significant change in thinking about sustainable coastal management and protection of delta and lowland coastlines (Figure 1). Impacts of climate change should be fully integrated into coastal planning and engineering and uncertainties factored in. Risk and vulnerability need to be assessed to decide on the most cost-effective approach for projects designed for the next 50 years, to 2100 and well beyond.

The changes in climate already have significant consequences with regard to precipitation (extreme rain events, flooding, droughts), temperature extremes, and ocean levels (storm surge, sea-level rise, tidal flooding). Sea-level rise, increasingly recognized as important on local and regional scales, is a dominant driver of coastal change, along with storms.

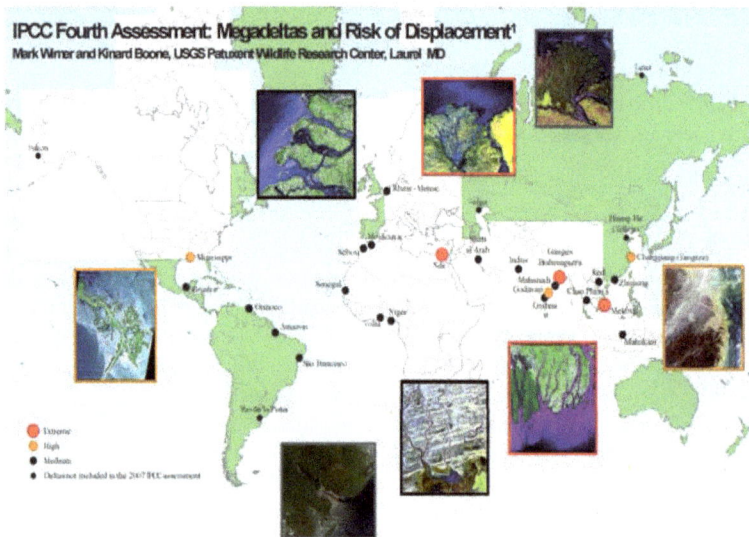

Figure 1. World Mega Deltas threatened by climate and human changes (open source poster modified by USGS from [1]).

J. Mar. Sci. Eng. **2015**, 3, 591–606

Global mean sea level was relatively stable for the past several thousand years under a mild climate until the mid-19th century, but during the 20th century sea level began rising due to global warming resulting from human activities at a global average rate of 1.2 mm/year. The current average global rise rate is 3.2 mm/year, 2.5 times faster [2,3]. Many coastal regions, particularly deltas, however, are experiencing much greater local or relative sea-level rise (LSLR), defined as global mean sea-level rise plus subsidence, plus sediment compaction, and in minor cases land emergence. Example rates are: Nile delta region 2–5 mm/year, Louisiana delta region ~4–10+ mm/year, and New York City area 4 mm/year. These higher than global rates have significant areal variations and are due to local geophysical (subsidence, faulting), oceanographic, and human factors (oil, gas, and ground water extraction, as well as wetland reclamation). Combined with storms, this rise is resulting in greater surge elevations, more frequent tidal nuisance flooding, and record increases in damage to coastal infrastructure and loss of life [2]. Projected GSLR by 2100 [4–7] is 0.2 m to 2 m and would be in addition to local rise factors such as subsidence and oceanographic processes (Figure 2). The purple and red color bars simply denote the most recent published projections. The range of projections is large due to the uncertainty of continued global warming, which in turn depends on the extent of continued man-made emissions of greenhouse gases into the atmosphere. Even greater global sea-level rise is possible (3 m+) over the next several hundred years [8,9], unless carbon emissions, global warming, and ice-sheet melting are greatly slowed or ideally reversed.

ETL 1100-2-1
30 Jun 14

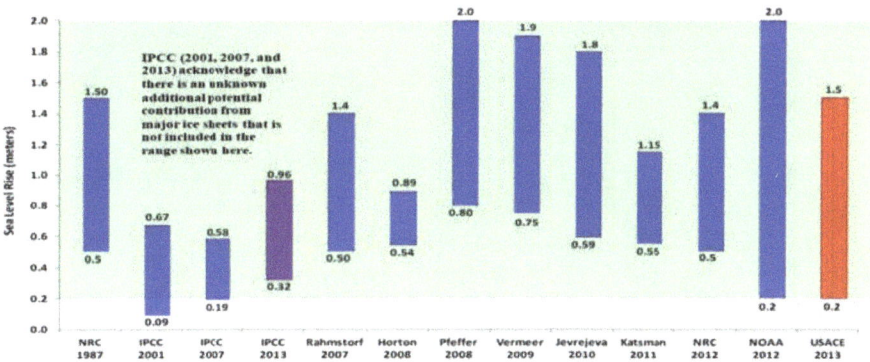

Figure 2. Comparison of maximum and minimum estimates of global sea level rise by 2100.

4. Recent World Catastrophic Storm Events

Hurricane Katrina, with a storm surge of up to 8.5 m and wave heights of 5.5 m (total 14 m), made landfall in eastern Louisiana and the Mississippi coast on 29 August 2005, resulting in major loss of lives from massive levee failures and flooding of New Orleans, loss of wetlands, and extensive erosion of barrier islands along the Gulf coast.

The Nile Delta Super Storm on 12 December 2010 is a striking example of the severity of more energetic storm events since 2003. The storm caused extensive flooding of the delta and coastal cities in Egypt. A surge of over 1.2 m and 7 m waves forced the closure of Alexandria's main harbor. The main damage resulted from wave and surge overtopping of coastal structures designed for smaller events.

Hurricane "Super storm" Sandy made landfall along the New Jersey–New York coast on 29 October 2012 at high tide after taking an unusual path at landfall. The largest storm on record for the Atlantic, Sandy, a combination tropical–extra tropical storm, caused catastrophic damage by overtopping barrier islands and coastal structures with damage throughout the region, including New York City, which experienced a 4.4 m storm surge.

5. Vulnerability of Deltas and Lowland Coastlines to Climate and Anthropogenic Changes

All the World's major deltas, including the Nile Delta and Louisiana's Mississippi River Delta, developed about seven to eight thousand years ago when Holocene sea-level rise slowed and rivers were able to transport sediment to the coast in volumes sufficient to maintain landforms acted on by marine processes. Deltas develop by lateral movement of channels that spread alluvial sediments while at the same time thick deposits undergo rapid subsidence. Acted on by marine processes over time, deltas become fronted by sandy barrier island shorelines, backed by vegetated wetlands and marshes [10].

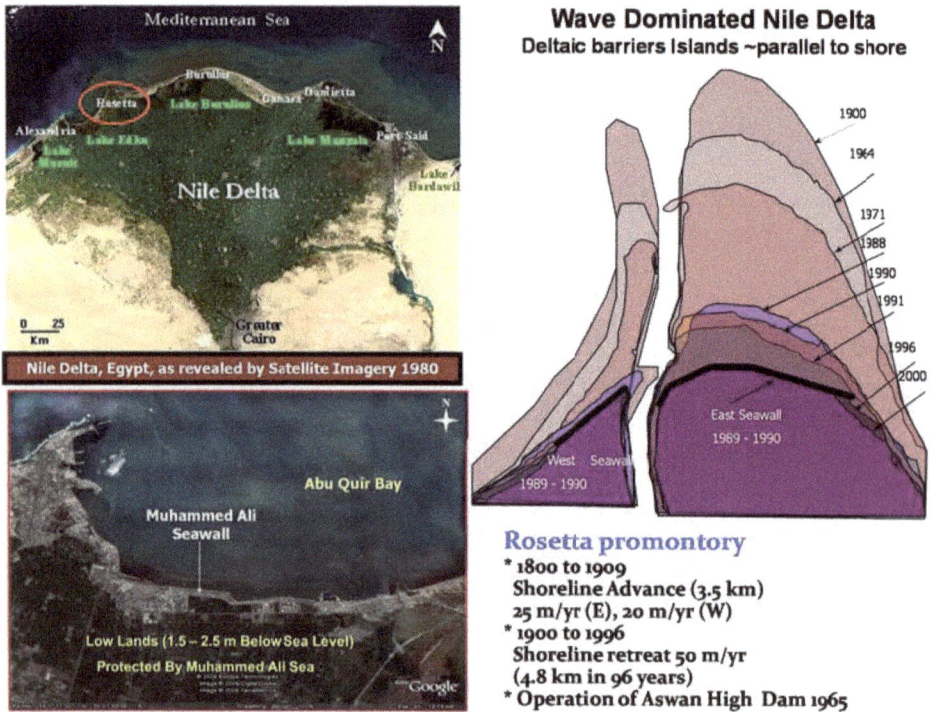

Figure 3. Coastal retreat of Rosetta Headland (1900 to 2010) and M. Ali Seawall [11] from open source maps.

Deltas have been subjected to significant coastal changes due to a variety of man-made factors over the past several hundred years, such as reduction in sediment load due to the construction of dams and levees, such as along the Mississippi River and the Nile River. In addition, the reduction of river currents modifies the near shore circulation, induced by wave–current interaction, which maintained the delta morphology for many centuries. Such effects of the combined wave–current flow, on nearshore circulation, are detailed in studies by Ismail [12,13]. For the Nile, the Aswan High Dam built in 1965 has caused significant coastal changes due to sediment depletion, exacerbating subsidence and shoreline erosion (Figure 3). For the USA, Louisiana's delta plain has the highest rates of coastal erosion and wetland loss of any region in the world due to a combination of complex natural processes and anthropogenic actions. Since 1900, about 4900 km^2 of wetlands in coastal Louisiana have been lost (Figure 4).

Figure 4. Map image of the Louisiana delta plain, Mississippi River delta, coastal barrier islands and wetlands, and New Orleans just south of Lake Pontchatrain (Google Earth 2014).

6. Nile Delta Coast and 2010 Storm Effects

The coastal flooding in Alexandria on 12 December 2010 is a striking example of the more progressive severity of events since 2003. Egypt was hit by strong winds, exacerbated by heavy precipitation, up to 60 km/h with 10 h duration. These weather conditions resulted in waves of ~7 m height with a surge of 1.2 m, which forced the closure of Alexandria harbor. The typical significant wave height is 1–1.5 m. Maximum wave height during storm conditions averages 4.5 m. Typical values of storm surge on the delta coast are 40–50 cm. The storm had profound destructive effects on Alexandria, Abu Qir Bay as well as on shorelines between the two river Nile promontories, Rosetta and Damietta.

A comprehensive program has been underway since 1970 to gather data related to coastal erosion along the Nile Delta shore. Review of major coastal problems and general description of the recommended protective measures to address these problems are in the coastal Master Plan for Phase I, as reported by [14]. Plans are underway to prepare Master Plan Phase II for coastal protection of the Nile Delta coast that will further address future climate change impacts and the need for adaptation alternatives (Figure 5).

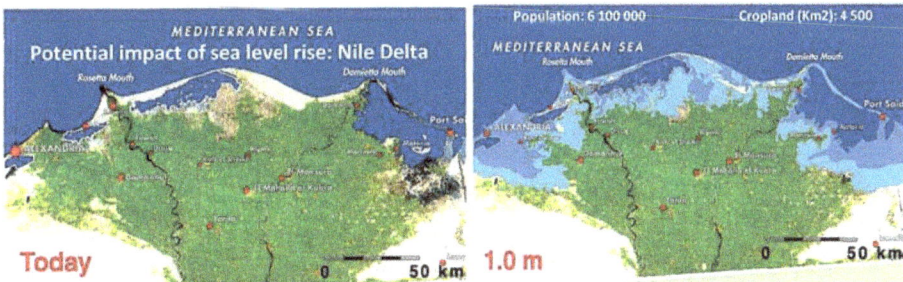

Figure 5. Maps of the Nile Delta showing potential impacts of a 1 m rise in sea level (Maps from open sources).

Mohamed Ali-Seawall Case Study

This project is the design-review and upgrade of Mohamed Ali seawall (revetment) in the partially protected Abu Qir Bay under scenarios of extreme storm conditions for the future 50 years return period. These design conditions include: storm wave height, storm surge, subsidence, and projected LSLR (Figure 6). Rosetta headland with its adjacent Abu Qir Bay form a littoral cell, in which sedimentation is controlled by a combination of waves, coastal currents, tides, and river discharge. The M.A. seawall located in Abu Qir Bay was constructed in 1830 to protect the agriculture lowland against sea flooding (Figure 3). The seawall was repaired and upgraded in 1981 and a beach segment was created for recreation. As in many coastal regions near major urban areas, the coast is used for a variety of purposes including, a power plant and a liquefied natural gas export terminal. The seawall was upgraded again in 2009, one year before the 2010 storm. Damages from the storm showed an urgency for further upgrades, as highlighted by [11].

Hydrodynamic analyses were conducted to provide design recommendations to increase the life of the structure, deter flooding of the lowlands, and increase the stability of the structure to impacts of climate change. Using a future LSLR value of 0.5 m in the next 50 years, resulted in a design wave run-up height of 5.4 m (Figure 6). Two design scenarios were considered to estimate wave height distributions within Abu Qir Bay and along M. Ali seawall.

I. Current morphology with the obtained maximum storm yearly wave condition (H_{max} = 3.5 m, T = 7.5 s, N direction, Water Depth = 14 m).

II. Future morphology (recent bathymetry +0.5 m sea-level rise within the next 50 years) with the maximum yearly wave condition in Abu-Qir Bay (4.8 m).

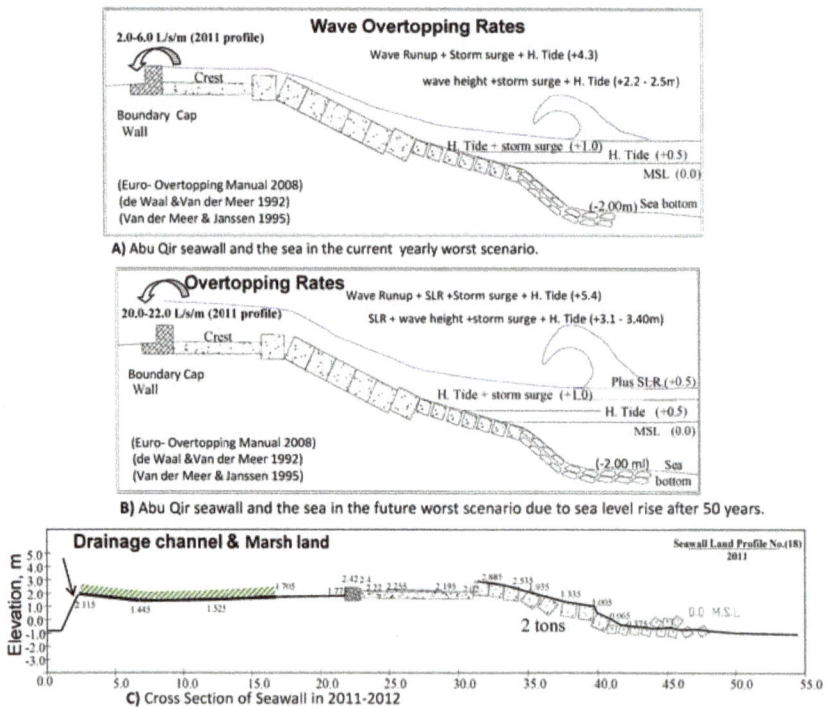

Figure 6. Impact of climate change on the the M.A. design and recommended drainage channel and creation of wetlands. The value of 0.5 m includes projected global sea-level rise and local subsidence [11].

The model results showed that sea-level rise has a noticeable effect for the case of maximum wave condition at the wall. The wave height in front of the coastal structure at the toe will increase by about 25% after 50 years. Further, based on the obtained wave run-up of 5.4 m and subsequent overtopping, recommendations were made to increase the weight of the armor layer from 0.5 t to 2 t, increase the seawall top elevation to 3 m, and reinforce the revetment toe. Further recommendations are given to increase the height of the seawall cap, and to strengthen the beach top and back slope with a facility to drain the overtopped storm water behind the seawall (Figure 6).

7. Hurricane Katrina, Storm Protection, and Louisiana Coastal Restorations Plans

Hurricane Katrina (2005) was the fourth most powerful storm to strike Louisiana since 1893 with respect to maximum wind speed and surge elevations at landfall. As Katrina progressed across Breton Sound and Lake Borgne, it generated storm surge of 8.5 m, and 5.5 m wave run-up (total 14 m) on the Mississippi coast, and a 6 m surge southeast of New Orleans, and up to 2 m of additional wave run-up, a total of 8 m. In addition, Katrina dropped 20–30 cm of rain across southeastern Louisiana. In southeastern Louisiana, communities unprotected by levees were flooded; the surge overtopped and destroyed levees protecting eastern New Orleans as well as parishes (i.e., counties) to the south and east, flooding the city and killing an estimated 1400 people. Floodwalls failed along drainage and navigation canals connected to Lakes Pontchartrain and Borgne with significant coastal erosion and wetland loss.

In response to Hurricane Katrina, the Louisiana Coastal Protection and Restoration (LACPR) technical report was prepared by [15]. In collaboration with the State, the USACE developed and analyzed a range of alternatives, based on a number of structural, nonstructural, and coastal restoration measures, to reduce storm surge risk in south Louisiana. As a representation of Category 5 storm risk reduction, the report presents alternatives for the 100-year, 400-year, and 1000-year design levels. The 400-year flood event is an approximation of Hurricane Katrina. The philosophy of integrated defense for reducing risk from hurricane surge was used based on a "Multiple Lines of Defense" strategy that no single measure or approach will be sufficient for achieving the risk reduction objectives. This integrated defense solution has been the Hurricane and Storm Damage Risk Reduction System (HSDRRS), with a budget of $14.5 billon. It includes five parishes and consists of 560 km of levees and floodwalls, 73 non-federal pumping stations, three canal closure structures with pumps, and four gated outlets. One major structural component of the system is the Inner Harbor Navigation Canal—Lake Borgne Surge Barrier (IHNC-LBSB) [16–18].

7.1. IHNC Lake Borgne Storm Surge Barrier—Sea Level, Subsidence and Barrier Design Height

The 2.9 km-long IHNC-LBSB is located at the confluence of the Gulf Intracoastal Waterway (GIWW) and the Mississippi River Gulf Outlet (MRGO), about 19 km east of New Orleans (Figure 7). The surge barrier, completed in 2011, works in tandem with the Seabrook Floodgate Complex constructed at the north end of the IHNC at Lake Pontchartrain, (Figure 8) [18,19].

The IHNC-LBSB has three gates that allow vessel passage through the barrier and provides a complete closure of the MRGO navigation channel.

The surge barrier design was based on detailed modeling of wind, surge, waves and rainfall for 152 synthetic storms that were selected to span the historical record of storms that have impacted southeastern Louisiana. Responses from this limited storm set were then interpolated to a much larger storm set to represent the full range of possible storm conditions and to enable recurrence analysis. The final 8 m design height (selected to minimize the impact of a 100-year, 1% chance annual occurrence event and provide resilience to a 500-year, 0.2% chance annual occurrence event) included 0.3 m of LSLR over the 50-year design life [20]. Given the magnitude of the IHNC-LBSB project, the significance of the urban area that is being protected, the global sea level rise predicted by [15], (i.e., [15] Table 4.1, 0.64–1 m by 2100) and the substantial subsidence that has been observed in parts of southern Louisiana, (e.g., Figure 9), one might question why a more conservative design height was not selected (e.g.,

~10 m) that would minimize the effects of a 500-year event over a design lifetime reaching until 2100. An alternative means of increasing the design capacity of the IHNC-LBSB is to expand the storage capacity for water that overtops the barrier. Currently there is a modest storage volume inside the IHNC for water to accumulate before exceeding the IHNC floodwalls if it overtops the IHNC-LBSB barrier. This storage volume is critical to providing system resilience to the 500-year event. However, the elevated water levels in the IHNC would bring any vessel that remained inside and broke its mooring to a level where they could directly impact and damage the IHNC floodwalls. The IHNC storage volume could be expanded by more than ten-fold, and the water level minimized, if water is allowed to flow through the Bayou Bienvenue Sector Gate (Figure 8) and enter a marsh area called the Central Wetlands (Figure 8). While the Central Wetlands are actually located inside the HSDRRS, the inhabited areas of Orleans and St. Bernard Parishes are protected by an interior "back" levee system, as shown in Figure 8. We strongly encourage investigation of this option as a critical, long term means of lowering the risk of overtopping or failure of the floodwalls in the IHNC and another catastrophic flooding of New Orleans.

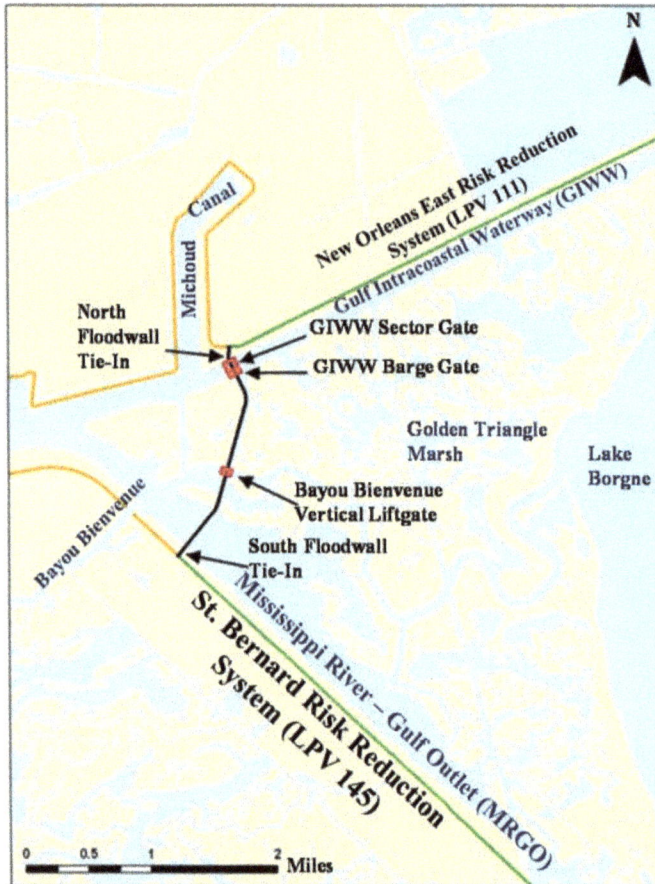

Figure 7. Map from open source of the Lake Borgne surge barrier [15].

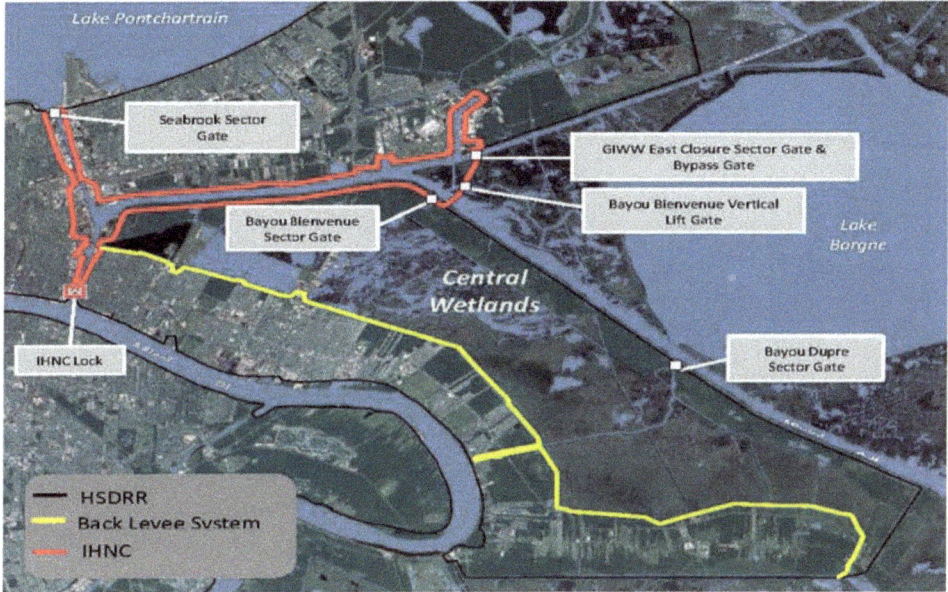

Figure 8. Location map of the IHNC-LBSB, IHNC floodwalls and the Central Wetlands.

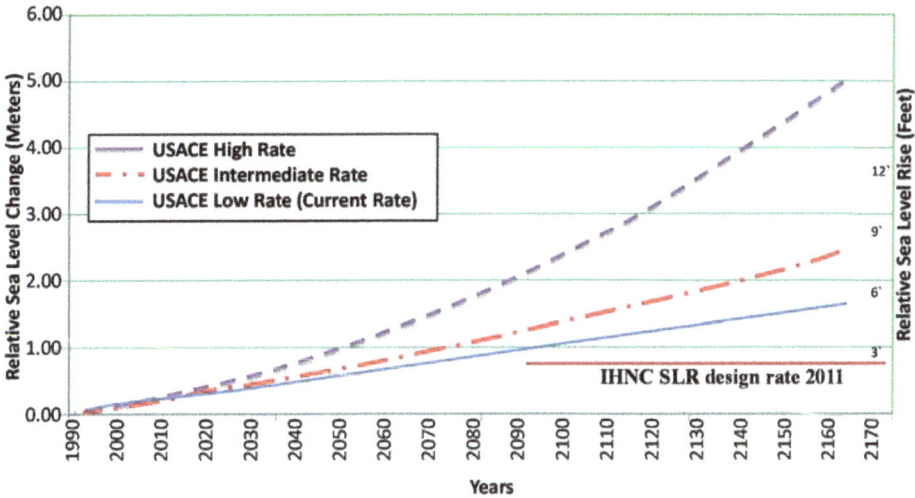

Figure 9. USACE projected sea-level rise curves for Grand Isle, Louisiana. These curves include eustatic (global) sea-level rise values and subsidence rates [7].

7.2. West Closure Pump Station

On the opposite side of the Mississippi River, lies the other main perimeter defense against flooding from storm surge (Figure 10). The West Closure Complex must operate in a fail-safe mode whenever severe storms threaten New Orleans, to prevent interior flooding of the west basin. When the navigable sector gate is closed to block storm surges, catastrophic flooding can be mitigated by pump station operation. Similar concerns to those raised for the Lake Borgne surge barrier exist about the design parameters (including the design storm recurrence and the design lifetime) and the accompanying adequacy of LSLR (subsidence and sea-level rise) assumptions in the design of this facility. In this case, the situation is further exacerbated by the potential impact to pumping operations if the closure is overtopped by surge and waves or experiences an extended power outage that is likely to accompany such an event.

Figure 10. Image of the GIWW—West Closure Pump station.

7.3. Louisiana Comprehensive Master Plan 2012

Over the past 30 years, a series of plans for addressing land loss have been developed, the LACPR is described above, but the most recent and ambitious is the Comprehensive Master Plan completed in 2012 [21]. This plan builds on earlier plans and discusses the importance of trying to deal with land loss by means of a variety of structural, nonstructural, and even "voluntary acquisition". The time frame is limited to 50 years into the future with a projected budget of $50+ billion. Restoration efforts aim at reconnecting the Mississippi River to the deltaic plain. While the objective of the plan is to ultimately have a healthy and sustainable coast, the lack of discussion of conditions likely to prevail beyond the next 50 years is an unfortunate omission.

Future changes to the delta plain over just the next 50 years were addressed in maps and figures showing predicted change, and rather than using the usual terms for sea-level rise projections due to global warming, the plan instead used "moderate environmental scenarios" for the current situation and "less optimistic environmental scenarios" using the GSLR reported values of 0.12 to 0.65 m by 2050. But there is no discussion of climate change and its impacts as the central driver of the less optimistic environmental conditions. Land subsidence, a large component of LSLR (~75%), is highly variable across the delta plain (~2–35 mm/year) and was factored into the reported scenarios. However, discussion is lacking on the natural geologic forces and man-made causes of subsidence such as oil, gas, and ground-water extraction. Since LSLR is already 0.9 m over parts of the coast over the past century [22], the current global average is 3.2 mm/year, and recent climate literature projects GSLR to be 2 m+ by 2100, it appears the CPRA's plan scenarios understate the risk to New Orleans and the delta plain due to continued rapid land subsidence and projected sea-level rise to 2100 and for centuries into the future. More realistic scenario values of LSLR for planning and design might be 0.5 to 1 m by 2050 and 3 m by 2100.

8. Hurricane Sandy and New York City Coastal Protection Plans

The New York City (NYC) Metropolitan area is the largest of the 25 most densely populated areas in the east coast of the U.S., out of which 23 are highly vulnerable to storms and sea-level rise. Hurricane Sandy made landfall along the New Jersey coast on 29 October 2012. Sandy was the largest extra-tropical cyclone in recorded history for the north Atlantic region. It made landfall near high tide with maximum surge elevation of 4.4 m (NOAA Battery tide gauge station), while the average tide range is 1.4 m. Deaths totaled 72 people ; total damages were ~$50 Billion and New York City sustained $19 Billion in damages, mostly due to surge flooding and fires. Damages were also significant for the northern New Jersey region and extended as far east as Cape Cod, Massachusetts. While Hurricane Sandy was not "caused" by climate change, storm surge elevations were enhanced by LSLR over the past century, which is partly an impact of climate warming. The Battery gauge shows 28 cm of relative rise over the past century and of that total about 12.2 cm is due to land subsidence as a result of rebound of Earth's crust. Moreover, the Sandy Hook, New Jersey gauge located across Lower New York Harbor has even higher rates of 39 cm LSLR over the past century, of which ~22.7 cm is due to subsidence [22].

Integrated elements of plans to rebuild after Sandy include elevating structures, nourishing barrier islands and dunes, and flood-proofing major infrastructure. Another long-term infrastructure adaptation measure considered was installation of storm surge barriers across vulnerable openings to the sea, including the Verrazano Narrows, upper East River and the Arthur Kill channel.

New York City Climate Change Plans for Adaptation

Following Sandy, NYC undertook comprehensive studies of the effects of Sandy on the coast as well as the urban infrastructure and developed plans for making the city more resilient to future storms and the increasing impacts of climate warming [23]. The plans were intermediate in time span, looking to year 2080 (Figure 11). For projecting sea-level rise, two model scenarios were employed. One, using methods from the recent IPCC assessment resulted in GSLR projections of 20.3–58.4 cm by 2080. A second method that included local subsidence and uncertainty values, more rapid rates of ice sheet melt for Greenland and Antarctica and greater atmospheric GHG emissions resulted in LSLR projections of 1.0–1.4 m by 2080, which is in close agreement with [6]. Beyond 2080, GSLR very likely will continue, possibly at even higher rates. Levels of 3 m or higher are possible over the next several hundred years if current high GHG emissions, ice sheet melting and global warming continue.

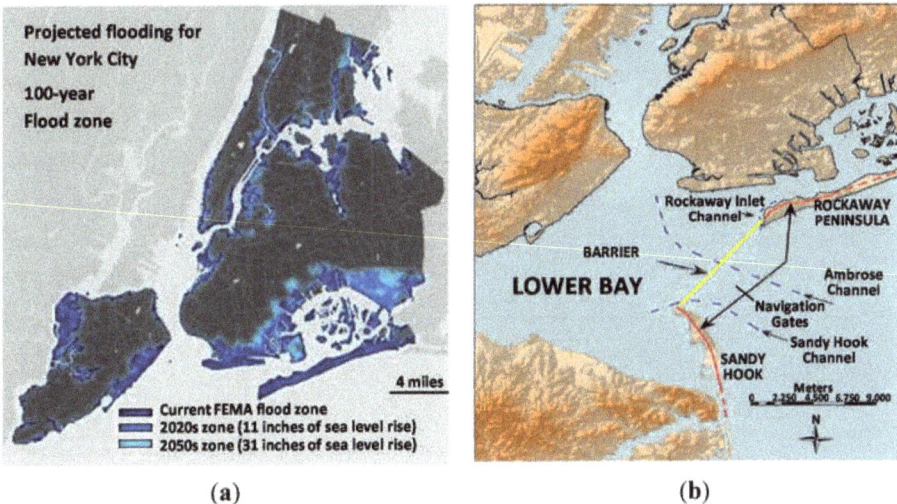

(a)   (b)

Figure 11. (**a**) NYC projected flooded areas for 100-year storm, SLR 79 cm by 2050 [23]. (**b**) Proposed surge barriers [24] and surge barrier with gates [24].

For many recent U.S. coastal protection plans, sea level change is factored in as a critical element from response to Hurricane Sandy, to the North Atlantic Coast Comprehensive Study, to analyses for projects currently in planning and engineering design. The planning process by NYC will be refined as climate science and predictive tools improve. New Jersey also sustained major flood damage and loss of life from Sandy and restoration is underway, mostly on barrier islands, but except for some floodplain "buy-outs", there appears to be less planning for future conditions and adaptation. Integrated elements of plans to rebuild after Sandy include elevating structures, nourishing barrier islands and dunes, and flood-proofing major infrastructure.

For New York, conceptual design of storm surge barriers was presented by Arcadis [24] across the vulnerable openings to the sea including the Verrazano Narrows, upper East River and Arthur Kill. Due to requirements of maintaining ship navigation, tidal exchange, and river discharge, barrier gates were proposed by Halcrow [24] within a surge barrier to be placed across Ambrose channel to accommodate passage of navigating vessels (Figure 9b). Due to high cost and complexity of such large structures, surge barriers are not in current plans.

9. Conclusions and Recommendations

Review of response and planning of coastal projects following major storms in Egypt, Louisiana, and the New York City region are presented. The review considers varying degrees of climate change impacts projected for the next 50–100 years. To reduce risk and vulnerability, [3,8] recommended that adaptation planning using projected GSLR of 0.5–2 m by 2100, plus local geophysical and man-made factors, is advisable. Planning should fit the time frame and degree of protection desired. These boundary conditions demand that the protection of coastal areas requires flexible designs that can be adapted in a phased manner to increasing sea-level rise and increased storm activity over the next 50 to 100 years and beyond [7]. The reality is that it is not economically feasible to protect all coastlines from high rates of sea-level rise and catastrophicsuper-storms. Structures built for short term (~50 years) and modest sea-level rise (~0.5 m) should be designed such that their height can be raised and size increased in the event that even higher sea levels prevail. New planning protocols need to employ cost-effective coastal protection technologies affordable by developing countries. Engineering plans and design need to incorporate alternatives such as beach nourishment, dune stabilization, and wetland restoration, which could serve as buffers protecting mainland areas from full effects of storm surge and wave action as well as provide environmental benefits. High-risk urban areas will likely require hard structures (e.g., seawalls, revetments, and flood barriers) and pump station infrastructure. Pump operations, however, need to be assured during extended storms. Strategic development of new urban planning policies, including relocation, should be considered when risk is great and in-place adaptation is no longer economically feasible.

Acknowledgments: We would like to thank three anonymous reviewers and the editor for their constructive and helpful edits and suggestions that improved the quality of the paper.

Author Contributions: S.J.W. wrote parts of the manuscript on climate change, sea-level rise impacts, Louisiana management plans, and parts of the New York City case study. N. I. had the lead writing on Egypt and parts on the Louisiana Lake Borgne Surge Barrier project.

Conflicts of Interest: The authors declare no conflicts of interest.

References

1. IPCC. *Contribution of Working Group II to the Fourth Assessment Report of Intergovernmental Panel on Climate Change 2007*; Parry, M.L., Canziani, O.F., Palutikof, J.P., van der Linden, P.J., Hanson, C.E., Eds.; Cambridge University Press: Cambridge, UK; New York, NY, USA, 2007; p. 976.
2. Sweet, W.V.; Park, J. From the extreme to the mean: Acceleration and tipping points of coastal inundation from sea-level rise. *Earth's Future* **2014**, *2*, 579–600. [CrossRef]

3. Williams, S.J. Sea-level rise implications for coastal regions. In *Understanding and Predicting Change in the Coastal Ecosystems of the Northern Gulf of Mexico*; Brock, J.C., Barras, J.A., Williams, S.J., Eds.; Coastal Education and Research Foundation, Inc.: Coconut Creek, FL, USA, 2013; pp. 184–196.

4. Parris, A.; Bromirskiet, P.; Burkett, V.; Cayan, D.; Culver, M.; Hall, J.; Horton, R.; Knuuti, K.; Moss, R.; Obeysekera, J.; *et al. Global Sea-Level Rise Scenarios for the U.S. National Climate Assessment*; NOAA Tech Memo OAR CPO-1; Climate Program Office (CPO): Silver Spring, MD, USA, 2012; p. 37.

5. IPCC. *Summary for Policy Makers, Climate Change: 5th Assessment Report of the Intergovernmental Panel on Climate Change 2013*; Stocker, T.F., Qin, D., Plattner, G.-K., Tignor, M.M.B., Allen, S.K., Boschung, J., Nauels, A., Xia, Y., Bex, V., Midgley, P.M., Eds.; Cambridge University Press: Cambridge, UK; p. 28.

6. Kopp, R.E.; Horton, R.M.; Little, C.M.; Mitrovica, J.X.; Oppenheimer, M.; Rasmussen, D.J.; Strauss, B.H.; Tebaldi, C. Probabilistic 21st and 22nd century sea-level projections at a global network of tide-gauge sites. *Earth's Future* **2014**, *2*, 383–406. [CrossRef]

7. U.S. Army Corps of Engineers (USACE). *Procedures to Evaluate Sea Level Change: Impacts, Responses, and Adaptation, ETL 1100-2-1*; USACE: Washington, DC, USA, 2014; p. 254.

8. Moser, S.C.; Williams, S.J.; Boesch, D.F. Wicked Challenges at Land's End: Managing Coastal Vulnerability under Climate Change. *Annual. Rev. Environ. Res.* **2012**, *37*, 51–78. [CrossRef]

9. Rignot, E.; Mouginot, J.; Morlighem, M.; Seroussi, H.; Scheuchl, B. Widespread, rapid grounding line retreat of Pine Island, Thwaites, Smith, and Kohler glaciers, West Antarctica, from 1992 to 2011. *Geophys. Res. Lett.* **2014**, *41*, 3502–3509. [CrossRef]

10. Williams, S.J.; Kulp, M.; Penland, S.; Kindinger, J.L.; Flocks, J.G. Mississippi River Delta Plain, Louisiana coast and inner shelf: Holocene geologic framework and processes, and resources. In *Gulf of Mexico Origin, Water and Biota: Vol. 3, Geology*; Buster, N.A., Holmes, C.W., Eds.; Texas A & M University Press: College Station, TX, USA, 2011; pp. 175–193.

11. Ismail, N.; Iskander, M.; El-Sayed, W. Assessment of coastal flooding at southern Mediterranean with global outlook for lowland coastal zones. In Proceeding of International Conference on Coastal Engineering, ASCE, Santander, Spain, 1–6 July 2012.

12. Ismail, N.M. Effect of wave-current interaction on littoral drifts. *J. Am. Shore Beach Preserv. Assoc.* **1982**, *50*, 35–38.

13. Ismail, N.M.; Wiegel, R.L. Opposing waves effect on momentum jets spreading rate. *J. Waterw. Port. Coast. Ocean. Div.* **1983**, *109*, 465–483. [CrossRef]

14. Kadib, A.L.; Shak, A.T.; Mazen, A.; Nadar, M.K. Shore protection plan for the Nile Delta coastline. In Proceeding of 20th International Conference on Coastal Engineering, Taipei, Taiwan, 9–14 November 1986; Volume III, pp. 2530–2544.

15. U.S. Army Corps of Engineers (USACE). *Louisiana Coastal Protection and Restoration (LACPR), Final Technical Report*; USACE: New Orleans District, MVD, USA, 2009; p. 293.

16. Tetra Tech. Inner Harbor Navigation Canal (IHNC) Hurricane Protection Barrier. Avaiable online: http://www.incainc.com/flood-control/inner-harbor-navigation-canal-ihnc-hurricane-protection-barrier.html (accessed on 20 January 2015).

17. U.S. Army Corps of Engineers (USACE). Climate change adaptation plan. Available online: http://www.iwr.usace.army.mil/Media/NewsStories/tabid/11418/Article/494304/procedures to-evaluate-sea-level-change-impacts-responses-and-adaptation.aspx (accessed on 15 April 2015).

18. Huntsman, S.R. Design and construction of the lake Borgne surge barrier in response to Hurricane Katrina. In Proceeding of Conference on Coastal Engineering Practice 2011, San Diego, CA, USA, 21–24 August 2011.

19. DeSoto-Duncan, A.; Hess, C.; O'Sullivan, M. "The Great Wall of Louisiana" Protecting the Coastline from Extreme Storm Surge and Sea Level Rise. In Proceeding of the 2011 Solutions to Coastal Disasters Conference, Anchorage, AK, USA, 26–29 June 2011; pp. 690–701.

20. Grieshaber, J.; USACE/Hurricane Protection Office. *Inner Harbor Navigation Canal (IHNC) Basin 1% (100-year) and 0.2% (500-year) Surge and Wave Event Water Levels*; SE Louisiana Flood Protection Authority: New Orleans, LA, USA, 2010.

21. Coastal Protection and Restoration Authority of Louisiana (CPRA). *Louisiana's Comprehensive Master Plan for a Sustainable Coast*; Coastal Protection and Restoration Authority of Louisiana: Baton Rouge, LA, USA, 2012; p. 190.

22. Zervas, C.; Gill, S.; Sweet, W. *Estimating Vertical Land Motion from Long-Term Tide Gauge Records*; Tech Report NOS CO-OPS 065; NOAA: Silver Spring, MD, USA, 2013; p. 22.
23. NYCPCC. Climate risk information 2013: Observations, climate change projections, and maps. In *New York City Panel on Climate Change*; Rosenzweig, C., Solecki, W., Eds.; NYCPCC: New York, NY, USA, 2013; p. 38.
24. Hill, D.; Bowman, M.J.; Khinda, J.S. Storm surge barriers to protect New York against the deluge. In Proceeding of COPRI Committee Report, ASCE Conference, Brooklyn, NY, USA, 30–31 March 2009.

Journal of
Marine Science and Engineering

MDPI

Article

Coastal Flood Assessment Based on Field Debris Measurements and Wave Runup Empirical Model

David Didier *, Pascal Bernatchez, Geneviève Boucher-Brossard, Adrien Lambert,
Christian Fraser, Robert L. Barnett and Stefanie Van-Wierts

Coastal Geoscience, Centre for Northern Studies, University of Quebec in Rimouski,
Rimouski, QC G5L3A1, Canada
* Author to whom correspondence should be addressed; David_Didier@uqar.ca; Tel.: +1-418-723-1986
(ext. 1364).

Academic Editor: Rick Luettich
Received: 22 April 2015; Accepted: 23 June 2015; Published: 15 July 2015

Abstract: On 6 December 2010, an extra-tropical storm reached Atlantic Canada, causing coastal flooding due to high water levels being driven toward the north shore of Chaleur Bay. The extent of flooding was identified in the field along the coastline at Maria using DGPS. Using the assumption that the maximum elevation of flooded areas represents the combination of astronomical tide, storm surge and wave runup, which is the maximum elevation reached by the breaking waves on the beach, all flood limits were identified. A flood-zone delineation was performed using GIS and LiDAR data. An empirical formula was used to estimate runup elevation during the flood event. A coastal flood map of the 6 December flood event was made using empirical data and runup calculations according to offshore wave climate simulations. Along the natural beach, results show that estimating runup based on offshore wave data and upper foreshore beach slope represents well the observed flood extent. Where a seawall occupies the beach, wave breaking occurs at the toe of the structure and wave height needs to be considered independently of runup. In both cases (artificial and natural), flood risk is underestimated if storm surge height alone is considered. There is a need to incorporate wave characteristics in order to adequately model potential flood extent. A coastal flooding projection is proposed for Pointe Verte based on total water levels estimated according to wave climate simulation return periods and relative sea-level rise for the Chaleur Bay.

Keywords: storm surge; wave runup; coastal flood mapping; coastal defense

---

1. Introduction

Sea-level rise, as a consequence of climate change, is a primary concern to coastal communities due to the increased threat of storm surges and coastal flooding [1,2]. The rate of global mean sea-level rise increased from 1.9 mm/year to 3.2 mm/year between the periods 1961 to 2009 and 1993 to 2011 [3,4]. This acceleration, combined with future projections of sea level and extreme water level change [5–11] has led the scientific community to pay greater attention to the development of vulnerability assessment methods for coastal communities and infrastructure to flooding phenomenon [12]. Previously, the vulnerability of coastal communities to future sea-level rise has been estimated by applying hypothetical sea-level increases (e.g. often a hypothetical 1 m rise is used, [13–15]) or predictions from SRES and IPCC scenarios [16,17] to topographic maps or digital elevation models at national or greater scales [18,19]. However, adaptation approaches to coastal flooding require the development of local to regional level assessments [20]. In this context, the regional distribution and amplitude of sea-level rise remains a significant uncertainty when forecasting future trends [6]. In addition, the isostatic component (vertical land motion) is still rarely accounted for in coastal flooding assessment methods.

It is evidently essential, therefore, to assess future *relative* sea-level rise [21], especially in areas experiencing crustal subsidence due to postglacial isostasy [22,23].

Coastal flooding occurs when water levels overtop the first natural ridge or flood defense crest landward of the beach, generating landward flow and sediment transport [24]. Alongshore water level variability, relative to the crest elevation, further affects overwash potential [25] and can lead to an increase in the risk of flooding. During coastal storms, resulting water level maxima are a product of astronomical tide, barometric pressure and wind-induced surge [26,27]. In addition, landward propagating waves in shallow waters are affected by beach morphology due to the effects of, e.g., shoaling, wave breaking and energy dissipation [28]. Such interactions can increase water levels in the onshore direction through the process of wave runup following breaking in the surf zone [29]. Runup (*R*) is defined as the vertical elevation difference between the maximum shoreward location of water and still water level (predicted tide + surge) [30–32] and includes two components: the setup and swash [33]. Setup is defined as a mean water surface elevation in response to wave breaking, which is a superelevation, whereas swash refers to oscillations across the water-land interface around setup. The sum of tide prediction (T_{pred}), surge (S_{surge}) and runup on natural beaches correspond to total water levels (TWL) [34,35]. Runup is generally calculated in terms of the two percent exceedance value of wave runup ($R_{2\%}$) for the TWL [34].

Runup (*R*) on a natural beach mainly depends on offshore significant wave height (*H0*) and wavelength ($L_0 = gT^2/2\pi$; where g is the acceleration due to gravity (981m/s), *T* is the wave period (s)) and the beach slope [29,36–38] grouped under the non-dimensional Iribarren number ξ. The Iribarren number, also referred to surf zone similarity parameter, is widely used in the computation of runup and overtopping discharge [37] and in near-shore process assessments [28]. Beach slope is therefore an important parameter for determining maximum wave runup height along coasts. However, beach surfaces used to calculate slopes, which can differ significantly between case studies, are not always clearly identified or described in the literature [29,32,34,39].

Wave runup on structures is typically calculated using a design wave for an expected water level in order to determine the required crest elevation for defense structures [40]. Types of structures range from low sloping coastal defenses to vertical seawalls. The main factors that influence flooding over sloping structures are wave runup and overtopping [41,42]. However, for vertical seawalls where $\xi = \infty$, wave runup is minimal and is limited to a theoretical value of $R_{2\%}/H_0 = 1.4$ [43]. In such cases, where waves frequently impact the structure, wave height is usually twice the incident wave amplitude and can produce a vertical plume of *circa* 5.5 times the wave amplitude [44]. On sloping coastal defenses, such as dykes, wave overtopping occurs when wave runup exceeds the height of defense structures [45].

It is a challenge to accurately model the landward flow of water following overtopping of defense structures due to complex interactions between wave morphology and the structures themselves [46]. Studies on flow characteristics and discharge rates following wave overtopping have highlighted the importance of the roughness factor and permeability associated with different defense structures [47,48]. Whereas overtopping associated with different coastal structure types has been evaluated [47,49], less is known on the spatial propagation of water levels in backshore zones following overtopping of the structures [46], which is of key relevance to coastal managers responsible for land use planning and flood risk management [50]. Moreover, the establishment of rigid defense structures in response to ongoing erosion processes often results in decreasing beach and intertidal zone widths [51]. This can lead to increased wave overtopping of defense structures during storms [52,53]. Such modifications to natural beach states can increase flood risk [51,54]. The alongshore variability in flood elevation as a result of human intervention in response to coastal hazard represents a threat to coastal communities.

Studies on runup have typically been carried out along stretches of natural beaches (e.g. [29,38,55–57]) or in areas of armoured coastline (e.g. [41,42]). In reality, coastal zones prone to flood risk will contain a combination of both natural and artificial sections, which highlights the need for a local to regional approach when considering coastal flooding vulnerability.

J. Mar. Sci. Eng. **2015**, *3*, 560–590

To date, the majority of coastal flooding studies have been carried out along coastlines exposed to ocean swell with long fetch and rarely on sheltered embayments [58]. In sheltered coastal environments, it is generally considered that the waves have less influence on coastal flooding and the tide and storm surge are the main drivers [58]. In Eastern Canada, the entire area of the Chaleur Bay (Figure 1) has been hit by two major and destructive storms recorded within a five year interval (December 2005 to December 2010), despite the fact that the Acadian peninsula partially protect the bay to wave agitation coming from the Gulf of St. Lawrence. However, the recent storm events have raised questions with regards to the influence of wave effects along coastlines in sheltered bays with short fetch. There is a critical need to understand wave influence in these environments in order to adapt and improve coastal zone management in Eastern Canada.

Mapping coastal flood risk areas is an important step for risk management planning and for guiding adaptation procedures necessary to reduce vulnerability to flooding. The classical method of flood mapping, which typically integrates extreme water levels, derived from historic tide gauge data, with a digital elevation model ("bathtub model") to identify areas at risk [31,59–61], has been criticized in recent years [62]. Recent storm events along North American (e.g. Katrina, Wilma, Sandy) and European (e.g. Xynthia) coastlines have highlighted the need to understand the effects of beach geomorphology and hydrodynamic conditions close to the coast in order to accurately determine backshore zones at risk of flooding [20,31,34]. Due to a paucity of wave data at local to regional scales, wave runup is not always included in coastal flood mapping. This is especially prevalent in Atlantic Canada [63,64]. Another challenge to flood risk mapping derives from forecasting extreme water level return periods based only on short tide gauge series [65,66]. It is also important to integrate the probability of extreme tidal water levels with runup data [34] for future coastal management and robust flood hazard mapping. Recent studies have shown that airborne LiDAR data, widely used for digital elevation models in flood hazard mapping, are not precise enough to accurately identify crest elevations of the first line of defense and to represent adequately elevation changes of flood defenses (embankment crests and walls) or beach berms at the coastal front or along river banks, which leads to over- or under-estimating of risk areas [67–70].

The development of methods for flood hazard mapping remains a challenge, especially at the local to regional scale, in order to be applicable and easy to use by local authorities responsible for land use and risk management. Therefore, only few approaches integrate the different components required to effectively map the exposed areas to flood hazards for the future.

The aim of this paper is to develop a coastal flood mapping approach based on *in situ* measurements, modelled wave characteristics, sea-level rise, estimated extreme water levels return periods, and wave runup estimations using empirical formula. This method is compared and validated from field surveys during and immediately after the 6 December 2010 storm event that caused extensive damage to the coast of Eastern Canada. We compare the altitude reached by the total water level during the storm based on the presence or not of coastal defenses and assess the importance of the waves in the flood pattern for sheltered coasts. Since slope determination is often of primary concern in runup parameterization, a focus was made on beach morphology with a mobile terrestrial LiDAR system (MTLS) to calculate beach slopes. The use of terrestrial LiDAR system also aims to solve the problem of identifying the elevation crest of the first coastal defense line. Finally, the proposed approach addresses future coastal flood scenarios at the community scale that can be easy to apply for coastal managers.

2. Study Area

In Atlantic Canada, common extra-tropical cyclones typically generate northwestern and northeastern winds toward the shores of the Gulf of St. Lawrence, creating storm surges and high waves [71,72]. As a result of these low-pressure systems, coastal flooding and erosion have increased in recent years throughout the Chaleur Bay [73]. The study area is located in the municipality of Maria (Figure 1) on the north shore of the Chaleur Bay in the province of Quebec, Eastern Canada.

Maria represents a typical coastal community for Eastern Canada, with a population of 2500 and an established mix of infrastructure (such as residential, recreational, commercial and tourism). The national road, 132, which links coastal towns throughout the Gaspé Peninsula with eastern Quebec, is particularly vulnerable to flooding at Maria [74]. Despite damage caused by several flooding events [75], there is currently no coastal flood hazard mapping in the province of Quebec.

Flood hazard has been assessed for the southeast part of Maria where there are more than 40 houses built on the sandy low-lying coast (Figure 1). This area was flooded both on 2 December 2005 and 6 December 2010. This stretch of coastline contains a uniform sandy and gravel beach to the east (~22 m width), which is in a quasi-natural state (Figure 2a) and a protected zone to the west, which contains a wooden seawall protecting different properties [52]. The eastern section contains some low lying coastal defenses, which are situated landward of the beach or on the upper beach, with a structure toe above HAT. We therefore assume an alongshore and natural wave energy dissipation across the beach in this area, where runup can occur on the sloping beach, as underlined by the Mase et al. [41] study on wave runup and overtopping.

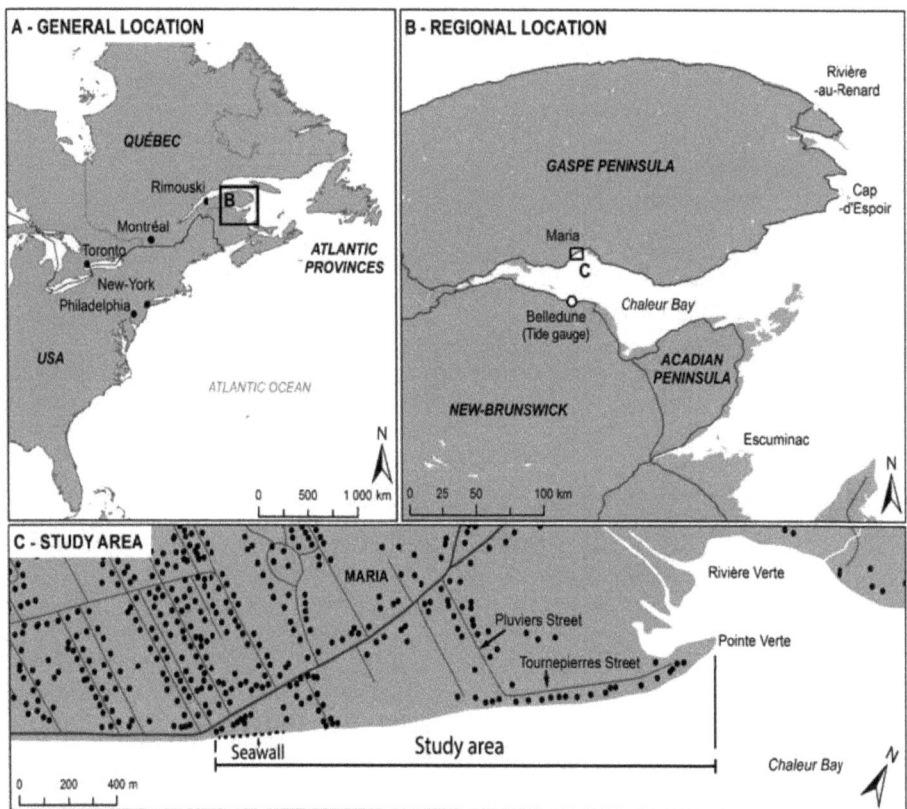

Figure 1. Location map of the study area in Maria (Quebec) on the north shore of the Chaleur Bay.

At the western zone of the study site, almost the entire beach has been eroded as a result of wave reflection and scouring at the wall toe [51,52] (Figure 2b). Beach erosion has exposed a substrate of pebbles and gravel (Figure 2b). Backshore elevation in this area is higher than in the eastern part of the site. The region experiences a semi-diurnal and meso-tidal regime, with a mean sea level (MSL) of 1.33 m relative to Chart Datum [76] (Table 1). During highest astronomical tide (HAT) (1.66 m), the seawall toe is submerged and oscillating water levels occur on the seaward face of the wall. As observed during the 2005 flood event [52], continuous wave overtopping can occur during periods of high water levels and large waves.

Figure 2. The beach of the study area is mainly natural on the eastern part (**a**) and characterized by a seawall on the western coastline (**b**).

Table 1. Mean estimated tidal values in 2010 according to chart and geodetic datum, Belledune station (NB) (Canadian Hydrographic Service).

Water Levels	Mean Estimated Value (2010) Chart Datum (CD)	Canadian Geodetic Vertical Datum 1928 (CGVD28)
Extreme level (tide + storm surge)	3.64	2.46
Highest Astronomical Tide (HAT)	2.84	1.66
Mean Sea Level (MSL)	1.33	0.15
Lowest Astronomical tide (LAT)	0.1	−1.08

3. Methodology

3.1. Post-Storm Surveys

Storm debris [77,78] and tidal deposits [79] can be used as water level limit and maximum runup extent indicators. A field survey was conducted on the day of, and immediately following, the 6 December 2010 flood event, in order to identify maximum flood water levels reached at Maria (Figure 3a).

Storm debris lines were located *in situ* using a DGPS Thales ProMark3 system (Figure 3b), with horizontal (x, y) and vertical (z) uncertainties of ± 1 cm. A total of 188 data points were collected for the 2010 event along debris lines, which were located along 2.2 km of coastline. A database of georeferenced points was then superimposed over a 2007 airborne LiDAR raster elevation data map (spatial resolution of 1 m, vertical accuracy of 20 cm) in ArcMAP. The points were connected together to form a continuous line indicating the landward limits of the water during the storm. Points were then generated every 5 m along the line and elevation values were extracted from the raster providing mean flood elevations for the (protected) western and (unprotected) eastern zones, relative to geodetic datum CGVD28.

3.2. Mobile Terrestrial LiDAR Data Acquisition

The alongshore and cross-shore variability in beach morphology at Maria has been studied by Bernatchez et al. [52] using DGPS profiles. Even if the natural beach is mainly homogenous in terms of sediments and morphological patterns (width, volume), morphological variability exists between areas with defense structures (west zone) and those in a quasi-natural state (east zone). We initially used airborne LiDAR data to map the first coastal defense line, but the spatial resolution and accuracy in z does not allow us to determine with sufficient precision the height of the crest elevation. We therefore used mobile terrestrial LiDAR system (MTLS) on the entire beach of the study area on 28 September 2011 (Figure 4a) to improve on spatial resolution and vertical precision of previous studies [51,52]. The survey was conducted at low tide and covered the entire foreshore to a minimum seaward elevation of 1 m below CGVD28 (close to LAT).

Figure 3. Field survey after the flood showing (**a**) flood generated overwash fan and (**b**) the debris line delineation. Limits were located using DGPS on the day following the storm event (Maria, 7 December 2010).

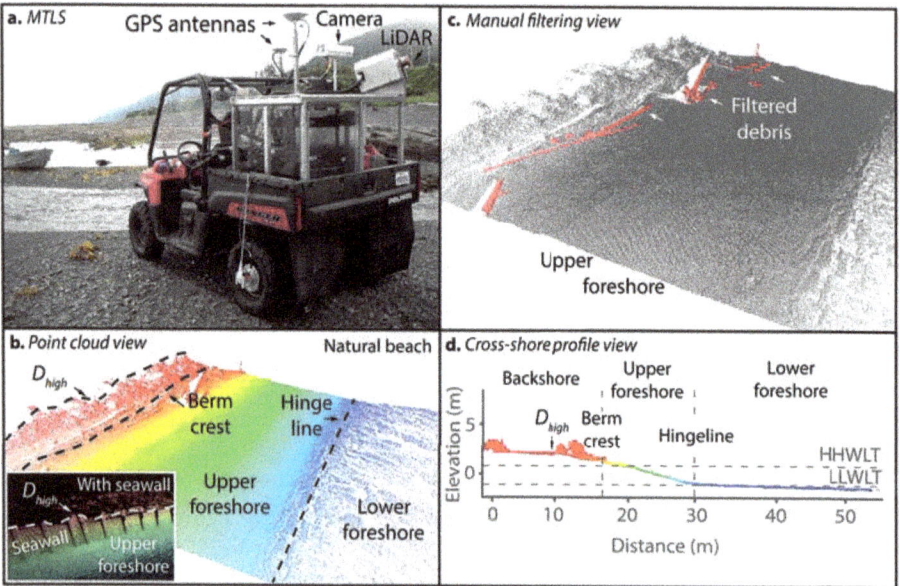

Figure 4. Mobile terrestrial LiDAR system (**a**) with point cloud and filtering view examples (**b**,**c**). Resulting data are used in cross-shore view to localize the first ridge on the natural beach or the coastal structure crest elevation (D_{high}), the berm crest and the hinge line (**d**).

The MTLS is a multi-sensor system mounted on a utility side by side vehicle (Figure 4). The multi-sensor system includes a laser scanner Riegl VQ-250, a camera PointGrey Grasshopper GRAS-50S5C-C and a GPS inertial navigation system Applanix POS-LV 220 (Figure 4a) [80]. Post-processing was performed using PosPAC and Trimble Trident Analyst software. The projection used was NAD83 MTM5 with geoid HT2.0. The LiDAR point cloud has a spatial resolution of less than 0.05 m × 0.05 m (Figure 4b). Different types of beach debris were filtered manually in LP360-QCoherent software (Figure 4c). A grid elevation with a spatial resolution of 0.5 m was created in LP360-QCoherent software in order to identify the beach slope with high accuracy.

The first line of defense (D_{high}), corresponding to the defense structure crest in the west zone, or to the crest elevation of the first ridge landward of the berm crest in the east zone, was manually digitized in ArcMAP. To calculate the accuracy of the laser data, a total of 23 control points were measured using a D-GPS Trimble R8 over the entire study site. The error was calculated by the root-mean-square (RMS) of the difference between control points and the adjacent laser points. Calculated horizontal (x,y) and vertical (z) errors are ± 0.03 m (RMS).

3.3. Hydrodynamic Data Analysis

3.3.1. Wave Dataset

No measured wave data are available for the Chaleur Bay. We therefore used a modelled wave dataset from Environment Canada's (EC) operational wave forecasting model based on WAM cycle 4 MW3 "PROMISE" [81–84]. Archived outputs cover a 10-year period, from 1 January 2003 to 31 December 2013 with a data hiatus for 2006. Time series are available at three-hour increments.

EC's implementation of WAM extends throughout the St. Lawrence Gulf and Estuary, from Montmagny to Cabot Strait and Belle-Isle Strait, at a spatial grid resolution of 0.04° latitude and 0.06° longitude (approximatively 5 by 5 km at 45° north). Selected source terms include formulation for exponential wave growth [85], formulation for linear wave generation [86], formulations for deep and intermediate water non-linear wave–wave interactions (quadruplets) and dissipation due to whitecapping [81], JONSWAP formulation for dissipation due to bottom friction [87], and formulation for dissipation due to depth-induced breaking [88].

Wave propagation schemes include refraction over bottom and non-stationary ambient currents. Non-stationary boundary conditions are based on 40 m tri-hourly winds from EC's Global Environmental Multiscale Model (GEM) and surface currents (at 5 m depth) from EC's oceanographic model of the Estuary and Gulf of St. Lawrence (MOR). Output wave parameters derived from spectra are significant spectral wave height (H_{m0}), mean spectral wave period (T_{m02}) and mean wave direction (Dir).Wave data quality has been assessed by Jacob et al. [84] and Lambert et al. [89–91] for the Estuary and Gulf of St. Lawrence domain. Selected data for this study were extracted from EC's model at grid point i = 67, j = 69 (−65.950 O, 48.1 N), located 7 km east south-eastward from Maria's shoreline, to a depth of −18.2 m. Given the limited fetch length in front of Maria (70 km), most simulated wave periods were under 5 s. We therefore selected a position for wave parameter extraction close to the shore yet remaining in deep-water conditions ($d > 1/2 L_0$) for most wave periods. Selected wave records were linearly interpolated to an hourly resolution in order to match temporal resolution with the available tidal dataset.

3.3.2. Determination of Near-Shore Wave Parameters

Runup values, as a result of wave setup and uprush, are determined by deep-water significant wave conditions (H_0 and L_0) and beach slope with the Stockdon et al. equation [29]. Wave shoaling and refraction over coastal waters can be empirically determined using weighting coefficients rather than physical computations. We will refer to H_0 in the paper as a reference to H_{m0}. The depth of wave breaking and the location of the break point on the sloping beach or relatively to a defense structure is an important factor contributing to erosion and flooding. Depth of breaking d_b is computed from (1) [92]:

$$d_b = \frac{1}{g^{\frac{1}{5}} \gamma^{\frac{4}{5}}} \left(\frac{H_0^2 C_0 \cos \theta_0}{2} \right)^{\frac{2}{5}}$$

(1)

where $\gamma = 0.8$, θ_0 = deep-water wave angle relative to shore normal direction (incident wave = $0°$), and deep-water phase velocity $C_0 = \frac{gT}{2\pi}$.

3.4. Wave Runup Estimation and Coastal Morphology

The flood assessment applied in this paper estimates wave runup with the empirical formulation proposed by Stockdon et al. [29] based on *in situ* measured runup values via extensive research conducted on ten sandy field experiments in the USA and Netherlands. The following equation has been applied because it is valid for a broad range of reflective to dissipative beaches:

$$R_{2\%} = 1.1 \left(0.35 \tan \beta (H_0 L_0)^{1/2} + \frac{\left[H_0 L_0 \left(0.563 \tan \beta^2 + 0.004 \right) \right]^{1/2}}{2} \right)$$

(2)

where $R_{2\%}$ is the elevation above the still water level that is exceeded by 2% of the wave runups [53] and $\tan \beta$ is the beach slope. This empirical runup formulation (RMS error = 0.38 m), hereinafter referred to as the RS06 model, is considered a good predictor of runup [25,33,93] and is often used in coastal flood and vulnerability assessment [67,94–96]. The static flood model applied in this paper consists of a summation of observed tidal level and storm surge, wave characteristics and wave runup according to the RS06 empirical model. It is based on the comparison of total water level (TWL) against ground elevation assuming an instant and constant flood surface on the area [67]. Despite its simplicity, the RS06 formulation requires a beach morphology assessment. In natural settings, the beach slope is not a straightforward calculation [32,39,97]. Some authors suggest using the foreshore beach slope in the computation of the empirical formula [29,30,63,96]. Cariolet and Suanez [32] obtained a good fit relationship between measured and predicted runup using the slope of the active section of the beach in a macrotidal environment, which corresponds to the upper foreshore slope. This is in accordance with Stockdon et al. [25] who used the mean beach slope between the berm crest and mean high water level to define $\tan \beta$ in storm conditions where the swash is moved up the beach. Two beach slopes on the natural beach were compared in our approach (Figure 5).

Beach slope at high tide during the flood event corresponded to the slope between the point of maximum runup elevation on the beach (R_{max}) and $T_{obs} - 1.5H_0$ [43], where T_{obs} is the still water level without the wave component. This slope, referred to as $\tan \beta_{1.5H0}$, was compared to the upper foreshore slope ($\tan \beta_{ufs}$). The upper foreshore slope is relatively located upward of mean sea level (CGVD = 0.15 m) and is located between the point of maximum runup elevation and the hinge line. Considering the maximum flood height along the coast on 6 December 2010, R_{max} is assumed to be equal to the berm crest elevation in both situations. Both berm crest and hinge line were manually delineated for the entire beach as a line feature in LP360 using a 3D visualization window (Figure 4b,d). The width of the beach Y_1 was calculated between beach berm crest and $T_{obs} - 1.5H_0$. Using the MTLS derived raster, a contour line equal to $T_{obs} - 1.5H_0$ was automatically digitised in ArcMAP. The mean beach slope, $\tan \beta_{1.5H0}$, was then defined $(Z_{crest} - Z_{Tobs-1.5H0})/Y_1$. The mean beach slope, $\tan \beta_{ufs}$, given a beach width Y_2 between the berm crest and the hinge line, equals to $(Z_{crest} - Z_{Hinge line}/Y_2)$. Beach slopes were calculated every 50 meters along the coast, for a total of 28 cross-shore profiles.

In the present study, where high water levels and wave conditions led to overtopping and flooding, bottom induced wave breaking was located close to the seawall. Since radiation stress induced wave set-down is at its maximum at the breaking point and static wave setup starts at the breaking point [98], we can assume that there is no contribution of wave setup to the total water level on the wall. As a consequence, neither runup nor setup were calculated in front of the seawall in the west zone. During the 6 December event, the cumulative offshore wave height plus storm water level elevation overtopped the defense crest by several tens centimeters. Thus, an approach using only the wave height as a criterion for flood assessment has been proposed (Figure 6).

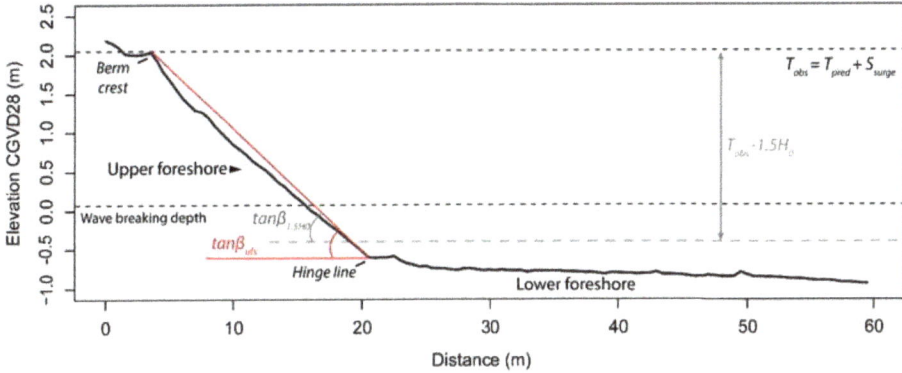

Figure 5. Natural beach slopes definitions used in the empirical runup formula. The slope $\tan\beta_{1.5H0}$ is located between the point of maximum runup elevation on the beach (R_{max}) and $T_{obs}-1.5H_0$. The upper foreshore slope ($\tan\beta_{ufs}$) is located between the point of maximum runup elevation and the hinge line. See the profile location in Figure 10, on the beach only (black line).

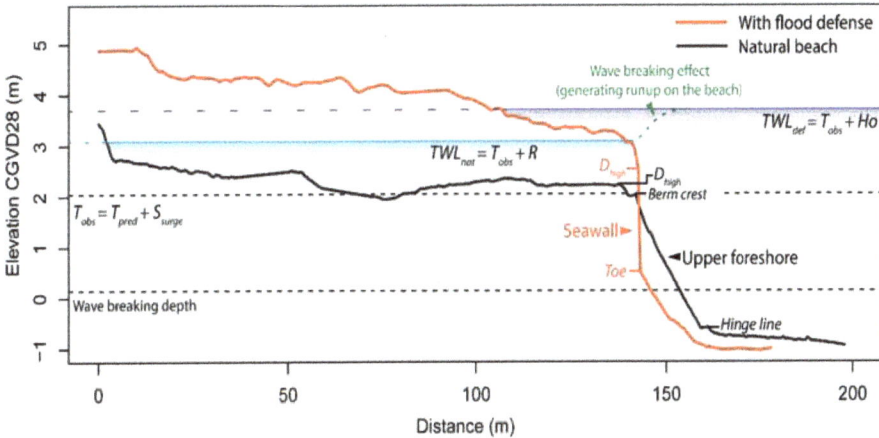

Figure 6. Sketch of the morphological difference between the defended (red) and natural (black) coastline. Modelled TWL are shown for both the natural beach (light blue) and defended coastline (dark blue). Profile locations are shown in Figure 10.

3.5. Sea-Level Rise Projection and Flood Mapping

3.5.1. Sea-Level Analysis

Relative sea-level rise, as a result of regional glacio-isostatic adjustment and eustatic contributions, is a concern along Atlantic Canada in terms of future coastal flood risk assessment [59,63,71,99]. In the southern Gulf of St. Lawrence, the isostatic crustal movement ranges from −1 to −4 mm/year [22,23]. In Maria, the rate is estimated to be −1.78 mm/year [100]. The nearby tide gauge at Belledune (Figure 1) provides a local relative sea-level trend of 4.06 mm/year between 1964 and 2014 for the study site, which is located 36 km northwest of the tide gauge. This tide gauge recorded data in 1964 and between 2000 and 2014. Hourly observed water levels were imported from the Canadian Tides and Water Levels Data Archives, available online [101]. The hourly predicted water levels were provided by the Canadian Hydrographic Service and were used to calculate the magnitude of surges [102]. Original tidal data are expressed relative to Chart Datum, which is 1.18 m below geodetic datum (CGVD28). All tidal and elevation references are expressed relative to CGVD28. Hourly data were smoothed to obtain monthly mean tidal levels from which the sea-level rise (SLR) trend was extrapolated. The averaged seasonal cycle and months containing less than 90% of data were removed following Daigle [103]. The SLR trend was obtained using linear regression following [104]:

$$Y = b_0 + b_1 X + \varepsilon \tag{3}$$

where b_1 is the regression coefficient estimating the SLR trend. An estimation of future water levels with the associated confidence intervals was also performed following the approach by Boon [104].

3.5.2. Estimation of Return Periods for Extreme Water Levels

For the natural beach at Pointe Verte, hourly wave runup time series were combined with measured sea-level (T_{pred} + surge) providing total water level (TWL$_{nat}$), following Ruggiero et al. [34]. For coastal flooding assessment, extreme water levels are most relevant. Different approaches based on observed water levels exist to calculated return periods [65,105–107] but all of them except the peak over threshold model [106] are time-series length dependent. Since the joint times series of wave runup data and water level data is relatively short (eight years), the peak over threshold model with the generalized Pareto distribution was used to compute return periods associated with extreme water levels [106]. This method is widely used elsewhere in the northern Atlantic for extreme water level predictions [108–110]. The analysis was performed on daily maximum levels (high water tide + sea level rise + surge + wave runup) rather than hourly values to avoid temporal dependence. A flood map assessment was performed for the 2010 event based on identified field debris and the RS06 model. The associated return period of this event was calculated with the Pareto analysis. Considering the relatively short time series of tidal data available from Belledune station, we limit the time horizon for flood mapping to 2030. Adding the sea level projection to tidal, barometric surge and runup components gives a future scenario of the same return period for any given year. For the natural beach:

$$TWL = T_{pred} + S_{surge} + R + SLR \tag{4}$$

where SLR is the sea-level rise between 2010 (T_1) and any future year T_2. For the artificially protected area, the runup component R is replaced by H_0 (see Section 3.4) thence, for defense structures:

$$TWL = T_{pred} + S_{surge} + H_0 + SLR \tag{5}$$

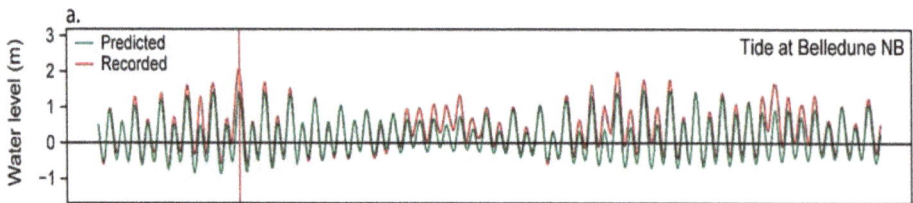

Figure 7. *Cont.*

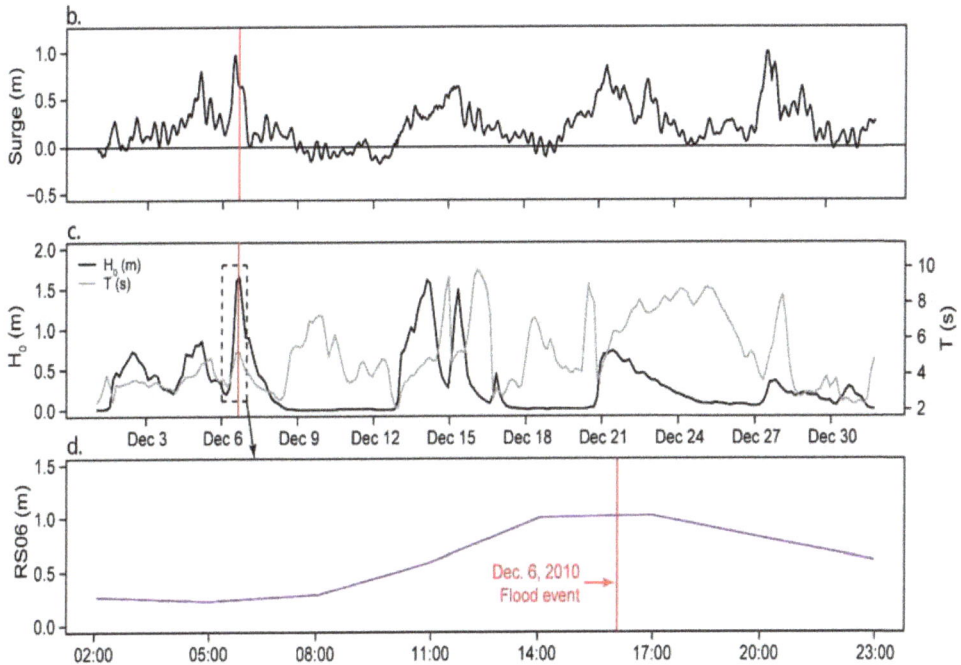

Figure 7. Tidal and hydrodynamic data on 6 December 2010. (a) Represents predicted and observed tide (m) at Belledune tide gauge and (b) The residual surge (m). The storm surge occurred during high waves reaching 1.63 m (c). (c,d) show the flooding event on 6 December, where a 1.02 m wave runup was estimated at 4:00 p.m. local time (8:00 p.m. UTC).

4. Results

4.1. Coastal Flood Mapping for the 6 December Event

4.1.1. Flood Event Analysis

During the day of 6 December 2010, the Belledune tide station recorded a high water level of 2.05 m at 4:00 p.m. local time (20:00 p.m. UTC), which included a storm surge of 0.63 m combined with an astronomical tide of 1.42 m (Figure 7a,b). Despite the fact that this water level was below the beach crest along most of the coastline (Figure 8), a significant area of the municipality of Maria was flooded by extreme water levels (Figure 9).

Mean observed flood limits reached 3.68 m and 3.07 m for areas with and without defense structures, respectively, therefore exceeding the observed tidal level (T_{obs}) by 1.63 m. The vertical seawall in the west zone increased total water elevations along this part of the coastline by 0.61 m in comparison to the natural zone. This is at least partly a result of the proximity of the wall to the still water level. The wall faces frequent wave impacts during such tidal conditions and the wall toe is completely submerged by more than 1.7 m of water depth, regardless of the wave component. Focusing on the wave breaking depth (D_b) enables a better understanding of the wave-breaking pattern both near the wall and on the natural beach (Figure 6). According to Equation (1), $D_b = 1.96$ m, providing a geodetic elevation of 0.09 m. Waves therefore break close to the toe of the wall, whilst in natural areas, breaking occurs on the foreshore slope (Figure 6). Where the seawall is built, no wave runup occurs. Instead, waves reach the defense almost at the breaking point.

J. Mar. Sci. Eng. **2015**, 3, 560–590

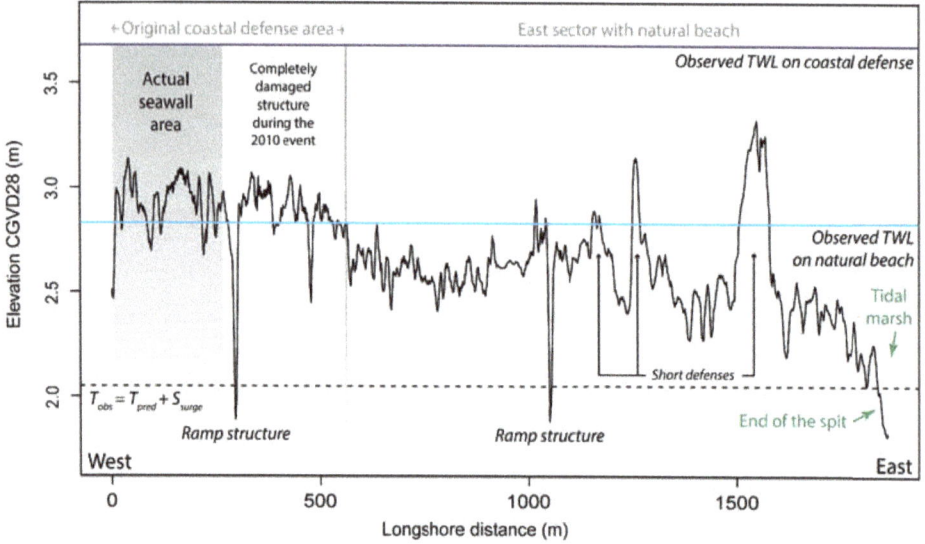

Figure 8. Recorded tide level (dashed black line) and mean observed flood limits along the coast. Longshore coastline crest elevation (D_{high}) variability between the natural zone and areas with defense structures (solid black line). Observed total water levels for the defended zone (dark blue line) exceeded those for the natural areas (light blue line).

Figure 9. (a) Flood extent as identified during the post-storm survey and location of the flow directions in response to land drainage; (b) The east zone, the landward side of the spit base acts as a natural outlet for the incoming drained seawater toward the tidal marsh. The picture was taken at 3:00 p.m. (local time) on 6 December, one hour before the flood peak.

During the 6 December 2010 event, unidirectional and landward flow submerged 0.18 km^2 of the coastline at Maria. Overwashed sediments were observed (see Figure 3a) to extend up to 20 m landward in the east and 30 m landward in the central part of the study site. Following north-westward landward propagation, flood waters receded in the opposite direction following land slope, drainage channels and pipes. Much of the floodwater drained along Tournepierres Street and through the tidal marsh (Figure 9b). As a result of this eastwards drainage via the tidal marsh, the landward flow of water in this zone was limited (Figure 9a). Houses located on embankments landwards of short defense structures (Figure 8) were spared from flooding. Results show that, despite the observed tide not reaching the beach or the defense crest on 6 December 2010, heavy flooding occurred along the entire Pointe Verte coastline in Maria due to wave effects. In the western zone, almost half of the original wooden seawall was destroyed during the event, probably due to different material robustness along the defense structure (see Figure 8). Where the wall material was older and thus the protection less sturdy, much of the beach was returned to a natural state in front of many properties.

4.1.2. Wave Runup Implications in Modelling Total Water Level for Flood Mapping

Calculated beach slopes show minimal variability and have a mean value of 0.14, with a mean standard deviation of 0.013 for $\tan\beta_{ufs}$ and 0.017 for $\tan\beta_{1.5Ho}$. We considered this mean value representative of the study area and valid for the cross-shore transect approach assuming an inshore wave direction perpendicular to the shoreline. The linear regression between the upper foreshore beach slope and the slope located between the berm crest and T_{obs}-$1.5H_0$ shows a good correlation ($r^2 = 0.87$, p-value < 0.001). Therefore, the upper foreshore beach slope was used in the RS06 formulation associated with the 2010 event.

The RS06 modelled runup (Figure 7d) and TWL for the natural beach was 1.02 m and 3.07 m, respectively. Observed total water level on the natural beach was 2.83 m (Table 2), thus resulting in a coastal inundation throughout Pointe Verte. The RS06 model predicts TWL to within uncertainty ranges (RMS = 0.38 m) of the RS06. Considering the relatively constant coastline elevation (D_{high}) (2.66 ± 0.28 m standard deviation), observed tidal levels fail to overtop the coastline crest (D_{high}) without the influence of wave effects. Wave breaking effects transfer wave energy into potential energy in the form of wave runup and thus create the potential for flooding to occur. Along the coastal defense, however, results show a higher observed TWL by at least 0.86 m compared to the natural beach (Figure 6 and Table 2), giving a water level of 3.69 m. With a 0.01 m underestimation difference, the estimated TWL on coastal defense gives 3.68 m, suggesting that adding the wave height to the still water level is a good approach to predict TWL in this case.

Table 2. Flood characteristics according to observed and predicted total water levels along natural and defended areas during the 2010 event.

2010 flood characteristics		
Coastline state	Natural beach	Coastal defense
Wave components (m)	Runup	H_0
	1.02	1.63
Time (UTC)	20:00	
Predicted tide	1.42	
Observed tide + Surge	2.05	
Surge only	0.63	
Estimated TWL (m)	3.07	3.68
Mean observed TWL (m) *(post-storm survey)*	2.83	3.69
	Estimated flood area (km^2)	Observed flood area (km^2)
Entire study area	0.267	0.179
West of Pluviers Street only	0.096	0.092
East of Pluviers Street only	0.171	0.087
West area with coastal structure	0.005	0.006

Figure 10. Modelled total water levels for the 2010 flood extent. The dashed line represents the observed flood extent during the post-storm survey. Red and black lines show profiles illustrated in Figure 6.

Estimated total water levels obtained from Equations (4) and (5) for the 2010 event were mapped over the 2007 LiDAR grid elevation whilst the data from the mobile terrestrial LiDAR system were used for defining crest elevations (D_{high}) (Figure 8). For the west and east areas, the predicted TWL estimates well the flooded surface. The static approach, however, does not take into account the flow pattern landward of the spit and therefore overestimates the overall flood extent by 0.088 km^2 (Table 2). The overestimation is mainly attributed to the area eastward of Pluviers Street, where the model overestimates TWL by 0.084 km^2. Considering westward of Pluviers Street, modelled and measured flooded surfaces are strongly correlated with a difference of only 0.004 km^2. The area protected by the seawall also offers good correlation between both surfaces where only 0.001 km^2 is underpredicted by the model. The static approach thus represents well the flood extent when a landward driven flow occurs compared with a drained area. The results also show that Environment Canada's (EC) operational wave forecasting model based on WAM cycle 4 MW3 "PROMISE" produced reliable data for the mapping of coastal flooding at Maria. This model was used for the first time to validate coastal flooding mapping applications.

4.2. Tidal Analysis and Sea-Level Trend

According to the Belledune tide station, the measured relative SLR trend was 4.06 mm/year between 1964 and 2014 (Figure 11). Once modelled with the peak over threshold method with a generalized Pareto distribution, the estimated total water level (including tides, surge and wave runup estimations) during the 6 December 2010 event exceeded the 100 years return period (Figure 12a) and corresponds to a level of 3.01 ± 0.78 m in 2010 and 3.09 ± 0.78 m in 2030. The observed water level on the natural beach (2.83 m) did not exceed the modelled water level and corresponds to a return period of 46 years. In the zone with vertical defense structures, the total water level (including tides, surge and deep-water significant wave height) reached 3.69 m during the 6 December 2010 event. This value has an annual probability of 0.0005 (Figure 12b).

Return periods (Figure 12) are particularly high considering only eight years of data are available. The confidence intervals suggest that, for the natural beach, the level reached during the December 2010 event could have a return period from 30 to over 200 years. An analysis of the extreme water levels return periods for the December 2010 event was conducted in the St. Lawrence maritime estuary from the longest available time series (1897–2010) at Pointe-au-Père-Rimouski station and indicates a return period of 150 years [75], which is closer to our upper interval.

Moreover, rising sea levels are likely to cause an increase in flood events and a shortening of return periods. A hypothetical 1 in 100 year event at Maria may become a 1 in 70 year event by 2030 (Table 3). This represents a 14% increase in frequency.

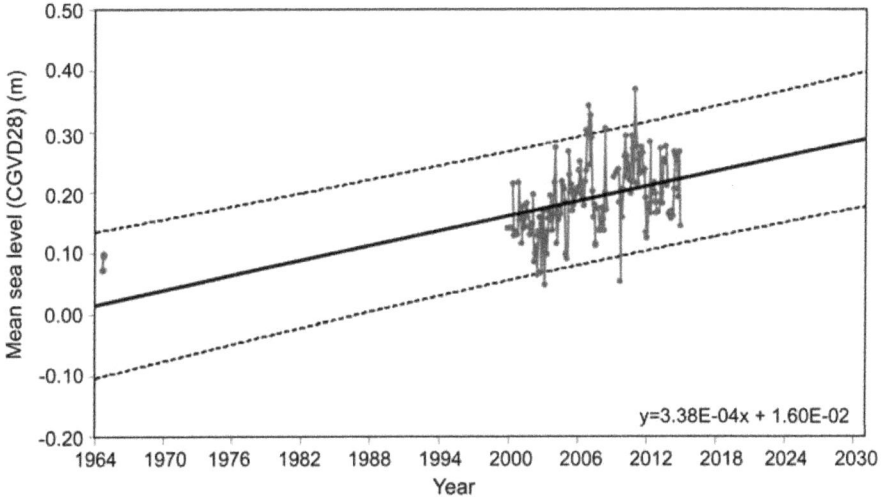

Figure 11. Linear model (solid line) fitted to 1964–2014 monthly mean sea levels at Belledune with 95% confidence bands. Linear term is statistically significant (99%).

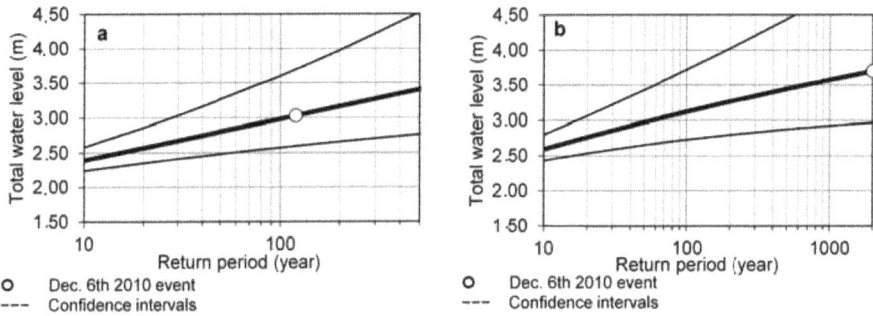

Figure 12. Estimated total water levels return periods at Belledune tide station according to the peak over threshold method including (**a**) wave runup for the natural beach on the eastern part of the study site and (**b**) significant wave height for the defended area on the west zone.

Table 3. Estimated extreme total sea levels for (**a**) (high water tide + sea-level rise + storm surge + wave runup) and (**b**) (high water tide + sea-level rise + storm surge + significant wave height) for different return periods for years 2005, 2010, 2020 and 2030 at Belledune tide station.

Estimated total water levels return periods			
(a)			
Return Period	Level 2005	Level 2010	Level 2020	Level 2030
---	---	---	---	---
1 year	1.78 ± 0.11	1.80 ± 0.11	1.84 ± 0.11	1.89 ± 0.11
2 years	1.95 ± 0.14	1.97 ± 0.14	2.02 ± 0.14	2.06 ± 0.15
10 years	2.36 ± 0.30	2.38 ± 0.30	2.42 ± 0.30	2.46 ± 0.30
30 years	2.63 ± 0.47	2.65 ± 0.47	2.69 ± 0.47	2.74 ± 0.47
50 years	2.76 ± 0.57	2.79 ± 0.57	2.83 ± 0.57	2.87 ± 0.57
100 years	2.94 ± 0.73	2.96 ± 0.73	3.00 ± 0.73	3.05 ± 0.73

Table 3. *Cont.*

Estimated total water levels return periods			
(b)			
Return Period	Level 2005	Level 2010	Level 2020	Level 2030
---	---	---	---	---
1 year	1.95 ± 0.11	1.97 ± 0.11	2.01 ± 0.11	2.05 ± 0.11
2 years	2.14 ± 0.15	2.16 ± 0.15	2.20 ± 0.15	2.24 ± 0.15
10 years	2.56 ± 0.31	2.58 ± 0.31	2.62 ± 0.31	2.66 ± 0.31
30 years	2.82 ± 0.47	2.84 ± 0.47	2.88 ± 0.48	2.92 ± 0.48
50 years	2.93 ± 0.57	2.95 ± 0.57	2.99 ± 0.57	3.03 ± 0.57
100 years	3.08 ± 0.71	3.10 ± 0.71	3.14 ± 0.71	3.18 ± 0.71

5. Discussion

Based on flood limits from post-storm survey, crest elevation and beach slope determination, wave climate modelling, wave runup estimations and sea-level rise projections, the proposed static flood approach presented here has emphasized the importance of integrating the wave component into flood assessments for low-lying coasts (Figure 13). Flood levels were identified with *in situ* measurements assuming that wave runup superimposes on tidal and surge components to increase water levels. Although runup values are mainly acquired by modelling, *in situ* measurements, or video monitoring systems [111–114], they can be estimated using measurements of field deposits [32,52,55,79]. The approach proposed here is based on deep-water wave modelling, as no wave buoy data were available for the study site during the 2010 event. The modelled results show good accordance with *in situ* measurements from the field. This is despite the empirical formula failing to account for physical processes in the surf zone [111] or the time-varying morphology of the beach in response to continuously changing wave characteristics [28].

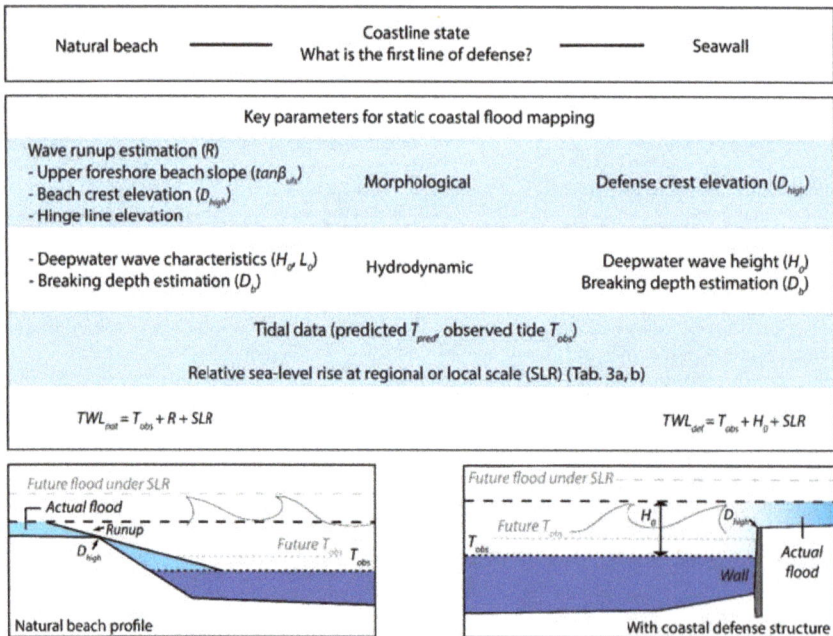

Figure 13. Conceptual and methodological approach for the static coastal flood assessment.

5.1. The Need for a Wave-Tide-Surge Integrated Approach

On 6 December 2010, no flood would have occurred in the area of Pointe Verte without a wave effect on the beach. In regards to the alongshore crest height variability and the water level observed at the tide gauge, only the concrete ramps, the end of the spit and the tidal marsh were affected by the tidal and surge components during the 2010 event. The crest elevation of both the natural beach and coastal structures was generally too high to enable a situation where the backshore would be entirely flooded by the storm surge alone [24]. Our results clearly show that the use of extreme water levels (tide + storm surge) alone, as often recommended and used for coastal flood mapping [59,61,63,64], is not necessarily sufficient for predicting coastal flooding, even in a fetch limited environment like the Chaleur Bay. The same site in 2005 experienced similar storm conditions, i.e. with upper foreshore limited tidal levels combined with high waves and overtopping of defense structures and overwashing of natural berms [52]. The flood extent elevation in 2005 was lower in comparison to the 2010 event (2.44 m on natural beach and 3.34 m on coastal structures), possibly due to a deep-water wave height of only 1.16 m in 2005, as obtained by the forecasting wave model in this study. Similar results have been observed on low-lying and rocky coasts of the St. Lawrence maritime estuary, where only 1% of the flooded zone between Rimouski and Sainte-Flavie would have been inundated by the tidal and surge components combined [77]. In all cases, total water levels, which caused major flooding, were increased by the wave effect. The results show the importance of integrating the wave component into the analysis of coastal flooding even for sheltered coastlines where their effects are often ignored or underestimated. Similar findings were observed in California where infrastructures at risk of flood are often concentrated around sheltered embayments. Originally, wave effects were ignored in flood analyses [58] yet subsequently have been integrated into more complete hydrodynamic flood models [67]. Other results were recently observed on the other side of the Atlantic, where in February 2010, Xynthia generated a large storm surge contributing to coastal flood in France. Bertin et al. [115] underlined the exceptional surface stress originating from young and steep waves propagating toward the shore, increasing water levels at coastline and contributing to flooding. For more than a decade, considering the wave component is also part of the FEMA's procedure for coastal flood mapping in the US [116–118]. Using computer programs such as Runup 2.0 and ACES (Automated Coastal Engineering System), numerical models like CSHORE, or empirical formulas, wave runup and wave overtopping are included in the flood map methodology for the National Flood Insurance Program [94].

For the natural beach of Pointe Verte, we modelled deep-water wave characteristics and integrated these into a static flood model under a wave runup empirical formulation [29]. The estimated wave implication for the 2010 event increased the still water level from 2.05 m to 3.07 m, resulting in a theoretical wave runup of 1.02 m in natural areas. This increase in water levels was also recorded by Bernatchez et al. [52], where an amplitude of 0.4 m over the still water level was recorded on the natural beach during the storm in December 2005.

Along the natural beach, wave runup occurs on the upper foreshore beach slope. Using this beach slope in the Stockdon runup formulation and adding this value to the observed tide and surge showed good correspondence with field debris elevation measurements. Although the foreshore slope has been used in previous works [29,32,34,67,96,119], the mean upper foreshore slope defined here was located between the beach crest and the hinge line, corresponding to the "slope of the active section of the beach", as suggested by Cariolet and Suanez [32] for macrotidal environments. Using this beach slope, the Stockdon model showed good results in estimating wave runup and mean flood limit elevations were within model errors (± 0.38 m). This is in accordance with authors who suggest using the Stockdon formula in order to estimate wave runup as a first step in coastal flood assessments [94,95]. Some studies show that the Stockdon formula tends to over-estimate the flooding extent [55,67] or occasionally underestimate it [111].

Other runup empirical formulas based on beach slope and deep-water wave characteristics exist [30,34,57,97,120] and demonstrated different results [32,77]. In our case, without precise field recordings of high frequency wave runup and near-shore hydrodynamics, we find the Stockdon model based on deep-water wave characteristics and upper foreshore beach slope suitable to our static flood model on the natural beach. The upper foreshore slope offers a practical advantage to coastal managers because it can be readily defined using GIS [51]. By digitizing the coastline and the hinge line from a series of aerial photographs and LiDAR products, one is able to estimate wave runup at high tide if wave climate data is available. This is a measureable morphological aspect that can be used in historic flood assessments and for predicting future extreme events.

As demonstrated along the seawall at Maria, wave crest elevation in relation to the height of the defense crest is also an important parameter in coastal flood management [121]. Upper foreshore sediment volumes are diminished in front of coastal structures in response to wall reflection and scouring at the toe [52,122]. Therefore, waves do not break on the foreshore until reaching the vicinity of the wall front. As a result, almost no dissipation occurs on the sloping beach in front of the wall and flood limits become higher landward of seawalls in comparison to behind natural beaches. No wave runup occurs in such cases [43] and flooding is due primarily to wave overtopping. Chini and Stansby [53] showed that the overtopping discharge was higher when beach level was lower at the toe of the seawall. This may explain the much higher TWL measured in front of defense structures compared to the natural beach in Maria. Since no wave overtopping discharge calculations were made in the present study, we propose adding significant deep-water wave heights to the still water level in order to estimate flood levels landward of the seawall. This approximation is a useful approach for coastal managers and showed good results in this study. Since coastal flood assessment and mapping for large areas can easily require time consuming and expensive studies [20], especially where complex coastal defense structures are present and require computationally demanding effort for bulk application [95], adding deep-water wave height to the observed tide (with surge) as a simple rule-of-thumb in order to obtain total water levels facing seawall can be suitable as a first approach to flood mapping in the present case where waves do not break before reaching the vicinity wall.

Another element vital to flood model robustness is the accuracy with which we can identify and map the first line of defense [25] both in natural environments and along defense structures. Recent studies show that airborne LiDAR digital elevation models, usually with a spatial resolution of 1 m and a vertical accuracy of ±15 cm [123,124] and sometimes up to 30 cm [125–127], are inadequate to determine this limit with sufficient precision as micro-scale topographic features relevant to hydraulic connectivity are overlooked [58]. The result is an over- or under-estimation of the flood extent. The mobile terrestrial LiDAR system (MTLS) that we used in this study produced a point cloud with a spatial resolution of less than 5 cm and vertical errors of ± 3 cm (RMS), which effectively solves the problem of having to determine defense crest elevation manually [25]. The MTLS also allows for much faster field surveys in comparison to fixed terrestrial laser scanning (TLS) or DGPS, which are sometimes used in lieu of airborne LiDAR data [128]. This is an important advantage in the coastal environment as the time available to map beach morphology can be short at low tide. Surveys with the MTLS also provide potential in urban flood modelling where the representation of micro-scale topographic effects and complex features (curbs, ditches roads, defense structure crests, walls, and/or buildings) or drainage systems have significant impact on flood propagation [67,69,70]. Airborne LiDAR data covering larges areas in a rapid survey can be combined with MTLS surveys along coastal borders to maximize the advantage of both platform to generate high-resolution digital elevation models for accurate coastal flood mapping.

5.2. Inland Drainage and Sea-Level Rise

During the 6 December 2010 event, water levels in the tidal marsh at Maria were lower than the predicted flood levels due to the protection offered by the coastal spit. Marshes attenuate wave energy and provide natural resilience to flooding exposure [129,130]. Due to a land gradient landward of the coastline, however, the Stockdon model did not predict well the flood limits. This model overestimation has been reported previously when applied to other flooded peninsula environments [67] and a decelerating landward flow with distance is normally seen in this particular situation [113]. Flood limits at the tidal marsh in east Maria were below the 2.83 m average for the region. Due to the morphology of the marsh, only the astronomical tide and storm surge components affected this flooding surface during the event, demonstrating the natural protection offered by this type of environment.

The relatively low 2.05 m still water level was too low to produce an inundation regime during the 2010 event. An inundation regime occurs when a constant inland flow of water under storm surge effects (tide + surge) is realized, causing continuous submersion of the coastline [24,25]. Under future sea-level rise, conditions where the still water line is raised, an inundation regime could occur over the Pointe Verte area, which has the potential to flood the entire spit and tidal marsh. In such a case, the land drainage effect would diminish and the static flood model presented here would likely give an accurate account of flooding for the area eastward of Pluviers Street.

According to our linear SLR prediction of 4.06 mm/year (1964–2014), the return period for an equivalent storm to generate a still water level (tide + surge) reaching the beach crest of 2.66 m causing an inundation situation is 150 years. This modelled linear trend is in accordance with the IPCC's RCP8.5 sea level rise scenario for at least the next two decades.

As an example, 20 years of SLR at this constant rate produces a net rise of 0.082 m from 2010 to 2030. Integrating the IPCC's RCP8.5 sea-level scenario [131] with current isostatic land motion (-1.78 mm/year; [100]) results in a net SLR of 0.122 m by 2030, relative to 2010 levels. By integrating the IPCC's projection and Koohzare's isostatic model, we can estimate the year when an inundation regime event (at least equal to the mean coastline elevation of 2.66 m) would happen. A hypothetical 2.66 m still water level could be reached by 2081, after 71 years. Moreover, in the context of climate change, the significant reduction of sea-ice cover will promote an increase in winter storm waves hitting the coast [132].

The trend of relative SLR has been 6.96 mm/y between 2000 and 2014 at the Belledune tide gauge. This evidence of acceleration in SLR is also observed at tide gauge stations in Eastern Canada and Northeastern United States [104]. The SLR acceleration since 1990 is a global response to the increased radiative forcing of the climate system [133]. Due to the short time series of tide gauge data, we opted for a conservative linear trend compared to the quadratic approach used by other authors to project sea-level rise [8,104].

The use of tide gauge data has the advantage of integrating both isostatic and eustatic components to project relative SLR at the regional scale [104]. However, due to the short time series and discontinuity of the tide gauge data from the Belledune station, the projection of relative SLR was made only for a short period. The recent acceleration in global SLR averages [3,4] and the large differences between scenarios of eustatic rise by 2100 [7] demonstrate the need to develop and use proxies to better establish regional relative sea-level trends [134–136]. These data will be needed to establish a more robust coastal flood risk mapping.

6. Conclusions

Wave runup estimation and modelling of water levels and deep-water waves have reproduced observed flood limit elevations during the storm event of 6 December 2010, as experienced along a natural and armoured coastline. Results clearly show the importance of integrating wave data and runup in assessing and mapping the risk of coastal flooding. They also demonstrate that the lowering of beach levels at a seawall toe increases total water level elevations in the backshore and consequently the risk of flooding due to a wave breaking occurring at the vicinity of the seawall. Our approach integrates runup, which takes into account beach slope (natural areas), or deep-water wave height (at seawalls), the return period of extreme water levels and future relative sea-level rise in order to establish, at the municipality scale, coastal flood risk maps. This approach, which takes into account several variables, is easy to use for coastal managers, which is an important factor in the process of adaptation. Such static approaches do not, however, account for land drainage. In the context of accelerating sea-level rise, it becomes necessary to establish, at the regional scale, a longer series of relative sea-level trends, a validation of wave, drainage, and numerical runup models in different geomorphological settings and hydrodynamic conditions. The next step will be to integrate wave overtopping discharge models to better represent overland flow patterns. In this perspective, the use of a mobile terrestrial laser system offers excellent potential to model and map more accurately coastal flooding. Integration of validated modelled data into a warning system offers a promising avenue for managing and mitigating coastal risks.

Acknowledgments: We thank the Quebec government for funding this project as part of its program for preventing the principal types of natural risks. Thanks as well to Tarik Toubal for the storm debris line mapping with DGPS and Susan Drejza, Patrick Bouchard and David Lacombe for the mobile terrestrial LiDAR survey. We thank Denis Jacob and Viateur Turcotte from Environment Canada for the wave data provided by the wave forecasting model, and to Phillip MacAulay (CHS) for the Belledune station data information. We also thank Azadeh Koohzare for sharing vertical crustal movement data. Thanks for the two anonymous reviewers and editors for their very pertinent and helpful comments.

Author Contributions: Field surveys (DGPS, LiDAR, flood limit): C.F., D.D., P.B., S.V.W. Hydrodynamic data analysis: D.D., A.L. Sea-level analysis: G.B.B., P.B., D.D. Flood mapping: D.D., C.F. Discussion on the results: D.D., P.B., A.L., G.B.B., C.F., R.L.B. Major contribution to the writing of the entire paper: D.D., P.B. with a main contribution of A.L. for introduction, discussion and wave analysis sections, G.B.B. for sea-level and return period for extreme water levels and S.V.W. for the methodology section of the mobile terrestrial LiDAR data. Edited English syntax: R.L.B.

Conflicts of Interest: The authors declare no conflict of interest.

References

1. Nicholls, R.J.; Tol, R.S.J. Impacts and responses to sea-level rise: A global analysis of the SRES scenarios over the twenty-first century. *Philos. Trans. A* **2006**, *364*, 1073–1095. [CrossRef] [PubMed]
2. Nicholls, R.J.; Cazenave, A. Sea-level rise and its impact on coastal zones. *Science* **2010**, *328*, 1517–1520. [CrossRef] [PubMed]
3. Church, J.A.; White, N.J. Sea-Level Rise from the Late 19th to the Early 21st Century. *Surv. Geophys.* **2011**, *32*, 585–602. [CrossRef]
4. Rahmstorf, S.; Foster, G.; Cazenave, A. Comparing climate projections to observations up to 2011. *Environ. Res. Lett.* **2012**, *7*, 044035. [CrossRef]
5. Pfeffer, W.T.; Harper, J.T.; O'Neel, S. Kinematic Constraints on Glacier Contributions to 21st-Century Sea-Level Rise. *Science* **2008**, *321*, 1340–1343. [CrossRef] [PubMed]
6. Church, J.A.; Clark, P.U.; Cazenave, A.; Gregory, J.M.; Jevrejava, S.; Levermann, A.; Merrifield, M.A.; Milne, G.A.; Nerem, R.S.; Nunn, P.D.; *et al.* Sea Level Change. In *IPCC. Climate Change 2013: The Physical Science Basis*; Stocker, T.F., Qin, D., Plattner, G.K., Tignor, M., Allen, S.K., Boschung, J., Nauels, A., Xia, Y., Bex, V., Midgley, P.M., Eds.; Cambridge University Press: Cambridge, UK; New York, NY, USA, 2013.
7. Horton, B.P.; Rahmstorf, S.; Engelhart, S.E.; Kemp, A.C. Expert assessment of sea-level rise by AD 2100 and AD 2300. *Quat. Sci. Rev.* **2014**, *84*, 1–6. [CrossRef]
8. Parris, A.; Bromirski, P.; Burkett, V.; Cayan, D.; Culver, M.; Hall, J.; Horton, R.; Knuuti, K.; Moss, R.; Obeysekera, J.; *et al. Global Sea Level Rise Scenarios for the US National Climate Assessment*; NOAA: Silver Spring, MD, USA, 2012.
9. Yin, J.; Schlesinger, M.E.; Stouffer, R.J. Model projections of rapid sea-level rise on the northeast coast of the United States. *Nat. Geosci.* **2009**, *462*, 262–266. [CrossRef]
10. Howard, T.; Pardaens, A.K.; Lowe, J.A.; Ridley, J.; Hurkmans, R.T.W.L.; Bamber, J.L.; Spada, G.; Vaughan, D. Sources of 21st century regional sea level rise along the coast of North-West Europe. *Ocean Sci. Discuss.* **2013**, *10*, 2433–2459. [CrossRef]
11. Obeysekera, J.; Park, J. Scenario-based projection of extreme sea levels. *J. Coast. Res.* **2012**, *29*, 1–7. [CrossRef]
12. Vafeidis, A.T.; Nicholls, R.J.; Mcfadden, L.; Tol, R.S.J.; Hinkel, J.; Spencer, T.; Grashoff, P.S.; Boot, G.; Klein, R.J.T. A new global coastal database for impact and vulnerability analysis to sea-level rise. *J. Coast. Res.* **2008**, *24*, 917–924. [CrossRef]
13. Gornitz, V.M.; Daniels, R.C.; White, T.W.; Birdwell, K.R. The Development of a Coastal Risk Assessment Database: Vulnerability to Sea-Level Rise in the U.S. Southeast. *J. Coast. Res.* **1994**, *371*, 327–338.

14. Jallow, B.P.; Toure, S.; Barrow, M.M.K.; Mathieu, A.A. Coastal zone of The Gambia and the Abidjan region in Cote d'Ivoire: Sea level rise vulnerability, response strategies, and adaptation options. *Clim. Res.* **1999**, *12*, 129–136. [CrossRef]

15. Titus, J.G.; Richman, C. Maps of lands vulnerble to sea level rise: Modeled elevations along the US Atlantic and Gulf coasts. *Clim. Res.* **2001**, *18*, 205–228. [CrossRef]

16. Solomon, S.; Qin, D.; Manning, M. *IPCC. Climate Change 2007. The Physical Science Basis*; Cambridge University Press: Cambridge, UK; New York, NY, USA, 2007.

17. Stocker, T.F.; Qin, D.; Plattner, G.K.; Tignor, M.; Allen, S.K.; Boschung, J.; Nauels, A.; Xia, Y.; Bex, V.; Midgley, P.M.; *et al.* IPCC. *Climate Change 2013: The Physical Science Basis*; Cambridge University Press: Cambridge, UK; New York, NY, USA, 2013.

18. Rao, K.N.; Subraelu, P.; Rao, T.V.; Malini, B.H.; Ratheesh, R.; Bhattacharya, S.; Rajawat, A.S. Sea-level rise and coastal vulnerability: An assessment of Andhra Pradesh coast, India through remote sensing and GIS. *J. Coast. Conserv.* **2008**, *12*, 195–207.

19. Nicholls, R.J.; Hoozemans, F.M.J.; Marchand, M. Increasing flood risk and wetland losses due to global sea-level rise: Regional and global analyses. *Glob. Environ. Chang.* **1999**, *9*, S69–S87. [CrossRef]

20. FEMA. *Atlantic Ocean and Gulf of Mexico coastal guidelines update. Final draft*; Federal Emergency Management Agency: Washington, DC, USA, 2007.

21. Walsh, K.J.E.; Betts, H.; Church, J.; Pittock, A.B.; McInnes, K.L.; Jackett, D.R.; McDougall, T.J. Using sea level rise projections for urban planning in Australia. *J. Coast. Res.* **2004**, *20*, 586–598. [CrossRef]

22. Gehrels, W.R.; Milne, G.A.; Kirby, J.R.; Patterson, R.T.; Belknap, D.F. Late Holocene sea-level changes and isostatic crustal movements in Atlantic Canada. *Quat. Int.* **2004**, *120*, 79–89. [CrossRef]

23. Koohzare, A.; Vaníček, P.; Santos, M. Pattern of recent vertical crustal movements in Canada. *J. Geodyn.* **2008**, *45*, 133–145. [CrossRef]

24. Sallenger, A.H. Storm impact scale for barrier islands. *J. Coast. Res.* **2000**, *16*, 890–895.

25. Stockdon, H.F.; Sallenger, A.H.; Holman, R.A.; Howd, P.A. A simple model for the spatially-variable coastal response to hurricanes. *Mar. Geol.* **2007**, *238*, 1–20. [CrossRef]

26. Benavente, J.; Del Río, L.; Gracia, F.J.; Martínez-del-Pozo, J.A. Coastal flooding hazard related to storms and coastal evolution in Valdelagrana spit (Cadiz Bay Natural Park, SW Spain). *Cont. Shelf Res.* **2006**, *2*, 1061–1076. [CrossRef]

27. Soldini, L.; Antuono, M.; Brocchini, M. Numerical Modeling of the Influence of the Beach Profile on Wave Run-Up. *J. Waterw. Port Coast. Ocean Eng.* **2012**, *139*, 115. [CrossRef]

28. Bauer, B.O.; Greenwood, B. Surf zone similarity. *Geogr. Rev.* **1988**, *78*, 137–147. [CrossRef]

29. Stockdon, H.; Holman, R.; Howd, P.; Sallenger, A. Empirical parameterization of setup, swash, and runup. *Coast. Eng.* **2006**, *53*, 573–588. [CrossRef]

30. Komar, P.D. *Beach processes and sedimentation*, 2nd ed.; Prentice Hall, Inglewood Cliffs: Upper Saddle River, NJ, USA, 1998.

31. Cariolet, J.M.; Suanez, S.; Meur-Férec, C.; Postec, A. Cartographie de l'aléa de submersion marine et PPR: Éléments de réflexion à partir de l'analyse de la commune de Guissény (Finistère, France). *Cybergeo* **2012**, *2012*, 1–21. [CrossRef]

32. Cariolet, J.M.; Suanez, S. Runup estimations on a macrotidal sandy beach. *Coast. Eng.* **2013**, *74*, 11–18. [CrossRef]

33. Ruggiero, P. Is the intensifying wave climate of the U.S. Pacific Northwest increasing flooding and erosion risk faster than sea level rise? *J. Waterw. Port Coast. Ocean Eng.* **2012**, *139*, 88–97. [CrossRef]

34. Ruggiero, P.; Komar, P.D.; McDougal, W.G.; Marra, J.J.; Reggie, A. Wave runup, extreme water levels and the erosion of properties backing beaches. *J. Coast. Res.* **2001**, *17*, 407–419.

35. Seabloom, E.W.; Ruggiero, P.; Hacker, S.D.; Mull, J.; Zarnetske, P. Invasive grasses, climate change, and exposure to storm-wave overtopping in coastal dune ecosystems. *Glob. Chang. Biol.* **2013**, *19*, 824–832. [CrossRef] [PubMed]

36. Battjes, J.A. Surf similarity. In Proceedings of the 14th International Conference on Coastal Engineering ASCE, Copenhagen, Denmark, 24–28 June 1974; pp. 466–480.

37. Cariolet, J.M. Quantification du runup sur une plage macrotidale à partir des conditions morphologiques et hydrodynamiques. *Geomorphol. Reli. Process. Environ.* **2011**, *1*, 95–108. [CrossRef]

38. Stockdon, H.F.; Thompson, D.M.; Plant, N.G.; Long, J.W. Evaluation of wave runup predictions from numerical and parametric models. *Coast. Eng.* **2014**, *92*, 1–11. [CrossRef]

39. Holman, R.A.; Sallenger, A.H. Setup and swash on a natural beach. *J. Geophys. Res.* **1985**, *90*, 945–953. [CrossRef]

40. Hughes, S.A. Wave momentum flux parameter: a descriptor for nearshore waves. *Coast. Eng.* **2004**, *51*, 1067–1084. [CrossRef]

41. Mase, H.; Asce, M.; Tamada, T.; Yasuda, T.; Hedges, T.S.; Reis, M.T. Wave runup and overtopping at seawalls built on land and in very shallow water. *J. Waterw. Port Coast. Ocean Eng.* **2013**, *139*, 346–357. [CrossRef]

42. Na, S.J.; Do, K.D.; Suh, K.-D. Forecast of wave run-up on coastal structure using offshore wave forecast data. *Coast. Eng.* **2011**, *58*, 739–748. [CrossRef]

43. Pullen, T.; Allsop, N.W.H.; Bruce, T.; Kirtenhaus, A.; Schüttrumpf, H.; van der Meer, J.W. *Eurotop: Wave overtopping of sea defences and related structures: assessment manual*; Environment Agency: Rotherham, UK; Expertise Netwerk Waterkeren: Utrecht, The Netherlands; Kuratorium fur Forschung im Kusteningenieurwesen: Hamburg, Germany, 2007.

44. Carbone, F.; Dutykh, D.; Dudley, J.M.; Dias, F. Extreme wave runup on a vertical cliff. *Geophys. Res. Lett.* **2013**, *40*, 3138–3143. [CrossRef]

45. Chen, X.; Hofand, B.; Altomare, C.; Suzuki, T.; Uijttewaal, W. Forces on a vertical wall on a dike crest due to overtopping flow. *Coast. Eng.* **2015**, *95*, 94–104. [CrossRef]

46. Peng, Z.; Zou, Q.-P. Spatial distribution of wave overtopping water behind coastal structures. *Coast. Eng.* **2011**, *58*, 489–498. [CrossRef]

47. Van der Meer, J.W.; Verhaeghe, H.; Steendam, G.J. The new wave overtopping database for coastal structures. *Coast. Eng.* **2009**, *56*, 108–120. [CrossRef]

48. Molines, J.; Medina, J.R. Calibration of overtopping roughness factors for concrete armor units in non-breaking conditions using the CLASH database. *Coast. Eng.* **2015**, *96*, 62–70. [CrossRef]

49. Van Gent, M.R.A.; Van den Boogaard, H.F.P.; Pozueta, B.; Medina, J.R. Neural network modelling of wave overtopping at coastal structures. *Coast. Eng.* **2007**, *54*, 586–593. [CrossRef]

50. Sabino, A.; Rodrigues, A.; Araujo, J.; Poseiro, P.; Reis, M.T.; Fortes, C.J. Wave overtopping analysis and early warning forecast system. In Proceedings of the ICCSA 2014, Le Havre, France, 23–26 June 2014; Murgante, B., Misra, S., Rocha, A.M.A.C., Torre, C., Rocha, J.G., Falcão, M.I., Taniar, D., Apduhan, B.O., Gervasi, O., Eds.; Springer: Cham, Switzerland, 2014.

51. Bernatchez, P.; Fraser, C. Evolution of coastal defence structures and consequences for beach width trends, Québec, Canada. *J. Coast. Res.* **2012**, *285*, 1550–1566. [CrossRef]

52. Bernatchez, P.; Fraser, C.; Lefaivre, D.; Dugas, S. Integrating anthropogenic factors, geomorphological indicators and local knowledge in the analysis of coastal flooding and erosion hazards. *Ocean Coast. Manag.* **2011**, *54*, 621–632. [CrossRef]

53. Chini, N.; Stansby, P.K. Coupling TOMAWAC and Eurotop for uncertainty estimation in wave overtopping predictions. In *Advances in Hydroinformatics, Springer Hydrogeology*; Gourbesville, P., Cunge, J., Caignaert, G., Eds.; Springer: Singapore, Singapore, 2014.

54. Dawson, R.J.; Dickson, M.E.; Nicholls, R.J.; Hall, J.W.; Walkden, M.J.A.; Stansby, P.K.; Mokrech, M.; Richards, J.; Zhou, J.; Milligan, J.; *et al.* Integrated analysis of risks of coastal flooding and cliff erosion under scenarios of long term change. *Clim. Chang.* **2009**, *95*, 249–288. [CrossRef]

55. Guimarães, P.V.; Farina, L.; Toldo, E.; Diaz-Hernandez, G.; Akhmatskaya, E. Numerical simulation of extreme wave runup during storm events in Tramandaí Beach, Rio Grande do Sul, Brazil. *Coast. Eng.* **2015**, *95*, 171–180. [CrossRef]

56. Ruggiero, P.; Holman, R.A.; Beach, R.A. Wave run-up on a high-energy dissipative beach. *J. Geophys. Res.* **2004**, *109*, 1–12. [CrossRef]

57. Holman, R. Extreme value statistics for wave run-up on a natural beach. *Coast. Eng.* **1986**, *9*, 527–544. [CrossRef]

58. Gallien, T.W.; Schubert, J.E.; Sanders, B.F. Predicting tidal flooding of urbanized embayments: A modeling framework and data requirements. *Coast. Eng.* **2011**, *58*, 567–577. [CrossRef]

59. Daigle, R.J. *Sea level rise estimates for New Brunswick municipalities: Saint John, Sackville, Richibucto, Shippagan, Caraquet, Le Goulet*; The Atlantic Climate Adaptation Solutions Association: New-Brunswick, QC, Canada, 2011.

60. Perherin, C.; Roche, A. Évolution des méthodes de caractérisation des aléas littoraux. In Proceedings of the XIèmes Journées Natl. Génie Côtier-Génie Civil, Les Sables d'Olonne, France, 22–25 June 2010; pp. 609–616.
61. Richards, W.; Daigle, R. *Scenarios and Guidance for Adaptation to Climate Change and Sea Level Rise- NS and PEI Municipalities*; Atlantic Climate Adaptation Solutions Association: Halifax, NS, Canada, 2011; p. 87.
62. Prime, T.; Brown, J.M.; Plater, A.J. Physical and economic impacts of sea-level rise and low probability flooding events on coastal communities. *PLoS ONE* **2015**, *10*, e0117030. [CrossRef] [PubMed]
63. Daigle, R. *Sea-Level Rise and Flooding Estimates for New Brunswick Coastal Sections*; The Atlantic Climate Adaption Solutions Association: Halifax, NS, Canada, 2012.
64. Webster, T.; McGuigan, K.; Collins, K.; MacDonald, C. Integrated river and coastal hydrodynamic flood risk mapping of the lahave river estuary and town of Bridgewater, Nova Scotia, Canada. Integrated river and coastal hydrodynamic flood risk mapping of the lahave river estuary and town of Bridgewater, Nova Scotia, Canada. *Water* **2014**, *6*, 517–546.
65. Pirazzoli, P.A.; Tomasin, A. Estimation of return periods for extreme sea levels: A simplified empirical correction of the joint probabilities method with examples from the French Atlantic coast and three ports in the southwest of the UK. *Ocean Dyn.* **2007**, *57*, 91–107. [CrossRef]
66. Thompson, K.R.; Bernier, N.B.; Chan, P. Extreme sea levels, coastal flooding and climate change with a focus on Atlantic Canada. *Nat. Hazards* **2009**, *51*, 139–150. [CrossRef]
67. Gallien, T.W.; Sanders, B.F.; Flick, R.E. Urban coastal flood prediction: Integrating wave overtopping, flood defenses and drainage. *Coast. Eng.* **2014**, *91*, 18–28. [CrossRef]
68. Webster, T.L.; Forbes, D.L.; Dickie, S.; Shreenan, R. Using topographic lidar to map flood risk from storm-surge events for Charlottetown, Prince Edward Island, Canada. *Can. J. Remote Sens.* **2004**, *30*, 64–76. [CrossRef]
69. Néelz, S.; Pender, G.; Villanueva, I.; Wilson, M.; Wright, N.G.; Bates, P.; Mason, D.; Whitlow, C. Using remotely sensed data to support flood modelling. *Water Management* **2006**, *159*, 35–43. [CrossRef]
70. Ozdemir, H.; Sampson, C.C.; de Almeida, G.A.M.; Bates, P.D. Evaluating scale and roughness effects in urban flood modelling using terrestrial LIDAR data. *Hydrol. Earth Syst. Sci.* **2013**, *17*, 4015–4030. [CrossRef]
71. Forbes, D.L.; Parkes, G.S.; Manson, G.K.; Ketch, L.A. Storms and shoreline retreat in the southern Gulf of St. Lawrence. *Mar. Geol.* **2004**, *210*, 169–204. [CrossRef]
72. Masson, A.; Catto, N. Extratropical Transitions in Atlantic Canada: Impacts and Adaptive Responses. *Geophy. Res. Abstr.* **2013**, *15*, 3149.
73. Boyer-Villemaire, U.; Bernatchez, P.; Benavente, J.; Cooper, J.A.G. Quantifying community's functional awareness of coastal changes and hazards from citizen perception analysis in Canada, UK and Spain. *Ocean Coast. Manag.* **2014**, *93*, 106–120. [CrossRef]
74. Drejza, S.; Frieseinger, S.; Bernatchez, P. Exposition des infrastructures routières de l'Est du Québec (Canada) à l'érosion et à la submersion. In Proceedings of the Actes du Colloque International Connaissance et Compréhension des Risques Côtiers: Aléas, Enjeux, Représentations, Gestion, Brest, France, 3–4 July 2014.
75. Bernatchez, P.; Boucher-Brossard, G.; Sigouin-Cantin, M. *Contribution des archives à l'étude des événements météorologiques et géomorphologiques causant des dommages aux côtes du Québec maritime et analyse des tendances, des fréquences et des temps de retour des conditions météo-marines extrêmes*; Chaire de recherche en géoscience côtière, Laboratoire de dynamique et de gestion intégrée des zones côtières, Université du Québec à Rimouski; Rapport remis au ministère de la Sécurité publique du Québec: Rimouski, QC, Canada, 2012; p. 140.
76. MacCaulay, P.; Canadian Hydrographic Service, Dartmouth, N.S., Canada. Personal communication, 2015.
77. Didier, D.; Bernatchez, P.; Marie, G. Évaluation de la submersion côtière grâce à l'estimation in situ du wave runup sur les côtes basses du Bas-Saint-Laurent, Canada (Québec). In Proceedings of the Actes du Colloque International Connaissance et Compréhension des Risques Côtiers: Aléas, Enjeux, Représentations, Gestion, Brest, France, 3–4 July 2014.
78. Ramana Murthy, M.V.; Reddy, N.T.; Pari, Y.; Usha, T.; Mishra, P. Mapping of seawater inundation along Nagapattinam based on field observations. *Nat. Hazards* **2012**, *60*, 161–179. [CrossRef]
79. Cariolet, J.-M. Use of high water marks and eyewitness accounts to delineate flooded coastal areas: The case of Storm Johanna (10 March 2008) in Brittany, France. *Ocean Coast. Management* **2010**, *53*, 679–690. [CrossRef]
80. Van-Wierts, S.; Bernatchez, P. *Relevé LiDAR terrestre à Sainte-Luce dans le secteur de l'Anse aux Coques dans le cadre d'une étude de recharge de plage en zone d'affouillement*; Chaire de recherche en géoscience côtière,

Laboratoire de dynamique et de gestion intégrée des zones côtières, Université du Québec à Rimouski; Rapport remis au ministère de la Sécurité publique du Québec: Rimouski, QC, Canada, 2012; p. 29.

81. WAMDI Group. The WAM Model: A Third Generation Ocean Wave Prediction Model. *J. Phys. Oceanogr.* **1988**, *18*, 1775–1810.

82. Komen, G.J.; Cavaleri, L.; Donelan, M.; Hasselmann, K.; Hasselmann, S.; Janssen, P.A.E.M. *Dynamic and Modelling of Ocean Waves*; Cambridge University Press: New York, NY, USA; p. 556.

83. Monbaliu, J.; Hargreaves, J.C.; Carretero, J.C.; Gerritsen, H.; Flather, R. Wave modelling in the PROMISE project. *Coast. Eng.* **1999**, *37*, 379–407. [CrossRef]

84. Jacob, D.; Perrie, W.; Toulany, B.; Saucier, F.; Lefaivre, D.; Turcotte, V. Wave model validation in the St. Lawrence river eastuary. In Proceedings of the 7th International Workshop on Wave Hindcasting and Forecasting, Banff, AB, Canada, 21–25 October 2004.

85. Janssen, P.A.E.M. Quasi-linear theory of wind–wave generation applied to wave forecasting. *J. Phys. Oceanogr.* **1991**, *21*, 1631–1642. [CrossRef]

86. Cavaleri, L.; Rizzoli, P.M. Wind wave prediction in shallow water: Theory and applications. *J. Geophys. Res.* **1981**, *86*, 10961–10973. [CrossRef]

87. Hasselmann, K.; Barnett, T.P.; Bouws, E.; Carlson, H.; Cartwright, D.E.; Enke, K.; Ewing, J.A.; Gienapp, H.; Hasselmann, D.E.; Kruseman, P.; *et al.* *Measurements of Wind-Wave Growth and Swell Decay During the Joint North Sea Wave Project (JONSWAP)*; Deutches Hydrographisches Institute: Delft, Netherlands, 1973.

88. Battjes, J.A.; Janssen, J.P.F.M. Energy loss and set-up due to breaking of random waves. In Proceedings of the 16th International Conference Coastal Engineering, Hamburg, Germany, 27 August–3 September 1978; pp. 569–587.

89. Lambert, A.; Neumeier, U.; Jacob, D. *Évaluation du modèle WAM opéré par Environnement Canada dans le Golfe du Saint-Laurent; résultats préliminaires pour les tempêtes de l'automne 2010; Institut des sciences de la mer de Rimouski, Université du Québec à Rimouski*; Rapport technique remis au Ministère des Transports du Québec: Rimouski, QC, Canada, 2012.

90. Lambert, A.P.; Neumeier, U.; Jacob, D.; Savard, J.-P. Are regional operational wind–waves models usable to predict coastal and nearshore wave climate? In Proceedings of the 2012 AGU Fall Meeting; San Francisco, CA, USA: 3–7 December 2012.

91. Lambert, A.; Neumeier, U.; Jacob, D.; Savard, J.-P. *Évaluation du modèle WAM opéré par Environnement Canada dans le Golfe du Saint-Laurent; résultats intermédiaires pour les années 2010-2011; Institut des sciences de la mer de Rimouski, Université du Québec à Rimouski*; Rapport technique remis au Ministère des Transports du Québec: Rimouski, QC, Canada, 2013.

92. Dalrymple, R.A.; Eubanks, R.A.; Birkemeier, W.A. Wave-induced circulation in shallow basins. *J. Waterw. Ports Coast. Ocean Div.* **1977**, *103*, 117–135.

93. Matias, A.; Williams, J.J.; Masselink, G.; Ferreira, Ó. Overwash threshold for gravel barriers. *Coast. Eng.* **2012**, *63*, 48–61. [CrossRef]

94. Melby, J.A.; Nadal-Caraballo, N.C.; Kobayashi, N. Wave runup prediction for flood mapping. *Coast. Eng. Proc.* **2012**, *33*, 1–15. [CrossRef]

95. Kergadallan, X. *Analyse statistique des niveaux d'eau extrêmes. Environnement maritime et estuarien*; CETMEF: Margny Lès Compiègne, France, 2013; p. 179.

96. Vousdoukas, M.I.; Wziatek, D.; Almeida, L.P. Coastal vulnerability assessment based on video wave run-up observations at a mesotidal, steep-sloped beach. *Ocean Dyn.* **2011**, *62*, 123–137. [CrossRef]

97. Nielsen, P.; Hanslow, D.J. Wave runup distributions on natural beaches. *J. Coast. Res.* **1991**, *7*, 1139–1152.

98. Dean, R.G.; Dalrymple, R.A. *Water waves mechanics for engineers and scientists. Advances Series on Ocean Engineering*; World Scientific: Singapore, Singapore, 1991; p. 368.

99. Shaw, J.; Taylor, R.B.; Solomon, S.; Christian, H.A.; Forbes, D. Potential impacts of sea-level rise on Canadian coasts. *Can. Geogr.* **1998**, *42*, 365–379. [CrossRef]

100. Koohzare, A.; University of New-Brunswick, Fredericton, NB, Canada. Personal communication, 2015.

101. Fisheries and Oceans Canada. Canadian Tides and Water Levels data Archives. Available online: http://www.isdm-gdsi.gc.ca/isdm-gdsi/twl-mne/index-eng.htm (accessed on 20 March 2015).

102. CHS–Canadian Hydrographic Service. Predicted Water Levels, numerical dataset. 2012. Available online: http://www.tides.gc.ca/eng/info/WebServicesWLD (accessed on 24 October 2012).

103. Garner, K.L.; Chang, M.Y.; Fulda, M.T.; Berlin, J.A.; Freed, R.E.; Soo-Hoo, M.M.; Revell, D.L.; Ikegami, M.; Flint, L.E.; Flint, A.L.; *et al.* Impacts of sea level rise and climate change on coastal plant species in the central California coast. *PeerJ* **2015**, *3*, e958. [CrossRef] [PubMed]

104. Boon, J.D. Evidence of sea level acceleration at U.S. and Canadian tide stations, Atlantic Coast, North America. *J. Coast. Res.* **2012**, *285*, 1437–1445. [CrossRef]

105. Gumbel, E.J. *Statistics of Extremes*; Columbia University Press: New York, NY, USA, 1958; p. 375.

106. Coles, S. *An Introduction to Statistical Modeling of Extreme Values*; Springer: London, UK, 2001.

107. Pugh, D.T.; Vassie, J.M. Extreme sea-levels from tide and surge probability. In Proceedings of the 16th on Coastal Engineering Conference, Hamburg, Germany, 27 August–3 September 1978; pp. 911–930.

108. Arns, A.; Wahl, T.; Haigh, I.D.; Jensen, J.; Pattiaratchi, C. Estimating extreme water level probabilities: A comparison of the direct methods and recommendations for best practice. *Coast. Eng.* **2013**, *81*, 51–66. [CrossRef]

109. Petrov, V.; Guedes Soares, C.; Gotovac, H. Prediction of extreme significant wave heights using maximum entropy. *Coast. Eng.* **2013**, *74*, 1–10. [CrossRef]

110. Tebaldi, C.; Strauss, B.H.; Zervas, C.E. Modelling sea level rise impacts on storm surges along US coasts. *Environ. Res. Lett.* **2012**, *7*, 014032. [CrossRef]

111. Laudier, N.A.; Thornton, E.B.; MacMahan, J. Measured and modeled wave overtopping on a natural beach. *Coast. Eng.* **2011**, *58*, 815–825. [CrossRef]

112. Holman, R.A.; Guza, R.T. Measuring run-up on a natural beach. *Coast. Eng.* **1984**, *8*, 129–140. [CrossRef]

113. Holland, K.; Holman, R.; Sallenger, A. Estimation of overwash bore velocities using video techniques. In Proceedings of 1991 coastal sediments conference, Seattle, WA, USA, 25–27 June 1991; pp. 489–497.

114. Holman, R.; Stanley, J. The history and technical capabilities of Argus. *Coast. Eng.* **2007**, *54*, 477–491. [CrossRef]

115. Bertin, X.; Li, K.; Roland, A.; Bidlot, J.-R. The contribution of short-waves in storm surges: Two case studies in the Bay of Biscay. *Cont. Shelf Res.* **2015**, *96*, 1–15. [CrossRef]

116. FEMA. *Guidelines Specifications for Flood Mapping Partners, Appendix D. Guidance for Coastal Flooding Analysis and Mapping*; Federal emergency management agency: Wichita, KS, USA, 2012.

117. FEMA. *Wave Runup and Overtopping-FEMA Coastal Flood Hazard Analysis and Mapping Guidelines*; Focused Study Report for FEMA: Seattle, WA, USA, February 2005.

118. FEMA. Great Lakes Coastal Guidelines, Appendix D.3. http://greatlakescoast.org/pubs/reports/Great_Lakes_Coastal_Guidelines_Update_Jan2014.png (accessed on 2 June 2015).

119. Holland, K.T.; Holman, R.A. The statistical distribution of swash maxima on natural beaches. *J. Geophys. Res.* **1993**, *98*, 10271. [CrossRef]

120. Mase, H. Random wave runup height on gentle slope. *J. Waterw. Port Coast. Ocean Eng.* **1989**, *115*, 649–661. [CrossRef]

121. Saitoh, T.; Kobayashi, N. Wave transformation and cross-shore sediment transport on sloping beach in front of vertical wall. *J. Coast. Res.* **2012**, *280*, 354–359. [CrossRef]

122. Sabatier, F.; Anthony, E.J.; Héquette, A.; Suanez, S.; Musereau, J.; Ruz, M.H.; Regnauld, H. Morphodynamics of beach/dune systems: Examples from the coast of France. *Géomorphol. Reli. Processes Environ.* **2009**, *1*, 3–22. [CrossRef]

123. Jones, A.F.; Brewer, P.A.; Johnstone, E.; Macklin, M.G. High-resolution interpretative geomorphological mapping of river valley environments using airborne LiDAR data. *Earth Surf. Processes Landforms* **2007**, *32*, 1574–1592. [CrossRef]

124. Leon, J.X.; Heuvelink, G.B.M.; Phinn, S.R. Incorporating DEM Uncertainty in Coastal Inundation Mapping. *PLoS ONE* **2014**, *9*, e108727. [CrossRef] [PubMed]

125. Hodgson, M.E.; Jensen, J.R.; Schmidt, L.; Schill, S.; Davis, B. An evaluation of LIDAR- and IFSAR-derived digital elevation models in leaf-on conditions with USGS Level 1 and Level 2 DEMs. *Remote Sens. Environ.* **2003**, *84*, 295–308. [CrossRef]

126. Aguilar, F.; Mills, J.P.; Delagado, J.; Aguilar, M.A.; Negreiros, J.G.; Pérez, J.L. Modelling vertical error in LiDAR-derived digital elevation models. *ISPRS J. Photogramm. Remote Sens.* **2010**, *65*, 103–110. [CrossRef]

127. Han, J.; Kim, S. Spatial zonation of storm surge hazardous area in the Nakdong Estuary of Korea using high precision terrain data acquired with airborne LiDAR system and geospatial analysis. *J. Coast. Res.* **2013**, *65*, 1385–1390.

128. Schubert, J.E.; Gallien, T.W.; Majd, M.S.; Sanders, B.F. Terrestrial Laser Scanning of Anthropogenic Beach Berm Erosion and Overtopping. *J. Coast. Res.* **2015**, *299*, 47–60. [CrossRef]
129. Arkema, K.K.; Guannel, G.; Verutes, G.; Wood, S.A.; Guerry, A.; Ruckelshaus, M.; Kareiva, P.; Lacayo, M.; Silver, J.M. Coastal habitats shield people and property from sea-level rise and storms. *Nat. Clim. Chang.* **2013**, *3*, 913–918. [CrossRef]
130. Gedan, K.B.; Kirwan, M.L.; Wolanski, E.; Barbier, E.B.; Silliman, B.R. The present and future role of coastal wetland vegetation in protecting shorelines: Answering recent challenges to the paradigm. *Clim. Chang.* **2011**, *106*, 7–29. [CrossRef]
131. Prather, M.; Flato, G.; Friedlingstein, P.; Jones, C.; Lamarque, J.-F.; Liao, H.; Rasch, P. Annex II: Climate System Scenario Tables. In *IPCC. Climate Change 2013: The Physical Science Basis*; Stocker, T.F., Qin, D., Plattner, G.-K., Tignor, M., Allen, S.K., Boschung, J., Nauels, A., Xia, Y., Bex, V., Midgley, P.M., Eds.; Cambridge University Press: Cambridge, UK; New York, NY, USA, 2013.
132. Senneville, S.; St-Onge Drouin, S.; Dumont, D.; Bihan-Poudec, A.-C.; Belemaalem, Z.; Corriveau, M.; Bernatchez, P.; Bélanger, S.; Tolszczuk-Leclerc, S.; Villeneuve, R. *Rapport final: Modélisation des glaces dans l'estuaire et le golfe du Saint-Laurent dans la perspective des changements climatiques*; ISMER-UQAR, Rapport final présenté au ministère des Transports du Québec: Rimouski, QC, Canada, 2014.
133. Church, J.A.; Monselesan, D.; Gregory, J.M.; Marzeion, B. Evaluating the ability of process based models to project sea-level change. *Environ. Res. Lett.* **2013**, *8*, 014051. [CrossRef]
134. Bittermann, K.; Rahmstorf, S.; Perrette, M.; Vermeer, M. Predictability of twentieth century sea-level rise from past data. *Environ. Res. Lett.* **2013**, *8*, 014013. [CrossRef]
135. Gehrels, W.R.; Woodworth, P.L. When did modern rates of sea-level rise start? *Glob. Planet. Chang.* **2013**, *100*, 263–277. [CrossRef]
136. Barnett, R.L.; Gehrels, W.R.; Charman, D.J.; Saher, M.H.; Marshall, W.A. Late Holocene sea-level change in Arctic Norway. *Quat. Sci. Rev.* **2015**, *107*, 214–230. [CrossRef]

Journal of
Marine Science and Engineering

MDPI

Article

Channel Shallowing as Mitigation of Coastal Flooding

Philip M. Orton [1,*], Stefan A. Talke [2], David A. Jay [2], Larry Yin [1], Alan F. Blumberg [1], Nickitas Georgas [1], Haihong Zhao [3], Hugh J. Roberts [3] and Kytt MacManus [4]

[1] Davidson Laboratory, Stevens Institute of Technology, Castle Point on Hudson, Hoboken, NJ 07030, USA; lyin1@stevens.edu (L.Y.); alan.blumberg@stevens.edu (A.F.B.); nickitas.georgas@stevens.edu (N.G.)
[2] Department of Civil and Environmental Engineering, Portland State University, Post Office Box 751, Portland, OR 97207-0751, USA; talke@pdx.edu (S.A.T.); djay@pdx.edu (D.A.J.)
[3] ARCADIS, 630 Plaza Drive, Suite 100, Highlands Ranch, CO 80129, USA; Haihong.Zhao@arcadis-us.com (H.Z.); Hugh.Roberts@arcadis-us.com (H.J.R.)
[4] Center for International Earth Science Information Networks, Columbia University, PO Box 1000, 61 Route 9W, Palisades, NY 10964, USA; kmacmanu@ciesin.columbia.edu
* Author to whom correspondence should be addressed; philip.orton@stevens.edu; Tel.: +1-201-216-8095.

Academic Editor: Rick Luettich
Received: 4 May 2015; Accepted: 6 July 2015; Published: 21 July 2015

Abstract: Here, we demonstrate that reductions in the depth of inlets or estuary channels can be used to reduce or prevent coastal flooding. A validated hydrodynamic model of Jamaica Bay, New York City (NYC), is used to test nature-based adaptation measures in ameliorating flooding for NYC's two largest historical coastal flood events. In addition to control runs with modern bathymetry, three altered landscape scenarios are tested: (1) increasing the area of wetlands to their 1879 footprint and bathymetry, but leaving deep shipping channels unaltered; (2) shallowing all areas deeper than 2 m in the bay to be 2 m below Mean Low Water; (3) shallowing only the narrowest part of the inlet to the bay. These three scenarios are deliberately extreme and designed to evaluate the leverage each approach exerts on water levels. They result in peak water level reductions of 0.3%, 15%, and 6.8% for Hurricane Sandy, and 2.4%, 46% and 30% for the Category-3 hurricane of 1821, respectively (bay-wide averages). These results suggest that shallowing can provide greater flood protection than wetland restoration, and it is particularly effective at reducing "fast-pulse" storm surges that rise and fall quickly over several hours, like that of the 1821 storm. Nonetheless, the goal of flood mitigation must be weighed against economic, navigation, and ecological needs, and practical concerns such as the availability of sediment.

Keywords: storm surge; flooding; tides; adaptation; wetlands; bathymetry; hurricane; Hurricane Sandy; Jamaica Bay; New York City

1. Introduction

Our coastlines provide many social, economic, and ecological benefits, but are increasingly vulnerable to damages from coastal storms and sea level rise. There is an increasing societal awareness and interest in the application of natural coastal systems for coastal risk reduction, through both reducing flood levels and avoiding increased development. In response to this trend, as well as the devastation caused by Hurricane Sandy, the US Army Corps of Engineers is evaluating the use of natural and nature-based approaches where possible to support coastal resilience and risk reduction [1]. In Northern Europe, beach nourishment projects, dunes and other natural features have long been used along with engineered infrastructure to reduce and mitigate against coastal flooding, e.g., [2].

J. Mar. Sci. Eng. **2015**, *3*, 654–673

Hurricane Sandy severely impacted New York City (NYC) in 2012, flooding a 132 km^2 area (17% of total land mass) with a population of 443,000 people [3]. Flooding was particularly widespread around Jamaica Bay, a large coastal embayment in NYC where surrounding neighborhoods include a large proportion of the 1,116,000 city residents (Figure 1) living on land within range of a 5 m coastal flood level (measured relative to the 1983–2001 mean sea level, MSL). The baseline FEMA 500-year coastal flood at Howard Beach, northern Jamaica Bay, is 3.8 m MSL, but this increases to 5.4 m with a high-end projection of sea level rise for the 2080s [4], raising the flood risk for this population.

The flood risk in Jamaica Bay has possibly been aggravated by widespread anthropogenic alterations over the past century. The entrance channel in Jamaica Bay was dredged to a depth of 5.5 m and width of 150 m, beginning about 1910, and eventually deepened to 9.1 m. Archival research has shown that the average depth of the bay has increased from 1 m in the mid-1800s to 5 m at the present [5]. As a result of this deepening and possibly also the smaller water depth increases due to sea level rise, tide ranges have increased ~0.5 m (25%–45%) and tides are amplified instead of being damped as they flow from the entrance to the inner reaches of the bay [6], though the contribution of sea-level rise to tidal range change is unknown. The estimated total loss of interior wetlands for the bay since the mid-1800s is 49 km^2 of the original 65 km^2 [7]. However, a successful Corps of Engineers pilot program rebuilt 0.64 km^2 of wetlands from 2006 to 2014 [8], with a goal of eventually restoring them to their 1974 footprint [9].

The scale of the wetland losses and depth increases in Jamaica Bay, combined with the at-risk population nearby, provide an important opportunity to test whether natural and nature-based strategies can protect against coastal floods and mitigate against local sea level rise. Furthermore, the bay has a large enough length (~10 km) to make wetland-based protection plausible, if one considers qualitative (order-of-magnitude) metrics such as the 15 km for 1 m storm surge reduction that was developed for historical Gulf of Mexico wetlands and hurricanes [10,11].

The purpose of this paper is to use hydrodynamic modeling to explore the potential for using nature-based coastal flood protection to reduce flood risk in the neighborhoods surrounding Jamaica Bay. We highlight and discuss the positives and negatives of a new concept for nature-based flood protection—Channel depth shallowing, the strategy of reducing estuary or inlet depths to reduce the inland penetration of a coastal flood. The efficacy of channel shallowing and wetland restoration are quantified using the simulations of the city's two highest known historical flood events—A Category-3 hurricane that passed over the city in 1821 [12,13] and Hurricane Sandy in 2012.

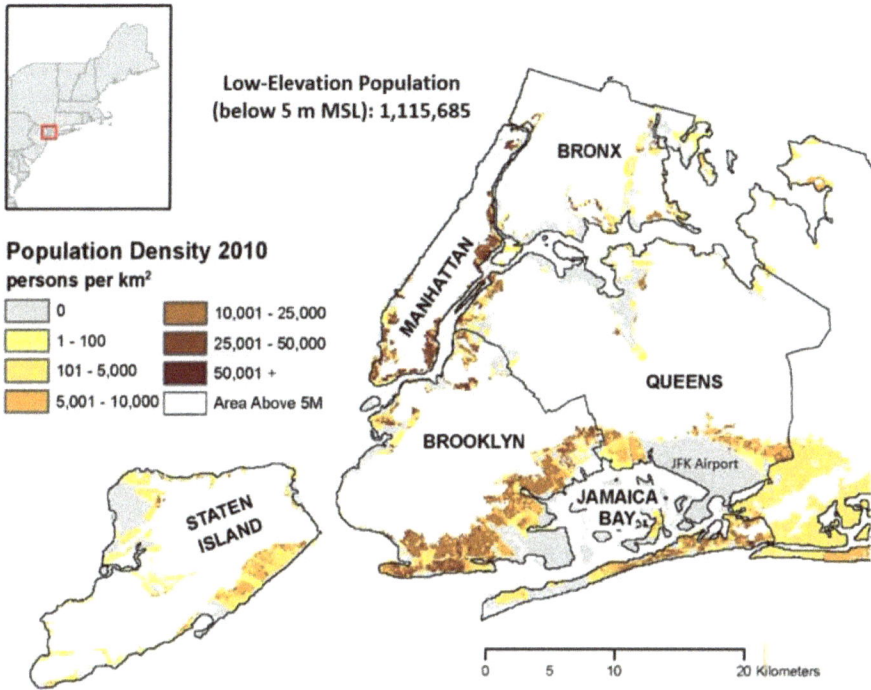

Figure 1. New York City boroughs and the western edge of Nassau County (right edge of map) showing the population density in contiguous low-elevation coastal zones below 5 m above mean sea level (MSL). The population density data are based on 100 m resolution 2010 population data [14]. LIDAR elevation data at 1 m resolution are used to filter for elevation; the map only shows population density data where >50% of a cell area is below 5 m elevation. Data available at [15].

2. Methods

2.1. Hydrodynamic Modeling

A validated hydrodynamic model and grid covering Jamaica Bay and surrounding neighborhoods is used in this study to investigate different nature-based mitigation scenarios. The Stevens ECOM hydrodynamic model (sECOM), e.g., [16,17], provides accurate coastal flood predictions as part of the NY Harbor Observation and Prediction System (NYHOPS) and the Stevens Institute of Technology Storm Surge Warning System, with water level root mean square errors (RMSE) of 0.10 m since 2007 [18]. The sECOM model includes a coupled wave model [18,19] and robust upland inundation capabilities, known as "wetting and drying" [20].

The Jamaica Bay grid is a 30 m × 30 m, 1069 by 550 square-cell grid covering the watershed up to an elevation of 6 m (Figure 2, top left panel), and the flood simulations on it are run in two-dimensional barotropic mode. The bathymetry and topography information in the model is based on the region's best available datasets, compiled by FEMA Region 2 for their recent coastal storm surge study [21]. The grid is doubly-nested, with a large, coarse grid modeling the Northwest Atlantic (the Stevens Northwest Atlantic Predictions grid, SNAP), a finer-resolution regional grid (the NYHOPS grid) that resolves New York Harbor regions at ~100 m [20,22], and a 30 m resolution grid that covers Jamaica Bay.

In storm surge modeling studies, a common simplified approach to representing the effects of wetlands is to treat them as enhanced landscape roughness features, through Mannings-*n*. While this neglects water—Vegetation interactions within the water column, it provides good model agreement for surges and flood recession processes at inland locations, e.g., [23]. Following this approach, Mannings-*n* is set via landscape types defined across Jamaica Bay using the National Land Cover Database (NLCD), as per [24]. These values are converted to drag coefficients (C_D) within the model through the standard formula $C_D = gn^2h^{-1/3}$ where g is gravitational acceleration, n is Mannings-*n*, and h is the time and space varying water depth. Mannings-*n* values for wetlands are 0.045, and those for other common land-cover types in the model are: 0.02 for open water, 0.09 for barren land (rock, sand or clay), and 0.10 and 0.13 for medium and high intensity developed land, respectively (Figure 2, right panels).

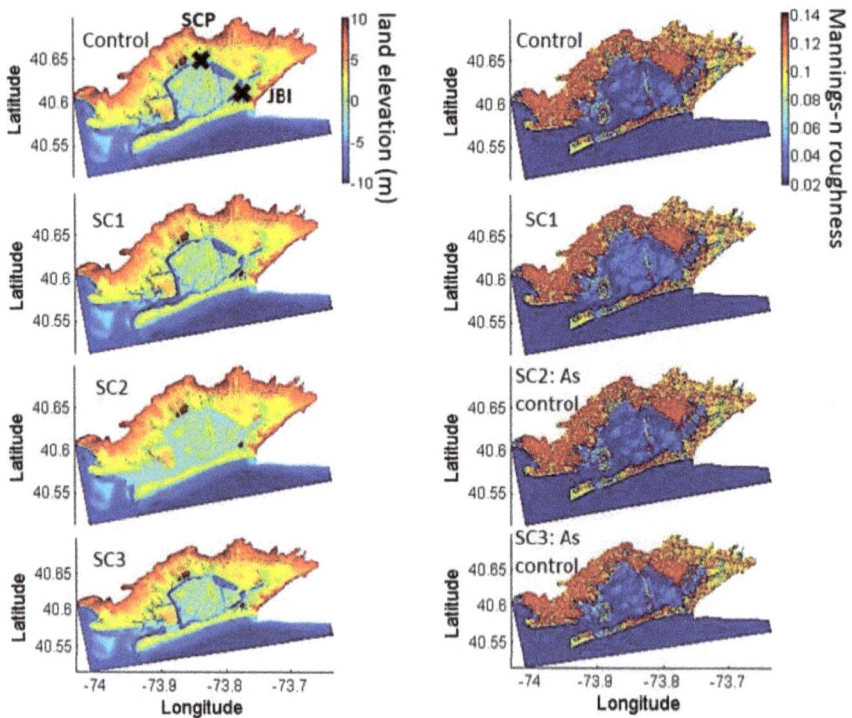

Figure 2. Maps of adaptation scenarios 1–3 and control—Land elevation (**left**) and Mannings-*n* (**right**). The Spring Creek Park (SCP) and Inwood (JBI) sites are also indicated (X) in the top left panel. Elevations are shown relative the NAVD88 datum.

2.2. Storms and Forcing Data

Storm surge is an increase in water level caused by wind and low atmospheric pressure, and combines with the tide to form a storm tide, which here we measure relative to the storm year's mean sea level, e.g., [25]. The two largest known historical hurricane storm tide events for New York Harbor (The Battery) are Hurricane Sandy, which caused a 3.4 m storm tide, and a Category-3 hurricane that passed over the city in 1821, which caused a storm tide estimated as 2.7–3.7 m [12], or 2.7–3.0 m [26] (Figure 3). Here, we use simulations of these two storms as tests of the efficacy of nature-based adaptation scenarios for Jamaica Bay. These two storms are highlighted here because of their magnitude and also because they represent two different types of event. The storm surge during Sandy rose slowly over three days, though it rose more rapidly in the final hours near the time of landfall. In contrast, the 1821 hurricane storm surge rose very rapidly over a period of a few hours [13].

The Sandy simulation on the SNAP and NYHOPS domains is forced by a meteorological reanalysis produced by Oceanweather, Inc., Cos Cob, CT, USA). The resulting water levels have an RMSE at The Battery of 0.16 m. The model results from the NYHOPS domain are used as a clamped offshore boundary condition on the nested Jamaica Bay grid. Local wind forcing for the Jamaica Bay grid is applied based on measurements at Breezy Point, and is assumed to be spatially homogeneous (data courtesy of Weatherflow, Inc., Scotts Valley, CA, USA). A model validation for the Hurricane Sandy control simulation on the Jamaica Bay grid is given below in Section 3.

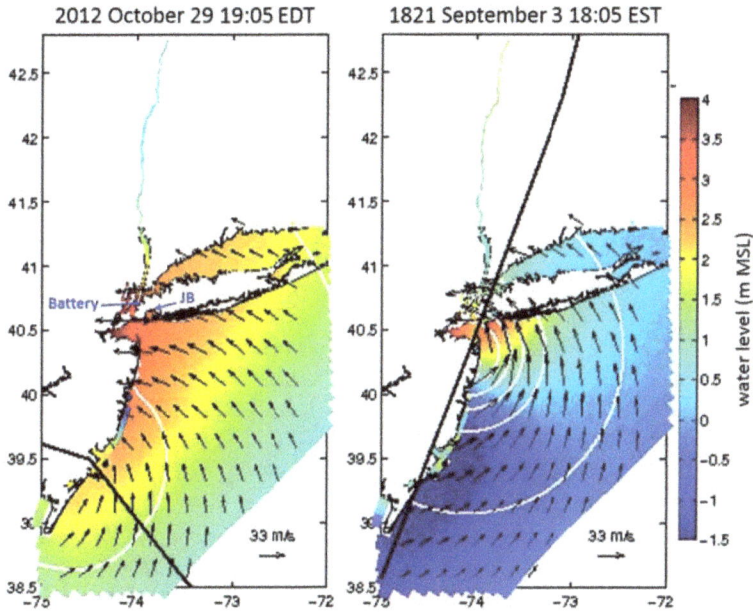

Figure 3. Hurricane Sandy (**left**) and the 1821 Hurricane (**right**) storm track (**black lines**), isobars (**white lines**), wind vectors, and simulated water levels relative to mean sea level near the time of final landfall. The 1821 simulation is performed on the mean sea level for the 1983–2001 epoch. Blue arrows in the left panel point to locations of the Battery tide gauge and Jamaica Bay (JB).

The hurricane of 1821 was simulated using synthetic tropical cyclone parameters based on work by Boose et al. [27] and Swiss Re [13]. The storm makes its first landfall near Cape May, New Jersey, tracks over land along the coast, crosses Raritan Bay, and makes a final landfall at NYC (Figure 3). Maximum sustained winds and radius of maximum winds at landfall are estimated to be 58 m·s^{-1} and 50 km, respectively. We utilize parametric equations to represent the storm's wind and pressure forcing for our ocean model—The Holland pressure model [28] and SLOSH wind model [29]. The simulation is run with the 1983–2001 mean sea level to represent the possible flooding for a similar type of storm if it were to occur in modern times.

2.3. Landscape Scenarios

Four experiments with differing landscape scenarios for Jamaica Bay are used for each storm, to test the effect of changing conditions. The landscape scenarios, defined as specific cases of land elevation (topography, bathymetry) and land cover, are:

CONTROL: Present-day landscape and land cover;

SC1: "Wetland restoration": Restoring the 1879 wetland footprint and bathymetry, while not altering areas of present-day deep channels or neighborhoods;

SC2: "Full shallowing": Shallowing deep (>2m) channels to 2 m depth below Mean Low Water;

SC3: "Inlet shallowing": As SC2, but shallowing only the narrowest region in the inlet;

The depths and Mannings-*n* roughness values for the control run and adaptation scenarios are shown in Figure 2. Scenario SC1 is a wetland restoration scenario for the bay interior, including both restoration of the wetlands and the eroded land beneath them (often referred to as "marsh islands"). The restoration is to the 1879 wetland footprint [30], but ignoring areas that are deep channels (>2 m depth) or upland neighborhoods in today's landscape. This experiment is intended to show how the present goal of restoring wetlands in the center of the bay [9] would affect extreme flooding events, if it were greatly expanded. Wetlands are represented with land elevations of 0.5 m above MSL and a characteristic wetland Mannings-*n* value of 0.045 [24]. Note that results with SC1 were insensitive to choosing a cutoff of 3 m instead of 2 m.

The shallowing experiments (SC2, SC3) are deliberately extreme, intended to demonstrate system sensitivities (leverage) and to provide process understanding. The bathymetry changes in each scenario would be unlikely to reflect equilibrium morphological conditions, and the inlet shallowing experiment (SC3) might induce water quality problems by creating stagnant, poorly flushed deep waters in the bay. The two shallowing scenarios have no changes to land cover; both are the same as the control, with present-day land cover.

2.4. Model Validation

The Hurricane Sandy control experiment using modern bathymetry was validated using water level data from the United States Geological Survey, including a pressure sensor near Spring Creek Park (SCP) and a tide gauge at Inwood (JBI; Figure 2). The SCP station is at the northern part of the bay, and is a rapid-deployment storm surge sensor (SSS-NY-QUE-002WL), a pressure gauge that is deployed when a tropical cyclone is approaching. This sensor elevation is at 0.2 m above NAVD88, so that it provides data only when the water level is higher than this elevation. Data are corrected for atmospheric pressure based on a nearby USGS barometer deployment. The JBI tide gauge is at the eastern end of the bay, and has been running since 2002. High water marks from Sandy within Jamaica Bay collected by USGS were also used to validate the model.

Because an early 19th century digital elevation model does not exist (at present), we make the assumption that the 1821 storm tide heights in NY Harbor can be modeled (where the validation data was observed), to a first approximation, using modern bathymetry. However, as noted by Talke et al. [25], changes to bathymetry at the entrance bar and within the harbor may have altered storm tide characteristics since the mid-19th century.

3. Results

Results for the Hurricane Sandy control simulation and validation are shown in time series form in Figure 4, demonstrating good agreement. The RMSEs between the model and observations are 0.22 and 0.18 m at SCP and JBI. Model skill, e.g., [31], is 0.99 for the tide gauge site JBI, but could not be applied for SCP due to obvious wave-driven fluctuations in the pressure sensor's data. A one-to-one plot of modeled and observed high water marks (HWMs) is shown in Figure 5, demonstrating an RMS difference of 0.19 m and r^2 value of 0.89. The HWM observations have moderate uncertainty, as shown by the error bars in the figure. These results are similar to other model studies of Sandy [32,33].

Maps of temporal maximum water level results for Hurricane Sandy for the control run and the three scenarios are shown in Figure 6. These results show peak water levels (averaged over the bay interior) decreasing by 0.01 m for SC1 (wetland restoration), 0.50 m for SC2 (full shallowing), and 0.23 m for SC3 (inlet shallowing). Table 1 shows detailed results for temporal maxima and their reductions.

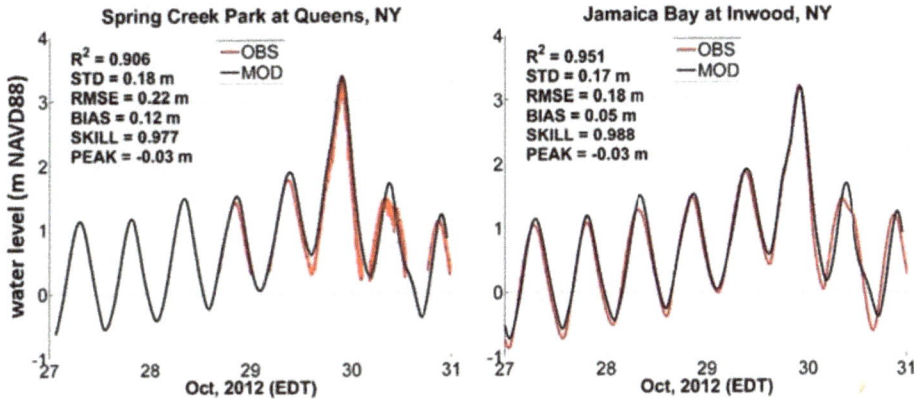

Figure 4. Hurricane Sandy model validation in Jamaica Bay, at (**left**) Spring Creek Park and (**right**) the Inwood tide gauge. Station locations are shown in the first panel of Figure 2.

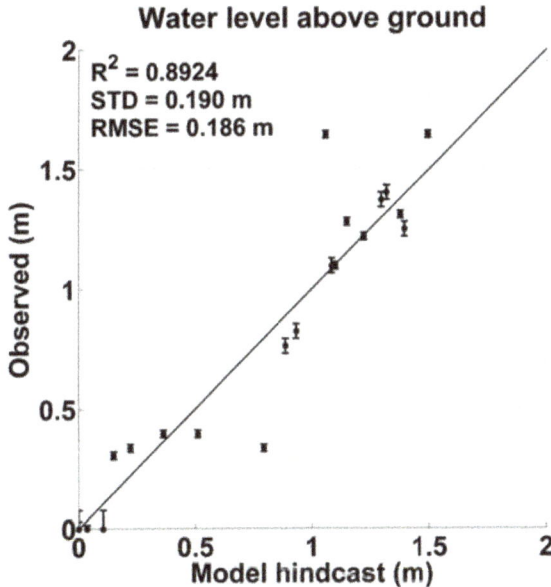

Figure 5. Sandy simulation *vs.* Jamaica Bay high water marks (HWMs). Error bars on the observed water levels come directly from the USGS field report for each HWM.

The control simulation run for the 1821 Hurricane on the NYHOPS grid (Figure 3) has a peak water level of 2.95 m over local mean sea level (MSL) at the Battery, within the range of historical estimates given in Section 2.2. Temporal-maximum water level results in Jamaica Bay for the 1821 Hurricane are shown in Figure 7. For the control run, the temporal maximum water level in the bay (2.54 m) is substantially lower than at The Battery (2.95 m), likely due to the strong (over 50 m·s^{-1}) east winds blowing against the water flowing in through the inlet prior to the final landfall at New York City (Figure 3). The results maps show peak water levels (averaged over the bay interior) decreasing by 0.06 m for SC1 (wetlands), 1.16 m for SC2 (full shallowing), and 0.75 m for SC3 (inlet shallowing) (Table 1).

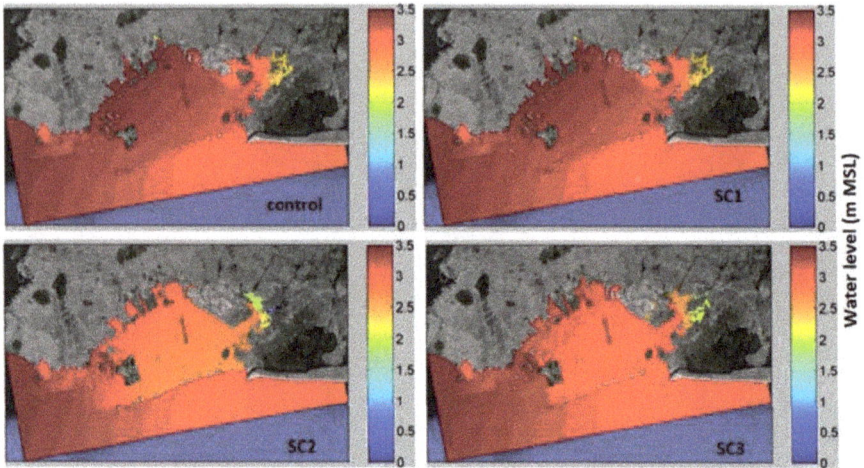

Figure 6. Results for Hurricane Sandy (peak water levels) in map view for control run and Scenarios 1, 2 and 3. Water levels are shown relative to local mean sea level (MSL) at JBI.

Table 1. Temporal maxima (peaks) and their reductions for each storm and scenario.

Storm Events	Scenario	Bay-Average			JBI Station		
		Peak Water Level (m [a])	Reduction (m)	Reduction Percentage (%)	Peak Water Level (m [a])	Peak Surge [b] (m)	Peak Tide [c] (m)
Hurricane Sandy 2012	Control	3.39	–	–	3.27	2.38	1.03
	SC1	3.38	0.01	0.3	3.25	2.33	1.01
	SC2	2.89	0.50	14.7	2.77	2.23	0.58
	SC3	3.16	0.23	6.8	3.06	2.32	0.76
Hurricane 1821	Control	2.54	–	–	2.60	2.87	0.78
	SC1	2.48	0.06	2.4	2.50	2.73	0.74
	SC2	1.38	1.16	45.7	1.27	1.46	0.49
	SC3	1.79	0.75	29.5	1.85	2.20	0.64

[a] Peak (temporal maximum) water levels are relative to the 1983–2001 mean sea level datum; [b] Peak surge was computed as the temporal maximum of the quantity (modeled storm tide-modeled tide); [c] Peak tide was computed as the temporal maximum in the tide-only model run (over the storm simulation time range).

Reductions in peak water level and changes to the flooded area are mapped in Figure 8. Compared to the control model runs, each scenario demonstrates a significant reduction in the flooded area (black dots). Each scenario also wets (floods) areas which remained dry in the control run, primarily near the entrance (red dots). On aggregate, the induced flooding for each scenario is small, and substantially more land is protected from flooding than is put at risk. For Sandy, the area of prevented flooding for each scenario is 0.4 km², 19.6 km², and 10.3 km², respectively, and induced flooding is negligible (below 0.1 km²). For 1821, the area of prevented flooding is 1.4 km², 18.1 km², and 11.4 km², respectively, and areas with induced flooding are 0.2, 0.1, and 0.8 km². The full shallowing scenario (SC2) for the 1821 hurricane prevents flooding of elevated areas inside the bay (above 1.5 m MSL) altogether (Figure 7). We also conducted experiments (not shown) where we combined the wetlands of SC1 with the shallowing of SC2, but we found that results were nearly indistinguishable from SC2. Therefore, adding the wetlands into the shallowing scenario did not further reduce water levels.

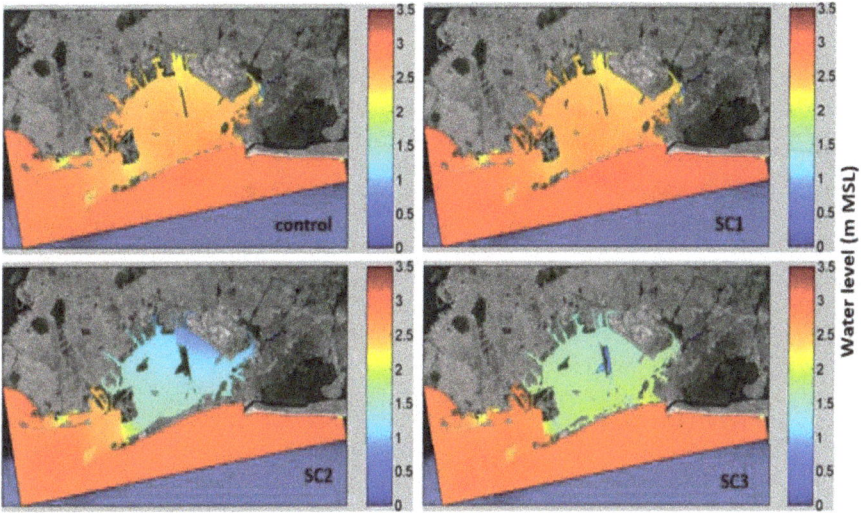

Figure 7. Results for the 1821 Hurricane (peak water level) in map view for control run and scenarios 1, 2 and 3, as with Figure 6.

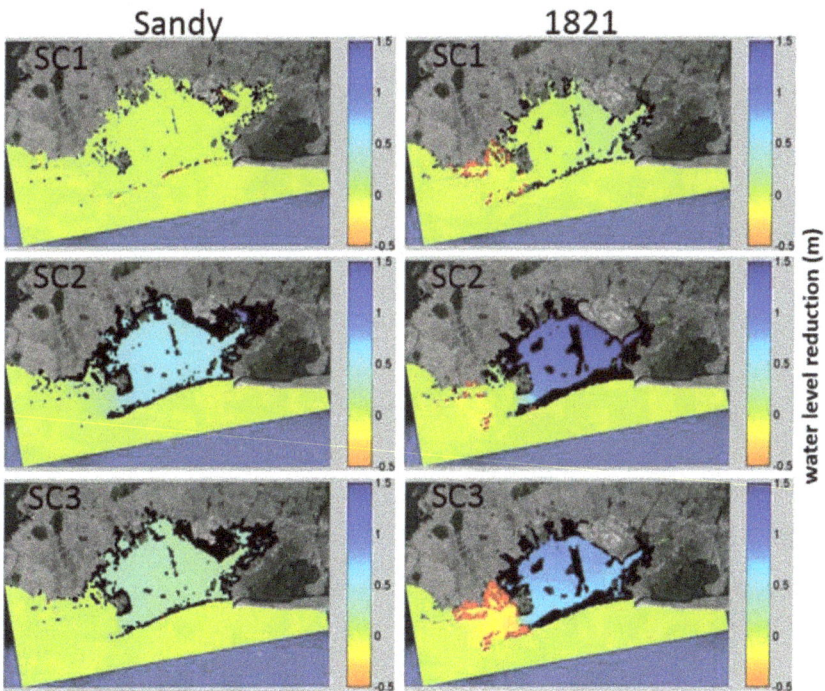

Figure 8. Reductions in peak water level with the three scenarios for Hurricane Sandy and the 1821 Hurricane. Black dots show areas where flooding was stopped (wet in control, dry in scenario), and red dots show areas where new flooding was induced (opposite).

Time series of water level at JBI show substantial reductions in peak water levels for SC2 and SC3 (Figure 9 top panels). Additional tide-only model runs were performed for the time period of each storm simulation, on each landscape scenario (and the control landscape), to separately elucidate effects of the adaptations on tide and surge (computed as storm tide minus tide) (Table 1; Figure 9 middle and bottom panels). In these tide-only runs, peak tide levels at JBI during 1821 and Sandy are reduced by 0.02–0.04 m (2%–4%), 0.29–0.45 m (31%–38%) and 0.14–0.27 m (15%–23%) respectively for the three experiments, relative to a tide-only run on the control landscape. Reductions in storm surge for Sandy are small (0.05, 0.15 and 0.06 m), whereas they are very large for the 1821 Hurricane (0.14, 1.41 and 0.67 m), respectively. Thus, the reductions in peak water levels and flooding for Sandy are mainly due to a reduction in the astronomical tide, whereas for 1821 the reduction in storm surge is dominant. Since the 1821 event occurred near tidal Low Water (LW), the decreased tidal range (smaller LW) actually increases the astronomical contribution to the peak water level for SC2.

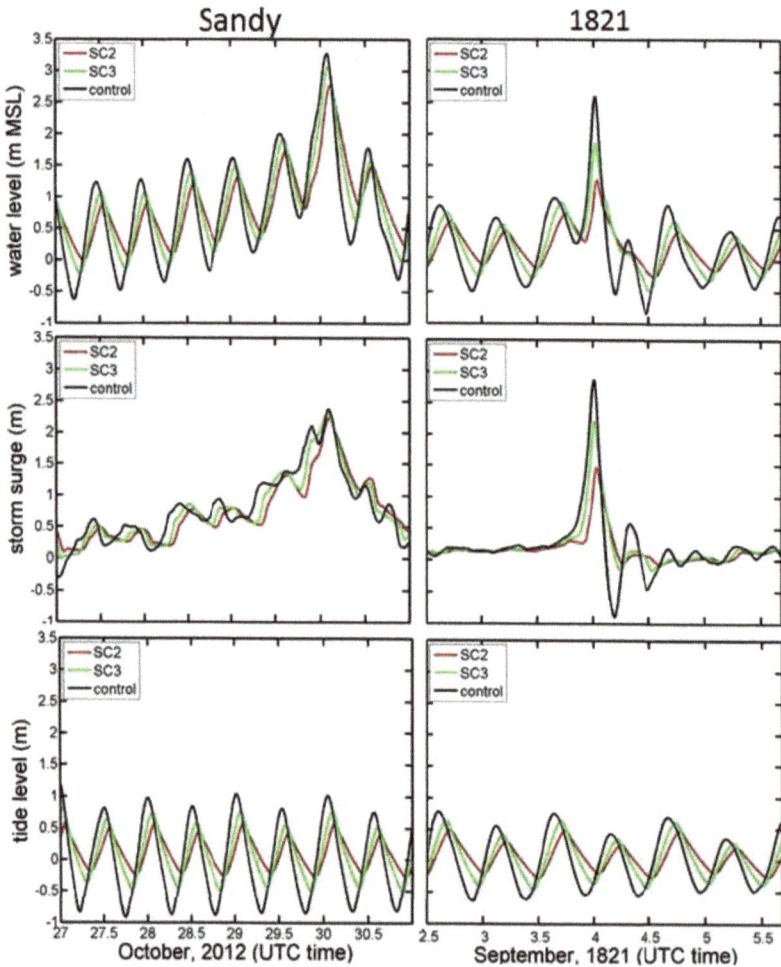

Figure 9. Time series of (**top**) water level, (**middle**) surge, and (**bottom**) tide level at JBI for Sandy (**left**) and the 1821 Hurricane (**right**), for the control runs and two shallowing scenarios.

4. Discussion

The primary result of the scenario experiments is that reductions of channel depth have a much stronger leverage on flood levels inside Jamaica Bay than wetland restoration, for these two extreme flood events. The full shallowing scenario (SC2) reduces peak water levels in the bay by 15% and 46%, for Sandy and 1821 respectively, and full wetland restoration only 0% and 2%, respectively. The narrow section of the inlet is demonstrated to exert strong leverage over water levels in the bay, as scenario SC3 shows reductions of 7% and 30%, respectively. This suggests that the inlet alone can provide about half (46% for Sandy and 65% for 1821) of the benefit observed in the full shallowing scenario (SC2), dividing the water level reduction for SC3 by the reduction for SC2. Thus, the channel shallowing scenarios demonstrate that reducing estuary or inlet depth can be a powerful option for reducing the inland penetration of coastal floods.

The utility of wetlands for flood protection is often cited as a reason to build or restore wetlands, and in some cases they can cause reductions in flooding [11]. These results suggest that the presence of deep shipping channels can eliminate this benefit, even with wetland restoration on a massive scale. Though 40% (24 km^2) of Jamaica Bay is wetlands in SC1 (Figure 2), the deep inlet and circumferential deep channels inside the bay efficiently deliver flood waters to the surrounding neighborhoods. A recent modeling study showed similar results for an idealized coastline that mimicked the Gulf Coast, demonstrating how increasing channel sizes lead to increasing storm surge penetration through a coastal wetland [34]. In spite of the minimal effect wetlands have on coastal flooding in this study, wetlands are also known to provide wave attenuation benefits, and they can reduce erosion by stabilizing sediments [35]; neither is quantified in this study. Moreover, they provide high-value habitat and ecosystem services, as summarized in the recent Corps of Engineers report on nature-based coastal resilience [1].

Storm characteristics such as track, speed, size, central pressure and wind speed (Figure 3) are well-known to influence the development of surge on the continental shelf [31,36], but these results also suggest that the characteristics of the hurricane affect the amount of flood-mitigation benefit of local changes to the bay, as has previously been observed with wetlands, e.g., [11]. Even though the overall peak water level in the 1821 control case was about 1 m less than the Sandy case in Jamaica Bay (Figure 9), significantly more benefit was derived thru shallowing in terms of reductions in peak water level and flood area (Figures 8 and 9). As discussed below in Section 4.1., tidal theory provides additional insights into the observed differences in the efficacy of shallowing for the two hurricanes. In Section 4.2, we discuss the broader considerations and timescale over which a shallowing strategy could take place.

4.1. Physics of Channel Depth and Storm Tides

Flood waters are widely known to be slowed by increased friction when passing through wetlands, and this can reduce flooding when it prevents inland propagation of a storm surge [36]. Moreover, the amplitude of the surge wave is reduced, much like a river flood wave is reduced when flowing through vegetation. In studies of tidal dynamics, the interplay between water depth, tidal frequency and friction has long been studied [37–40]. However, little attention has been paid to the influence of channel and inlet depths on storm surge propagation.

Both tides and storm surge are "sea-level events" causing anomalies in sea level of varying durations, although surges are atmospherically-forced waves and tides are gravitationally-forced waves composed of constituents with well-defined periods. Tides and surges in harbors and inlets are both defined as long, or shallow water, waves, in which the wavelength is much longer than the depth. Shallow (or narrow) channels at inlets and inside estuaries can reduce inland penetration of a sea level anomaly due to a "choking" effect of reducing the cross-sectional area of the channel and the amount of water transported in and out of a bay or estuary, as is the case in the NY Bight region at Barnegat Bay and Great South Bay [41]. This is illustrated by the relative effectiveness of the inlet shallowing scenario, SC3.

J. Mar. Sci. Eng. **2015**, *3*, 654–673

The time scale of the sea-level event, whether it be tides or a storm surge, can also influence its penetration into a bay or estuary. Compared to fast-pulse events such as the 1821 Hurricane, storm surges from cold-season storms (e.g., nor'easters) have a longer time scale, on the order of 1 day, and the choking and frictional effects of a shallow inlet or channel may be lessened. A more rapid sea-level event such as semi-diurnal tide may be attenuated more strongly by a shallow inlet or channel [41]. Physically, a fast moving wave produces stronger velocities, leading to greater dissipation of energy (which scales as U^3/H, where U is the velocity and H is the water depth) and attenuation. A sea-level event like Hurricane Sandy's storm tide can be a combination of both short- and long-timescale events. During the days preceding landfall, water levels in Jamaica Bay steadily increased and were 1.3 m above the predicted tidal levels 12 h before the peak. This "forerunner surge" [42] penetrated into coastal embayments with little attenuation, and was followed by a more rapid sea-level pulse in the final hours. The 1821 Hurricane, on the other hand, was a very fast-pulse flood event, with nearly all of its storm surge (82% or 2.3 m) arriving in the final four hours (Figure 9). The different storm sizes and tracks played a large role in creating these differences in surge time-scale; Sandy, with a more offshore track (Figure 3) and very large size (radius of maximum winds 220 km just before landfall), caused moderately strong northeast winds in NY Bight for over two days before landfall, whereas the 1821 storm had an along-coast track right over NYC and a much smaller size (estimated radius of maximum winds 40 km), causing the wind-blown surge to be driven by east winds over only a few hours.

The interacting roles of friction, channel depth and wave period have been considered for estuary channels before, in the context of tide dynamics. For an idealized, rectangular basin [39], the damping term in the equation of motion depends on rU/H, where r is the linearized friction coefficient, U in the tidal velocity, and H is the mean depth. Hence, as noted by Ianniello [37] and Burchard [43], decreasing the system's depth has a similar dynamical effect as increasing the friction or wave frequency. This is one reason why the attenuation of the diurnal tide can be different than the semi-diurnal tide in tidal rivers, though non-linear interaction between constituents and river flow must be considered [40]. Based on this linearized scaling, a 3-fold decrease in channel depth (e.g., 6 m to 2 m) has a similar dynamical effect as a 3-fold increase in drag coefficient. Similar results are found by Friedrichs and Aubrey [38] for tidal propagation in convergent estuaries.

The peak surge results for Sandy (Table 1) show a relatively small reduction of 0.15 m at JBI for the full shallowing scenario, indicating that the water level reduction of 0.49 m at that station was predominantly (69%) due to a reduction in the tide. On the other hand, the surge reduction of 1.41 m for the 1821 Hurricane at JBI represents slightly more than 100% of the reduction in total water level, 1.33 m. The forerunner surge, combined with higher mean depth due to the tidal stage, means that the mean water depth just before Sandy's landfall was ~2 m greater than during a similar phase of the 1821 event. Model results also demonstrate that flow velocities in the inlet were significantly larger during 1821 (1.94 m·s^{-1}) than in 2012 (1.03 m·s^{-1}), due to the smaller time scale over which water levels rose. Combined, the depth and velocity effects both act to attenuate and mitigate the 1821 event more than Sandy in SC2 and SC3.

Two primary processes reduce storm-tide magnitudes within Jamaica Bay. The reduction of depths with SC2 (full shallowing) and the greater resulting frictional drag increase the shallow water wave attenuation (SC2). The reduction of inlet depth only (SC3) constricts flow, producing only a choking effect. The reductions in peak water level for SC3 are about half of those for SC2, demonstrating that the effects are of roughly similar magnitude. Nonetheless, the increased damping in the SC2 scenario appears to produce the positive effect of reducing surge wave reflection, which likely causes increased flooding near the inlet in SC3 (Figure 8). Careful consideration must therefore be made to avoid unwanted side-effects when designing coastal flood mitigation strategies.

The shallowing experiments also induce a lag in the times of peak surge and peak water level, relative to the control runs, and the sizes of these lags typically show inverse relationships with water depth and sea level pulse period. Defining the "lag" as the time of the peak in the full shallowing (SC2) experiment minus the time of the peak in the control run, the lags of peak surge are 10 (Sandy) and 40 min (1821). The lags of peak water level are 60 (Sandy) and 30 min (1821). The lags in tide-only runs are substantially larger, at 110 (Sandy) and 90 min (1821) (Figure 9). The lags are usually larger for cases when there is shallower water (e.g., tide-only runs), consistent with stronger frictional effects. The reason for the small time lag for surge with Sandy SC2 (10 min) is because the water is deep when the final pulse arrives, even in the shallowing experiments, and the pulse has a long period (relative to tide or water level, which have lags of 110 and 60 min, respectively). The 1821 surge has a 40 min lag, and this moderately large lag can be explained by the surge pulse coming on top of a low tide, so in shallow water, and having a short pulse period. Lastly, the lags for SC3 (inlet shallowing) are about half those for SC2 (full shallowing), suggesting more shallow water area (and smaller bay-average depth) leads to larger lags. In conclusion, these results are generally consistent with the interpretation of frictional effects being the dominant reason that shallowing is effective at reducing peak storm-driven water levels, dependent upon both pulse period and water depth. It is also noteworthy that the lags of 30–60 min in peak water level could be viewed as additional protection against flooding, slowing arrival of the flooding in addition to reducing the water levels.

The results shown in this manuscript are broadly consistent with the results of Talke et al. [25], who hypothesized that local changes to bathymetry and channel depth in New York Harbor could help explain the long-term trends in storm tides measured at tide gauges. Nonetheless, the detailed hydrodynamic effects of channel shallowing require further study and may be highly site or storm specific. As shown in Figure 8, there are locations where a shallowed channel might worsen flooding. An analogy can be made to river channels, where constricting a channel cross-sectional area can produce flooding upstream.

4.2. Broader Considerations for Shallowing as Mitigation of Coastal Flooding

As a flood adaptation strategy, channel shallowing can be challenging if the channels were historically dredged for navigational purposes and there is an economic driver behind their maintenance. In the case of Jamaica Bay, the use of deep-hull shipping in the bay has decreased, but there are still important commercial shipping and municipal (e.g., sewage sludge transport, fire safety access) uses of the deep channels [44]. The use of a dredged channel as a public recreational amenity (e.g., fishing charters) can be extremely important. However, most fishing and recreational vessels in this region have draft depths of no more than 2 m, relative to the channel depth of 6 m, so some degree of shallowing may be possible. Looking more broadly around the world, the flood impacts of proposed channel deepening projects should be quantified using a range of realistic local storm tide events (e.g., fast-pulse and slow-pulse), to evaluate whether they would worsen flooding at inland locations.

The practicality of creating and maintaining shallow depths is also an open question. At one extreme, shallowing could be accomplished through direct sediment in-fill, similar to beach replenishment. However, for some systems like Jamaica Bay, this would require much larger volumes of sediment than beach replenishment. Moreover, designing an optimal in-fill strategy that would allow a bathymetric equilibrium to quickly develop is a challenging engineering problem.

At the other extreme, shallowing could simply be initiated by decommissioning a dredged channel and allowing natural sedimentation to occur over a longer period of time. This could be a simple and potentially cost-effective strategy, though the shallowing would need to occur at a rate faster than the sea level rise rate, to reduce flood levels. The cessation of maintenance dredging is referred to by the USACE as "de-authorization" and would require an economic assessment of tradeoffs. De-authorization is not likely viable for large, high-traffic channels. However, one recent example exists with the Mississippi River Gulf Outlet canal, a shipping channel that was de-authorized by the USACE in 2007, in part because of a debate over whether it caused increased storm surge penetration. The channel was physically blocked in 2009 [45].

These considerations suggest that further study is required to understand the morphodynamic processes and timescales that control long-term sedimentation in Jamaica Bay and the broader New York Harbor area. At least one formerly-dredged system within New York Harbor area, the Passaic River estuary, has recovered to pre-dredging depths over the time scale of decades [46]. There is some evidence that Jamaica Bay is sediment-starved [44], but a shallowing of the Jamaica Bay might naturally occur over many decades. More research is needed to better understand the sediment transport processes and budget for Jamaica Bay and the surrounding region, and the potential effects of channel-depth modification thereon. It is possible, for example, that channel shoaling would reduce sediment transport by reducing tidal exchange, thus exerting a self-reinforcing effect, once initiated.

A gradual channel shallowing over many decades could provide flood risk reduction on a timescale that is similar to that of the sea level rise. The full shallowing scenario results reported here show reductions in flood levels of 0.50 and 1.16 m for Sandy and the 1821 Hurricane, comparable to the central range of sea level rise estimates (25th–75th percentile) expected for the 2080s, 0.56–1.27 m [47]. It is also noteworthy that the reduction in peak tide levels for adaptation scenarios 2 and 3 (Figure 9) would also protect against nuisance-flooding from "king tides" or from rainfall at times when sewer drainage is blocked by high tides. These types of flooding already occur in the low-lying Jamaica Bay neighborhoods such as Old Howard Beach and Broad Channel.

The water-quality and ecosystem impacts of shallowing for a given site would also require considerable study. Creek and inlet bathymetric recontouring has been identified as a restoration technique appropriate for smaller-scale tidal tributaries of Jamaica Bay, to decrease residence time of water and improve water quality [48]. However, the shallowing of an inlet can reduce the water exchange between the bay area and the open ocean, which could adversely impact the water quality inside the bay. An extreme example of the relationship between inlets and flushing is the breach on Fire Island, which was left open because of its potential to improve water quality due to increased flushing of stagnant bay waters [49].

5. Summary and Conclusions

In this study, we have explored the potential for mitigation of coastal flooding by natural systems, focusing on New York City's Jamaica Bay. Water depths in the bay have increased from an average of 1 m to 5 m since the mid-1800s, due in part to extensive channel dredging, and tidal marsh islands in the Bay have been rapidly eroding. For these reasons, we conducted numerical model experiments into how much leverage extensive restoration of wetlands or shallowing of dredged channels can have in reducing coastal flooding from the city's two highest historical storm tide events, Hurricane Sandy and the 1821 Hurricane.

Results show that restoring wetlands in the center of the bay to their 1879 footprint results in relatively small reductions in peak water levels (0.3% and 2.4% for Sandy and 1821, respectively) and flood area, while shallowing of dredged channels leads to much greater reductions (15% and 46%). A third scenario that shallowed only the narrowest region of the bay's inlet found roughly half as much (46%–65%) reduction in flood levels in the bay, relative to the full shallowing scenario.

Flood waters are widely known to be slowed by increased friction when passing through wetlands, and this can in some cases reduce flooding when it prevents inland propagation of a storm surge. Water depths and friction have been shown to have parallel influences through estuarine channels when it comes to tide propagation, and here we have demonstrated similar effects for storm tides at Jamaica Bay. The relative efficacy of shallowing for the 1821 Hurricane demonstrates that channel shallowing is particularly effective at reducing peak water levels at inshore locations for fast-pulse sea level events with periods of several hours.

For Jamaica Bay, we have begun conducting deeper analyses into the impacts that a broader set of coastal adaptation measures would have on flooding, water quality and storm waves, as well as for a range of different storms. The realism of channel shallowing must be evaluated in terms of other societal interests (e.g., recreation or shipping), costs, time frame, and the availability of sediments or

J. Mar. Sci. Eng. **2015**, *3*, 654–673

other materials for in-fill. However, no method of coastal flood mitigation is without its limitations, and here we have demonstrated a novel and effective approach for reducing coastal flooding.

These results suggest that when looking at nature-based solutions to sea level rise, more focus needs to be placed on sedimentary systems and bathymetry, including deep channels and topography in wetland areas, and not simply on wetlands. Regardless of whether Jamaica Bay shallowing is ever practical or economically affordable, we hope that the modeling and physical conceptualization above will help inspire additional creative research into nature-based and gray-green solutions for coastal flooding.

Acknowledgments: The authors would like thank Gena Wirth, Philippa Brashear, Lauren Elachi, Emily Silber and Kate Orff (SCAPE design), and Eric Sanderson and Mario Giampieri (Wildlife Conservation Society) for their fruitful collaborations and important contributions to moving this research forward. We would also like to thank Megan Linkin (Swiss Re) for sharing data on the 1821 Hurricane. Funding for the Stevens Institute researchers conducting this research was provided by the NOAA-RISA project "Consortium for Climate Risk in the Urban Northeast" (Rosenzweig, PI), and the NOAA-CPO-CSI-COCA project "Quantifying the Value and Communicating the Protective Services of Living Shorelines Using Flood Risk Assessment", award NA13OAR4310144. Funding for Stefan Talke was under National Science Foundation (NSF) award OCE-1155610 and U.S. Army Corps of Engineers, sponsor award number W1927N-14-2-0015. Computer modeling was made possible, in part, by a grant of computer time from the City University of New York High Performance Computing Center under NSF Grants CNS-0855217, CNS-0958379 and ACI-1126113.

Author Contributions: P.O. designed the experiments, interpreted results, and wrote most the text; S.T. and D.J. helped interpret results and write text; L.Y. designed the model grid and ran the validation experiments; A.B. helped write the text and create the model; N.G. helped create the model and interpret results; H.Z. helped write text; H.R. helped interpret results; and K.M. developed a figure and helped write text.

Conflicts of Interest: The authors declare no conflicts of interest.

References

1. Bridges, T.S.; Wagner, P.W.; Burks-Copes, K.A.; Bates, M.E.; Collier, Z.A.; Fischenich, C.J.; Gailani, J.Z.; Leuck, L.D.; Piercy, C.D.; Rosati, J.D. *Use of Natural and Nature-Based Features (NNBF) for Coastal Resilience*; Engineer Research and Development Center, Vicksburg MS Environmental Lab: Vicksburg, MS, USA, 2015.

2. Hanson, H.; Brampton, A.; Capobianco, M.; Dette, H.; Hamm, L.; Laustrup, C.; Lechuga, A.; Spanhoff, R. Beach nourishment projects, practices, and objectives—A European overview. *Coast. Eng.* **2002**, *47*, 81–111. [CrossRef]

3. City of New York. *A Stronger, More Resilient New York*; City of New York: New York, NY, 2013; p. 445.

4. Orton, P.; Vinogradov, S.; Georgas, N.; Blumberg, A.; Lin, N.; Gornitz, V.; Little, C.; Jacob, K.; Horton, R. New York City Panel on Climate Change 2015 Report Chapter 4: Dynamic Coastal Flood Modeling. *Ann. N. Y. Acad. Sci.* **2015**, *1336*, 56–66. [CrossRef] [PubMed]

5. Swanson, R.; West-Valle, A.; Decker, C. Recreation *vs.* waste disposal: The use and management of Jamaica Bay. *Long Isl. Hist. J.* **1992**, *5*, 21–41.

6. Swanson, R.L.; Wilson, R.E. Increased tidal ranges coinciding with Jamaica Bay development contribute to marsh flooding. *J. Coast. Res.* **2008**, 1565–1569. [CrossRef]

7. NYC-DEP. *Jamaica Bay Watershed Protection Plan*; New York City Department of Environmental Protection (DEP): New York, NY, USA, 2007; Volume 1, p. 128.

8. USACE. Fact Sheet: Jamaica Bay Marsh Islands. Available online: http://www.nan.usace.army.mil/Missions/CivilWorks/ProjectsinNewYork/EldersPointJamaicaBaySaltMarshIslands.aspx (accessed on 13 July 2015).

9. NYC-DEP. *Jamaica Bay Watershed Protection Plan, One-Year Progress Report*; New York City Department of Environmental Protection: New York, NY, USA, 2008; p. 64.

10. Corps of Engineers. *Interim Survey Report, Morgan City, Louisiana and Vicinity: Serial no. 63*; U.S. Army Corps of Engineers District: New Orleans, LA, USA, 1963.

11. Wamsley, T.V.; Cialone, M.A.; Smith, J.M.; Atkinson, J.H.; Rosati, J.D. The potential of wetlands in reducing storm surge. *Ocean Eng.* **2010**, *37*, 59–68. [CrossRef]

12. Scileppi, E.; Donnelly, J.P. Sedimentary evidence of hurricane strikes in western Long Island, New York. *Geochem. Geophys. Geosyst.* **2007**, *8*. [CrossRef]

13. Swiss Re. *The big one: The East Coast's USD 100 billion hurricane event*; Swiss Re America Holding Corporation: Armonk, NY, USA, 2014; p. 21.

14. Doxsey-Whitfield, E.; MacManus, K.; Adamo, S.B.; Pistolesi, L.; Squires, J.; Borkovska, O.; Baptista, S.R. Taking advantage of the improved availability of census data: A first look at the Gridded Population of the World, Version 4 (GPWv4). *Appl. Geogr.* **2015**, in press.

15. Columbia University Center for International Earth Science Information Network (CIESIN). New York City, 5 meter Low Elevation Coastal Zone Population Estimates. Available online: http://www.ciesin.columbia.edu/data/nyc5mlecz2010/nyc5mlecz2010.zip (accessed on 13 July 2015).

16. Blumberg, A.F.; Georgas, N. Quantifying uncertainty in estuarine and coastal ocean circulation modeling. *J. Hydraul. Engin.* **2008**, *134*, 403–415. [CrossRef]

17. Blumberg, A.F.; Khan, L.A.; St John, J. Three-dimensional hydrodynamic model of New York Harbor region. *J. Hydraul. Eng.* **1999**, *125*, 799–816. [CrossRef]

18. Georgas, N.; Blumberg, A.F. Establishing Confidence in Marine Forecast Systems: The Design and Skill Assessment of the New York Harbor Observation and Prediction System, Version 3 (NYHOPS v3). In Proceedings of the Eleventh International Conference in Estuarine and Coastal Modeling (ECM11), Seattle, WA, USA, 4–6 November 2009.

19. Georgas, N.; Blumberg, A.; Herrington, T. An operational coastal wave forecasting model for New Jersey and Long Island waters. *Shore Beach* **2007**, *75*, 30–35.

20. Blumberg, A.; Georgas, N.; Yin, L.; Herrington, T.; Orton, P. Street scale modeling of storm surge inundation along the New Jersey Hudson River waterfront. *J. Atmos. Oceanic Technol.* **2015**, *32*. [CrossRef]

21. FEMA. *Region II Coastal Storm Surge Study: Overview*; Federal Emergency Management Agency: Washington, DC, USA, 2014; p. 15.

22. Georgas, N.; Orton, P.; Blumberg, A.; Cohen, L.; Zarrilli, D.; Yin, L. The impact of tidal phase on Hurricane Sandy's flooding around New York City and Long Island Sound. *J. Extreme Events* **2014**, *1*, 1450006. [CrossRef]

23. Bunya, S.; Dietrich, J.C.; Westerink, J.J.; Ebersole, B.A.; Smith, J.M.; Atkinson, J.H.; Jensen, R.; Resio, D.T.; Luettich, R.A.; Dawson, C.; *et al.* A high-resolution coupled riverine flow, tide, wind, wind wave, and storm surge model for Southern Louisiana and Mississippi. Part I: Model development and validation. *Mon. Weather. Rev.* **2010**, *138*, 345. [CrossRef]

24. Mattocks, C.; Forbes, C. A real-time, event-triggered storm surge forecasting system for the state of North Carolina. *Ocean Model.* **2008**, *25*, 95–119. [CrossRef]

25. Talke, S.; Orton, P.; Jay, D. Increasing storm tides at New York City. *Geophys. Res. Lett.* **2014**, *41*, 1844–2013. [CrossRef]

26. Kussman, A.S. *Report on the Hurricane of September 3, 1821*; U.S. Weather Bureau: New York, NY, USA, 1957; p. 54.

27. Boose, E.R.; Chamberlin, K.E.; Foster, D.R. Landscape and regional impacts of hurricanes in New England. *Ecol. Monogr.* **2001**, *71*, 27–48. [CrossRef]

28. Holland, G.J. An analytic model of the wind and pressure profiles in hurricanes. *Mon. Weather. Rev.* **1980**, *108*, 1212–1218. [CrossRef]

29. Jelesnianski, C.; Chen, J.; Shaffer, W.; Oceanic, U.S.N.; Administration, A.; Service, U.S.N.W. *SLOSH: Sea, lake, and overland surges from hurricanes*; US Dept. of Commerce, National Oceanic and Atmospheric Administration, National Weather Service: Silver Spring, MD, USA, 1992.

30. USC & GS. *Jamaica Bay and Rockaway Inlet, Long Island, New York*; United States Coast and Geodetic Survey: Washington, DC, USA, 1879.

31. Orton, P.; Georgas, N.; Blumberg, A.; Pullen, J. Detailed modeling of recent severe storm tides in estuaries of the New York City region. *J. Geophys. Res.* **2012**, *117*, C09030. [CrossRef]

32. Wang, H.V.; Loftis, J.D.; Liu, Z.; Forrest, D.; Zhang, J. The storm surge and sub-grid inundation modeling in New York City during Hurricane Sandy. *J. Mar. Sci. Eng.* **2014**, *2*, 226–246. [CrossRef]

33. Forbes, C.; Rhome, J.; Mattocks, C.; Taylor, A. Predicting the storm surge threat of Hurricane Sandy with the National Weather Service SLOSH model. *J. Mar. Sci. Eng.* **2014**, *2*, 437–476. [CrossRef]

34. Loder, N.; Irish, J.; Cialone, M.; Wamsley, T. Sensitivity of hurricane surge to morphological parameters of coastal wetlands. *Estuar. Coast. Shelf Sci.* **2009**, *84*, 625–636. [CrossRef]

35. Shepard, C.C.; Crain, C.M.; Beck, M.W. The protective role of coastal marshes: A systematic review and meta-analysis. *PLoS ONE* **2011**, *6*. [CrossRef] [PubMed]

36. Resio, D.T.; Westerink, J.J. Modeling the physics of storm surges. *Phys. Today* **2008**, *61*, 33–38. [CrossRef]
37. Ianniello, J. Tidally induced residual currents in estuaries of constant breadth and depth. *J. Mar. Res.* **1977**, *35*, 755–786.
38. Friedrichs, C.T.; Aubrey, D.G. Tidal propagation in strongly convergent channels. *J. Geophys. Res.* **1994**, *99*, 3321–3336. [CrossRef]
39. Dronkers, J.J. *Tidal computations in rivers and coastal waters*; North Holland Publishing: Amsterdam, The Netherlands, 1964; p. 296.
40. Jay, D.A. Green's law revisited: Tidal long-wave propagation in channels with strong topography. *J. Geophys. Res.* **1991**, *96*, 20585–20598. [CrossRef]
41. Aretxabaleta, A.L.; Butman, B.; Ganju, N.K. Water level response in back-barrier bays unchanged following Hurricane Sandy. *Geophys. Res. Lett.* **2014**, *41*, 3163–3171. [CrossRef]
42. Kennedy, A.B.; Gravois, U.; Zachry, B.C.; Westerink, J.J.; Hope, M.E.; Dietrich, J.C.; Powell, M.D.; Cox, A.T.; Luettich, R.A.; Dean, R.G. Origin of the Hurricane Ike forerunner surge. *Geophys. Res. Lett.* **2011**, *38*. [CrossRef]
43. Burchard, H. Combined effects of wind, tide, and horizontal density gradients on stratification in estuaries and coastal seas. *J. Phys. Oceanogr.* **2009**, *39*, 2117–2136. [CrossRef]
44. Seavitt, C.; Alexander, K.; Alessi, D.; Sands, E. *Shifting Sands: Sedimentary Cycles for Jamaica Bay, New York*; Catherine Seavitt Nordenson: New York, NY, USA, 2015; p. 218.
45. Shaffer, G.P.; Day, J.W., Jr.; Mack, S.; Kemp, G.P.; van Heerden, I.; Poirrier, M.A.; Westphal, K.A.; FitzGerald, D.; Milanes, A.; Morris, C.A. The MRGO Navigation Project: A massive human-induced environmental, economic, and storm disaster. *J. Coast. Res.* **2009**, *54*, 206–224. [CrossRef]
46. Chant, R.J.; Fugate, D.; Garvey, E. The shaping of an estuarine superfund site: Roles of evolving dynamics and geomorphology. *Estuar. Coasts* **2011**, *34*, 90–105. [CrossRef]
47. Horton, R.; Little, C.; Gornitz, V.; Bader, D.; Oppenheimer, M. New York City Panel on Climate Change 2015 report Chapter 2: Sea level rise and coastal storms. *Ann. N. Y. Acad. Sci.* **2015**, *1336*, 36–44. [CrossRef] [PubMed]
48. USACE. *Hudson-Raritan Estuary Comprehensive Restoration Plan Potential Restoration Opportunities Project Summary Sheets: Jamaica Bay, United States Army Corps of Engineers*; USCAE: New York, NY, USA, 2014; p. 138.
49. Hapke, C.J.; Stockdon, H.F.; Schwab, W.C.; Foley, M.K. Changing the paradigm of response to coastal storms. *Eos Trans. Am. Geophys. Union* **2013**, *94*, 189–190. [CrossRef]

Journal of
Marine Science and Engineering

MDPI

Article

The Measurement of Personal Self-Efficacy in Preparing for a Hurricane and Its Role in Modeling the Likelihood of Evacuation

Alan E. Stewart

CHDS Department, 402 Aderhold Hall, University of Georgia, Athens 30602, GA, USA; aeswx@uga.edu;
Tel.: +1-706-542-1812; Fax: +1-706-542-4130

Academic Editor: Rick Luettich
Received: 7 April 2015; Accepted: 1 July 2015; Published: 21 July 2015

Abstract: Storm surges require that coastal residents make necessary preparations and evacuate the coast prior to hurricane landfall. An important individual characteristic in preparing for tropical cyclones is hurricane personal self-efficacy. Coastal residents who believe that it is possible to prepare for and evacuate ahead of a hurricane (hurricane response possibilities) and, further, believe that they personally can prepare and evacuate (hurricane personal self-efficacy) will be better prepared for hurricanes. In this study the author used a sample of 334 people to evaluate an 8-item self-report measure, the Hurricane Personal Self-Efficacy Scale (HPSES). This measure can be used to assess beliefs that it is possible in general to prepare for a hurricane and that the respondent him or herself can make these preparations and evacuate ahead of a hurricane. A factor analysis confirmed that the items measured two characteristics: (1) beliefs that is it possible for people in general to prepare for a hurricane; and (2) beliefs that the respondent personally could prepare for a hurricane and evacuate. The author also examined the functionality of the measure within a framework that was informed by Protection Motivation Theory (PMT). Hurricane response possibility beliefs, prior experiences with hurricane evacuation and hurricane-related property damages, and a tendency for people to sense and observe the weather were all predictive of personal self-efficacy in preparing for hurricanes, R^2_{adj} = 0.36. In operationalizing other constructs associated with PMT using weather-related psychological measures in a path analysis model, it was found that personal self-efficacy, fear of consequences of the severe and extreme weather, and appraisal of the threats posed by behaviors that could result in injury or death during severe weather together predicted the self-reported likelihood of evacuating, R^2_{adj} = 0.26. The implications of the study for coastal engineers and planners, ways of increasing hurricane personal self-efficacy in preparing for hurricanes, and the study's limitations are discussed.

Keywords: hurricanes; storm surge; evacuation; efficacy; human behavior

1. Introduction

Approximately 2544 people died in the United States from 1963 to 2012 from tropical cyclones in the Atlantic Ocean [1]. Although hurricanes and tropical storms may bring a variety of severe weather (e.g., rain, wind, tornadoes) that results in property damage, injuries, or deaths, the storm surge from tropical cyclones has historically posed the most deadly hazard, accounting for 49% of the deaths from 1963 to 2012 [1]. Recent hurricanes have produced some noteworthy storm surges that rapidly affected widespread areas and caught coastal dwellers by surprise with respect to the surge depth and force of the water [2,3]. The storm surge of Hurricane Katrina in 2005 contributed to the collapse of the levee system and the ensuing fatalities from drowning when regions like the lower 9th district in New Orleans were suddenly inundated. Hurricane Katrina also produced a 24–28 foot (7.3 to 8.5 m) storm surge that was 20 miles (32.2 km) wide, centered on St. Louis Bay, Mississippi that killed at least 180

J. Mar. Sci. Eng. **2015**, *3*, 630–653

people in that state [4]. Hurricane Ike, a large category 2 storm when it made landfall, brought a 10–15 foot (3.0 to 4.6 m) storm surge to the Galveston Island and Galveston Bay area of Texas in 2008 [5]. Hurricane Sandy produced historic damages in New Jersey, New York, and Connecticut in 2012 when storm surges ranging from 3 to 9 feet (0.91 to 2.7 m) occurred in the New York City metropolitan area and along the central and north coasts of New Jersey [6]. The implications of these recent storms are clear: Tropical cyclones that bring the potential of storm surge to an area demand that coastal residents prepare ahead of the storm and then evacuate to safer areas inland.

There is evidence that coastal residents do not possess a full understanding or awareness of the dangers that storm surges can create [1–3,7–9]. One possible reason for this is that forecasters historically have used the Saffir-Simpson Hurricane Wind Scale (SSHWS) to convey the intensity of hurricanes [10,11]. This use of the SSHWS along with narratives that emphasize the maximum sustained winds in a tropical cyclone may have had the effect of communicating that the most potent threats come from high winds. In recognition of the need to communicate the dangers of storm surges, the National Hurricane Center (NHC), based upon a coupled social and natural systems research approach, deployed an experimental product, *P-Surge* (probability of surge), in the 2014 hurricane season [1,8,12].

The availability of additional information about the possible storm surge effects of a tropical cyclone can help coastal residents to prepare for a storm and to make decisions about whether they will evacuate the coast [8,12,13]. This is significant because in the face of increased societal vulnerability to hurricane impacts, hurricane preparedness and the compliance with orders to evacuate ahead of an approaching hurricane remain below desired levels [13–17]. Researchers have identified numerous variables such as the perceived strength of one's dwelling to withstand a hurricane, not having a hurricane preparedness or evacuation plan, confidence in facing subsequent hurricanes based upon prior storm experiences, concerns about caring for elderly family members or pets, and transportation problems, among others, that influence hurricane-related responses and that may be evaluated alongside the risks of an approaching hurricane [14,16–30].

Among these influences on hurricane preparation, the variable of personal self-efficacy plays noteworthy role. The construct of self-efficacy comes from the Social Cognitive Theory (SCT) of Albert Bandura [31–33]. Bandura [33], p. 3 defined efficacy as "beliefs in one's capabilities to organize and execute the courses of action required to produce given attainments." Personal self-efficacy pertains to the degree to which a person believes that he or she can personally perform a behavior or bring about an outcome. The sense of personal self-efficacy is highly specific to the realm of activity a person is trying to learn; efficacy is not a general or global characteristic that people possess. For example, researchers discussed and examined efficacy within the contexts of: (1) engineering design [34]; (2) scientific leadership in marine ecology [35]; (3) technology adoption [36]; (4) teaching science in the K-12 grades [37]; and (5) coping with the psychological aftermath of a hurricane [38], among many others.

Beliefs about the extent to which preparatory and adaptive responses are possible in general and that an individual actually can perform these responses (personal self-efficacy) are critical in preparing for a hurricane, especially in coastal areas that may experience a storm surge. Although sheltering-in-place for a minor hurricane may place fewer demands upon coastal residents to make hurricane preparations, this is not the case for major hurricanes (i.e., category 3 or higher) or smaller storms that pose a storm surge threat. In these cases residents will likely have to evacuate to safer locations further inland in addition to securing their property prior to evacuation. It is essential for survival for people to know what to do and where to go, generally, and also to believe that they can personally perform these preparatory and evacuation behaviors. Coastal residents who know that preparatory and preventive options are possible and, further, possess the personal self-efficacy to respond appropriately will realize the benefits of innovations in storm surge modeling and the engineered infrastructure that has been designed to make coastal communities resilient to tropical cyclones [39–41].

This author's aims in this article are to: (1) Introduce a brief, 8-item measure that assesses the extent to which people believe various hurricane preparation and evacuation behaviors are possible and the extent to which they personally could perform these behaviors (personal self-efficacy); (2) Examine the psychometric properties of the measure that are important for understanding its reliability and validity for practical uses; and (3) Model the extent to which peoples' personal self-efficacy in making hurricane preparations is related to their prior experiences with the severe weather that hurricanes bring, with their prior hurricane evacuation experiences, and with their likelihood of complying with a recommended evacuation. In the sections below the author describes the development of the Hurricane Personal Self-Efficacy Scale and then discusses the theoretical framework of Protection Motivation Theory (PMT, [42,43]) in which the measure was used to model peoples' self-reported likelihood of evacuation.

1.1. Development of the Hurricane Personal Self-Efficacy Scale (HPSES)

Bandura stipulated in his Social Cognitive Theory that people learn many behaviors by first attending carefully to other people and the situations in which the behaviors are performed [31–33] (e.g., people preparing for a hurricane). What behaviors appear to be possible and what behaviors are performed? What stimuli seem important in guiding behaviors? Next, the theory indicates that people, encode, remember, and organize what they have been observing others do. This may take the form of remembering verbal instructions that they hear or read or of recalling important features of the behaviors that were performed. Beyond these memory processes, people mentally and behaviorally rehearse what was observed, remembered, and organized. Practice is important because it can help the person to omit unnecessary or wrong behaviors while reinforcing the necessary and appropriate behavioral steps. Finally, Bandura suggested that once a behavior is learned, the anticipated benefits or reinforcements for performing it serve as a motive to respond proactively [33] (e.g., evacuating because it more fully ensures safety from a storm surge).

With these features of SCT in mind, the author created 8 items (i.e., verbal statements) with the purpose of assessing the personal self-efficacy of people regarding their abilities to: Develop a safety plan for a hurricane, prepare and protect property from the hurricane, protect themselves in a hurricane, and to evacuate ahead of a hurricane. The items for Hurricane Personal Self-Efficacy Scale (HPSES) appear in Table 1. The author designed the first four items to assess the extent to which respondents believed that the hurricane preparation or evacuation responses were possible in general for people to perform (response possibilities). The knowledge that particular preparation responses are possible for people in general may come from observing or socially comparing oneself with others in a variety of ways [33]. The knowledge of possible responses, along with other influences such as personal enactment and experience with the behavior, verbal information, and physiological and emotional states, contributes to a sense of personal self-efficacy [33]. The remaining four items in the HPSES were designed to elicit respondents' perceptions of the extent to which they could personally perform the responses (personal self-efficacy). The scale instructions asked respondents to indicate their degree of agreement with each item using a 1 (*Strongly Disagree*) to 5 (*Strongly Agree*) fully-anchored rating scale.

Table 1. Items in the Hurricane Personal Self-Efficacy Scale (HPSES).

Hurricane Response Possibilities
1. It is possible for people in general to prepare and to secure their property ahead of time for a hurricane.
2. It is possible for people in general to develop a safety plan for how to deal with a hurricane.
3. It is possible for people in general to protect themselves against a hurricane.
4. It is possible for people in general to evacuate when necessary ahead of a hurricane.

Hurricane Personal Self-Efficacy
5. I feel that I can prepare and secure my property ahead of time for a hurricane.
6. I have a safety plan for how to deal with a hurricane.
7. I can protect myself against a hurricane.
8. I can evacuate when necessary ahead of a hurricane.

The author chose to develop a brief measure that tapped the global, summary perceptions of people's hurricane self-efficacy, rather than a lengthier instrument that inventoried multiple and detailed components of hurricane preparedness and evacuation behavior. The author believed that a brief measure may find more uses in research and practical settings than a longer scale. In addition, Stein *et al.* [26] recently established a precedent for using such brief and summative measures in assessing hurricane risk perceptions; these researchers asked a limited number of general questions about the perceptions of risks posed by a hurricane rather than a more expanded list of the different hurricane attributes that could be harmful (e.g., wind, rain, tornadoes).

1.2. Protection Motivation Theory: An Organizing Framework

The author chose to evaluate the functionality of the HPSES in predicting peoples' self-reported likelihood of hurricane evacuation within a framework that was inspired by Protection Motivation Theory (PMT). Protection Motivation Theory (Figure 1) provides a valuable perspective for understanding the multiple inputs that can affect coastal residents' hurricane preparation and evacuation behavior and the role of personal self-efficacy in formulating a response to the hazard [44]. Researchers originally created PMT as a model of disease prevention and health promotion [42,43,45]. The author selected PMT from among other possible models (i.e., the Health Belief Model [46], Theory of Reasoned/Planned Behavior [47] Protective Action Decision Model [48,49]), because its emphasis on cognitive and emotional variables and their hypothesized inter-relationships possessed the greatest potential to extend the understanding of coastal residents' preparation and evacuation behavior. In addition to health-related behaviors, researchers have used PMT to study the behavior of people affected by flood events [50]. Two meta-analyses of studies that employed PMT have been largely supportive of the model [42,51].

The PMT model contains three major components: (1) sources of information; (2) psychological (cognitive and emotional) mediating processes; and (3) adaptation modes (see Figure 1). The author will describe PMT and illustrate its components using hurricane preparation as an example. As depicted in Figure 1, the PMT model processes proceed from left to right with respect to time. The arrows along the bottom reflect that a person may remember the results of prior experiences with model variables and that these may affect the values of variables in subsequent experiences with hurricanes. Regarding the first component, information sources reside within the person's environment and include directly-sensed inputs of the weather along with communications from various forms of media (broadcast, internet, and social media). Information may also exist within the person as personality traits or dispositions, memories of past hurricane events (intrapersonal), gathered through verbal interactions and relationships with members of one's neighborhood or community, or through nonverbal communications (e.g., observing others' behaviors in preparation for a hurricane).

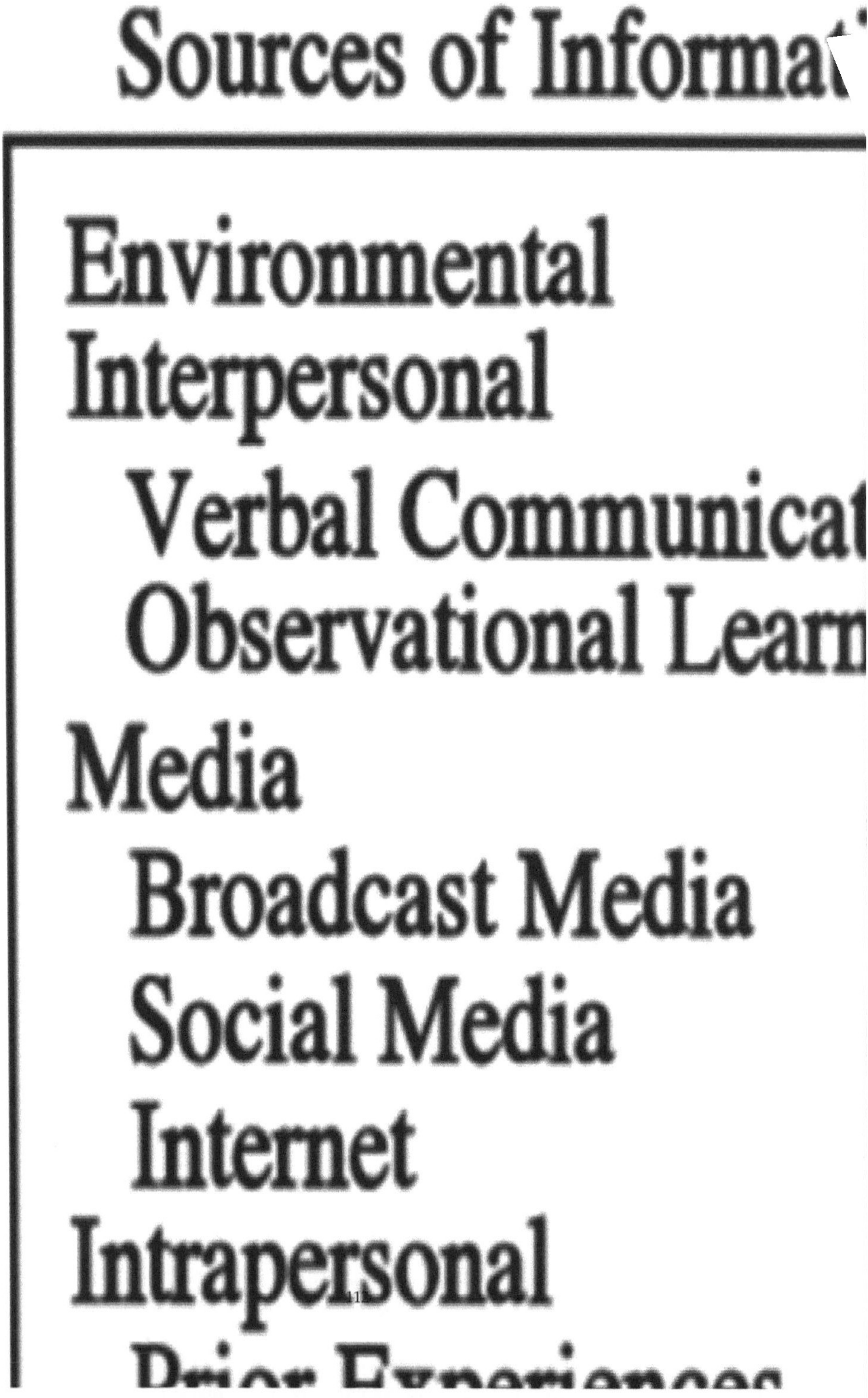

Sources of Informat

Environmental

Interpersonal

   Verbal Communicat

   Observational Learn

Media

   Broadcast Media

   Social Media

   Internet

Intrapersonal

   Prior Experiences

These sources inform the second PMT component, psychological mediating processes, which involve three components that jointly contribute to an evaluation of hurricane risks and culminate in protection motivation. First, threat and vulnerability appraisals pertain to the dangerousness of a hurricane and the specific weather-related risks that it is forecasted to bring. Given these risks and the person's intended or desired behaviors, how vulnerable is the person to the hurricane's impacts? What are the costs and benefits of preparing and evacuating (or not performing these actions)? Second, fear is an important component in the PMT model because it is a powerful motivator. Given the threats and vulnerabilities that may accompany a hurricane, to what extent do these beliefs give rise to fear? In this context, fear may have to do with the hurricane-related weather phenomena (heavy rain, high winds, lightning and thunder, storm surge, *etc.*). Fears also may stem from the anticipated impacts that the hurricane may cause (loss of power, damage or destruction of the home, flooding, *etc.*).

Hurricane personal self-efficacy and response efficacy form the third part of the psychological processes in the PMT model and represent the psychologically active ingredient for mobilizing an adaptive response. Response efficacy is defined as the belief that one's responses will be effective or successful in producing a desired outcome. The likelihood of an adaptive response is increased to the extent that the person knows about appropriate adaptive behaviors to perform ahead of a hurricane (response possibilities), believes that he or she personally can perform the behaviors (personal self-efficacy), and that these behaviors will effective in protecting oneself during a hurricane (response efficacy). The perceived costs in performing these preparatory behaviors (response costs) diminish the likelihood of an adaptive response. The level of personal self-efficacy in preparing for hurricanes and other natural hazards is not only an important component of PMT, but also play a significant role in the Protective Action Decision Model [48,49].

The appraisals of threat and vulnerability, of preparation and coping, along with the level of fear that is experienced give rise to protection motivation. Protection motivation, in turn, leads to PMT's third major component: Behavioral responses. These responses may be adaptive (dwellings are secured and then evacuated ahead of the hurricane) or maladaptive (e.g., people choose to shelter in place when a storm surge is forecasted).

In summary, the PMT model provides a meaningful account of how people receive environmental information (e.g., the forecast of a hurricane landfall with an accompanying storm surge), appraise their risks (of damage by the storm, the dangers of not preparing or evacuating), experience the feelings associated with a significant natural hazard and its possible consequences (i.e., fear), and then formulate decisions and enact behavioral responses. Hurricane personal self-efficacy is significant in PMT because it relates to knowledge of what response options are possible and what the person believes she or he can actually do to prepare for the storm. The section below describes the method used to develop the HPSES, to assess its measurement properties, and, to examine its functionality within a PMT-informed framework to model the likelihood of evacuation.

2. Method

2.1. Participants

The participants in this research were 334 university undergraduate students (76% men) from a large university in the southeastern United States. Approximately 90% of the sample reported that they were from one of the following states: North or South Carolina, Georgia, Florida, Alabama, Mississippi, Louisiana, Tennessee, or Texas. The participants' ages ranged from 18 to 42 years, $M = 21.3$ years, $SD = 1.92$ years. The sample's racial identifications were as follows: Caucasian 74.6%, Asian 9.4%, African American 7.6%, Hispanic/Latino/a 3.6%, and Other 4.9%. Given their residence in the southeastern United States, 15.5% of the sample reported that they or their families had sustained property damages from hurricanes previously and 13.6% of the sample indicated that they previously had evacuated their homes due to a hurricane. There were 5.8% of the participants that had both sustained hurricane-related property damage and had evacuated their homes previously.

The author relied upon a sample of university undergraduates to develop the HPSES for two reasons, the first of which pertained to the comparability of university undergraduates' responses with those of the general public with respect to weather-related attitudes, beliefs, preferences, and behaviors. In two previous investigations, the author observed that the responses of university undergraduates were very similar with those given by a representative sample of adults in the United States [52] and by those of adult residents who lived in coastal counties in the southeastern United States [17]. The second reason for relying upon an undergraduate sample stemmed from recent research reporting that 500 university students from the state of Florida lacked specific knowledge of hurricane risks and generally were unprepared for hurricanes that may affect that state [53]. Consequently, the use of an undergraduate same from the southeastern United States may be representative and instructive for developing a measure of hurricane response and self-efficacy.

The research protocol used in collecting the data for this article was reviewed and approved by the Institutional Review Board (IRB). All of the participants gave their informed consent to complete the research measures. The study participants completed the research to as part of their course requirements and received course credit for their participation.

2.2. Measures

The participants completed several self-report measures that were part of a larger research project whose goal was to examine the perceptions and psychological responses to weather-related risks. These measures included: (1) the HPSES (the primary focus of this article and whose items appear in Table 1); (2) the Weather Salience Questionnaire (WxSQ) [54]; (3) the Fear of Weather Scale (FOWS); (4) the Weather Risk-Taking Scale (WRTS); and (5) a demographic form assessing age, race, gender, and prior experiences with severe or extreme weather events that included experiencing property damages from hurricanes and previous experiences of hurricane evacuation. The HPSES and selected subscales of the remaining measures were used for the purposes of this article. The project measures are described more fully below.

2.2.1. Weather Salience Questionnaire (WxSQ)

The WxSQ is a 29-item self-report measure that assesses seven ways in which people may find weather and weather changes to be psychologically significant or salient. The respondents read each WxSQ statement and then use a five-point fully-anchored rating scale to indicate the frequency with which they engage in a weather-related behavior (1 = *Never* to 5 = *Always*) or the extent of their agreement with what the item describes (1 = *Strongly Disagree* to 5 = *Strongly Agree*). The author has previously established the measurement characteristics of the WxSQ and its seven subscales [51,53]. A description of the subscales follows, along with the calculated value of the Cronbach's coefficient alpha (α) for the respondents in this sample. Cronbach's coefficient alpha ranges from 0 to 1 and provides an estimate of the internal consistency reliability of the items, that is, the extent to which they consistently assess the construct for which they were designed [55,56]. In addition, 95% confidence intervals (CI) are provided for the alpha coefficients, based upon a Statistical Analysis System (SAS) procedure [57].

The WxSQ has yielded relatively stable scores over time, with a two-week test-retest coefficient of 0.91. People may experience the weather as psychologically significant by: (1) Seeking information about it in the media and online (9 items, $\alpha = 0.75$, CI: 0.70 to 0.80); (2) Sensing and observing the weather directly (5 items, $\alpha = 0.70$, CI: 0.65 to 0.75); (3) Experiencing different mood states that stem from weather or weather changes (6 items, $\alpha = 0.80$, CI: 0.76 to 0.83); (4) The impact of weather effects on activities of daily life (3 items, $\alpha = 0.45$, CI: 0.35 to 0.56); (5) Noticing the variability and changeability of the weather (4 items, $\alpha = 0.70$, CI: 0.65 to 0.76); (6) Experiencing psychological attachments for various kinds of synoptic weather (3 items, $\alpha = 0.89$, CI: 0.87 to 0.92); and (7) The disruption of daily routines (work or school) due to severe or extreme weather (3 items, $\alpha = 0.74$, CI: 0.67 to 0.80). Two of WxSQ items load on two subscales. The internal consistency reliability (Cronbach's α coefficient) of the 29 WxSQ items was 0.81, CI: 0.76 to 0.84.

J. Mar. Sci. Eng. **2015**, *3*, 630–653

The author used the first two WxSQ subscales (1) Seeking weather information; and (2) Sensing and observing the weather) to operationalize the *information sources* part of the PMT model that pertains to inputs from the natural environment.

2.2.2. Fear of Weather Scale (FOWS)

The author designed the remaining measures, the *Fear of Weather Scale* and the *Weather Risk-Taking Scale*, for the larger project to investigate the psychological experiences and correlates of routine, severe, and extreme weather events. The FOWS is an 87-item self-report measure whose purpose is to measure the fear of various components of severe weather (65 items) and of the effects and consequences of extreme weather events (22 items). For each weather component, people used a 0 (*No fear*) to 6 (*Terror*) fully-anchored rating scale to indicate their degree of fear of experiencing that type of weather. An exploratory factor analysis of the 65 severe weather items revealed 11 factors that corresponded to fears of storms and types of extreme weather: (1) Winter storms (8 items, $\alpha = 0.95$, CI: 0.94 to 0.95); (2) Hurricanes (7 items, $\alpha = 0.92$, CI: 0.91 to 0.94); (3) Floods (6 items, $\alpha = 0.91$, CI: 0.89 to 0.92); (4) Thunderstorms (7 items, $\alpha = 0.91$, CI: 0.90 to 0.93); (5) Tornadoes (5 items, $\alpha = 0.92$, CI: 0.90 to 0.93); (6) High winds/damaging wind storms (7 items, $\alpha = 0.91$, CI: 0.89 to 0.92); (7) Heavy rainfall, (5 items, $\alpha = 0.88$, CI: 0.85 to 0.90); (8) Foggy, overcast, and dark conditions (4 items, $\alpha = 0.78$, CI: 0.71 to 0.81); (9) Fire weather/drought conditions (4 items, $\alpha = 0.83$, CI: 0.80 to 0.86); and (10) Routine weather that may presage stormy conditions (e.g., humid, balmy, blustery conditions) (6 items, $\alpha = 0.80$, CI: 0.73 to 0.82). The last factor pertained to non-atmospheric natural hazards and involved fears of earthquakes, tremors, and tsunamis (5 items, $\alpha = 0.88$, CI: 0.86 to 0.90). One open-ended item was not factor analyzed.

The 22 items designed to assess peoples' fears of the consequences of extreme weather were associated with four factors in an exploratory factor analysis: (1) Loss of property (dwellings, belongings) and of what happens to the self and other people (loss of control over aspects of daily life) (8 items, $\alpha = 0.94$, CI: 0.93 to 0.95); (2) Damage to property and belongings (7 items, $\alpha = 0.94$, CI: 0.93 to 0.95); (3) Loss of utility infrastructure (e.g., power, water, natural gas, transportation systems) (5 items, $\alpha = 0.91$, CI: 0.88 to 0.92); and (4) Danger to the physical safety of self and of others (2 items, $\alpha = 0.92$, CI: 0.89 to 0.94). The Cronbach's alpha for the 22 items in the four subscales taken together was 0.97, CI: 0.96 to 0.97.

The scores of the items that compose each factor can be summed to create a subscale measure. The author used the FOWS Hurricane subscale and the four subscales pertaining to the fear of the consequences of extreme weather in the analyses to model hurricane evacuation likelihood. The four consequences subscales were summed to create a single composite score. The author reasoned that because hurricanes can produce such widespread and extensive damage (encompassing fear of loss and damage of property, loss of utilities, and danger to self and others), that it was acceptable to combine the subscales pertaining to fearful storm consequences.

The FOWS also contains an additional item that pertained to evacuation: *If severe weather was threatening and public officials were evacuating your area, how likely is it that you would evacuate your home and seek shelter elsewhere?* This question has a 6-point fully-anchored rating scale (1 = *Not likely at all (would stay at home)* to 6 = *Very likely to evacuate*). The author used responses to this item as a dependent variable in modeling the likelihood of evacuation as a function of the HPSES and other project variables within the PMT framework. Specifically, the likelihood of evacuation was taken to reflect the level of protection motivation that people possessed.

2.2.3. Weather Risk-Taking Scale (WRTS)

The WRTS is a self-report instrument that lists 32 behaviors that people might perform in severe or extreme weather situations that have the potential for injury or death (e.g., driving over inundated roadways or bridges, remaining outside during thunderstorms or extremes of temperature, swimming or surf-boarding in storm tides, *etc.*). The author developed the items based upon the categories of weather events that regularly result in accidents or fatalities as reported by the National Weather Service. Various weather safety resources (e.g., National Lightning Safety Institute, the National Weather Service, American Red Cross, *etc.*) were consulted to identify risky weather-related behaviors as well as responses that were safe and risk-averse.

Respondents to the WRTS evaluated the 32 behaviors under three instructional sets, the first of which was the rated likelihood of their performing the behavior. The respondents used a seven-point fully-anchored rating scale (1 = *Extremely Unlikely* to 7 = *Extremely Likely*) to evaluate each behavior. The second instructional set asked the respondents to evaluate the degree of riskiness associated with each behavior using a seven-point fully-anchored rating scale (1 = *Not at all Risky* to 7 = *Extremely Risky*). The third instructional set asked the respondents to consider the potential benefits of performing each of the 32 behaviors. Again, a seven-point fully-anchored rating scale was used (1 = *No Benefits at All* to 7 = *Great Benefits*). The responses to each of the 32 items for the three instructional sets are summed to create scores that summarize the respondents' overall degree of perceived riskiness of the behaviors, the likelihood of performing them, and the potential benefits of performing them.

The three WRTS subscales possessed good internal consistency: Rated riskiness of the weather-related behaviors ($\alpha = 0.87$, CI: 0.85 to 0.89), perceived benefits of performing the behaviors ($\alpha = 0.89$, CI: 0.87 to 0.92), likelihood of performing the behavior s ($\alpha = 0.84$, CI: 0.81 to 0.86). PMT specifies that the appraisal of threat comes from differencing the intrinsic and extrinsic rewards of a potentially maladaptive response from the severity and vulnerability posed by the response [42]. Consequently, the author operationalized threat appraisal for the purposes of this study by subtracting the perceived benefits of performing the behaviors from their rated degree of riskiness for each respondent. The order of the subtraction was chosen so that the resulting numbers would be positive, which in this study conveyed a greater degree of perceived risk (or appraised threat) compared to the benefits of performing the risky behaviors. Because the WRTS subscale assessing the likelihood of performing the behaviors was highly correlated with the Risk-Benefits difference score ($r = 0.89$, $p < 0.0001$) and because the difference score appeared to capture more fully the threat appraisal construct from PMT, the author chose not to use the WRTS likelihood subscale in the modeling analyses of this project.

2.3. Procedure

The participants completed the measures via networked computers in groups of 10–15 in a supervised and controlled laboratory setting. The order in which the measures were administered was varied throughout the data collection to avoid artifactual responses based upon the order in which the measures were completed. The project measures typically required 45 to 60 min of time for completion.

2.4. Data Analysis

The author used the Statistical Analysis System (SAS [58]) to calculate the descriptive statistics, and the Cronbach (α) coefficients. The author also used the SAS procedure known as the Covariance Analysis of Linear Structural Equations (CALIS) to assess the measurement model of the HPSES (i.e., to examine the degree to which the eight items measured hurricane response possibilities and hurricane personal self-efficacy) and to examine the contribution of the participants' demographic data, HPSES, FOWS, and WRTS to the likelihood of evacuation. Because the HPSES used an ordinal (rather than continuous) level of measurement, the relationships among the items were assessed with polychoric correlation coefficients. As recommended by several methodologists, this polychoric correlation matrix served as input to the CALIS procedure for evaluating the HPSES measurement model [59,60]. The author fit the both the HPSES measurement model and the path model of the likelihood of evacuation using weighted least squares estimation because some of the predictor variables deviated slightly from the standard normal distribution. In addition, because of missing data for three respondents, the sample size for the path analysis was 331 people.

3. Results

3.1. Properties of the Hurricane Personal Self-Efficacy Measure

The descriptive statistics for the eight HPSES items appear in Table 2. The mean values for each item revealed a response tendency that ranged from *uncertain* (numerical value of 3) to *agree* (4) on the five-point rating scale. This result also was evident in the generally small and negative values of the skewness for each item. The second HPSES item, *It is possible for people in general to develop a safety plan for how to deal with a hurricane*, had higher values for skewness and kurtosis given that many people generally indicated that they agreed with the item.

Table 2. Descriptive statistics and item correlations of the HPSES items.

	Descriptive Statistics				Polychoric Correlation Coefficients							
Item	Mean	SD	Skew.	Kurt.	1	2	3	4	5	6	7	8
1. It is possible for people in general to prepare and to secure their property ahead of time for a hurricane.	3.81	0.74	−0.75	0.90	–							
2. It is possible for people in general to develop a safety plan for how to deal with a hurricane.	4.28	0.65	−1.02	3.33	0.43	–						
3. It is possible for people in general to protect themselves against a hurricane.	3.65	0.88	−0.69	0.29	0.57	0.42	–					
4. It is possible for people in general to evacuate when necessary ahead of a	4.10	0.72	−0.87	1.86	0.30	0.35	0.39	–				

118

Table 2 also shows the matrix of polychoric correlations among the HPSES items. The correlations ranged from 0.06 (items 6 and 8) to 0.57 (items 1 and 3). Item 6, *I have a safety plan for how to deal with a hurricane*, generally exhibited correlations that were of lower magnitude than the inter-correlations of the other items. This result may have occurred because this item assesses a self-reported behavior and because the sample of university undergraduates were living at the time of their research participation at a location that was far inland and that did not necessitate the creation of a hurricane safety plan.

The Covariance Analysis of Linear Structural Equations (i.e., the SAS CALIS procedure) was used next to evaluate the extent to which the four HPSES items (#1 to #4) were related to the latent variable (factor) of hurricane response possibilities and the extent to which items (#5 to #8) were related to the latent variable of hurricane personal self-efficacy. It was expected that because all of the HPSES items inquired about capabilities to prepare, to evacuate and to remain safe ahead of a hurricane, but differed in focus (i.e., people in general for items #1 to #4 and for the self in items #5–#8), that the two latent variables would be correlated with each other.

This HPSES measurement model demonstrated a good fit to the data when interpreting the statistics used to assess model performance [61]. The residuals between the observed correlations and those predicted by the measurement model were generally small as given by the standardized root mean residual (SRMR = 0.04). Similarly, the root mean square error of approximation (RMSEA), another absolute indicator of model fit, was acceptable at 0.06; the 90% confidence interval for the RMSEA was 0.03 to 0.09. The probability of close fit index (*p*-close) was 0.06, a value that is associated with a good degree of fit between the model and the data. The Bentler Comparative Fit Index was 0.96 and the Bentler-Bonett Non-normed Index was 0.91. These latter fit indices range from 0 to 1, with higher values indicating a greater degree of fit between the measurement model and the data. In addition, the present model with two subscales exhibited significantly better fit than a model possessing a single, general scale, X^2 (df = 2) = 19.02, $p < 0.0001$.

Table 3 shows the correlations of each item with the latent variables corresponding to hurricane response possibilities and to hurricane personal self-efficacy; all of the correlations were statistically significant. The two latent variables (hereafter referred to as subscales) were significantly correlated, as would be expected, $r = 0.52$, $p < 0.0001$.

The Cronbach's alpha for the eight item HPSES scale was 0.80, CI: 0.71 to 0.82. The alpha coefficients were somewhat lower for the subscales because of the smaller number of items, which can attenuate the coefficient magnitude. For hurricane response possibility, $\alpha = 0.67$, CI: 0.58 to 0.76 and for personal self-efficacy, $\alpha = 0.63$, CI: 0.55 to 0.70.

The first four HPSES items can be summed to obtain an indication of the extent to which the respondent believed that making hurricane preparations, evacuating, and remaining safe were possible responses for people in general. Similarly, the sum of the last four items provides an indication of the respondent's sense of personal self-efficacy in preparing or evacuating before the hurricane and remaining safe. Table 2 provides the descriptive statistics for the response possibility and the personal self-efficacy subscales. The mean value of the response possibility subscale, 15.84 (*SD* = 2.14), corresponded qualitatively with *Agree* with respect to knowledge of hurricane preparation and evacuation response options for people in general. The mean value for the personal self-efficacy subscale, 13.23 (*SD* = 2.59), was somewhat lower than for response possibility and corresponded qualitatively with *Uncertain* to *Agree*. The mean scores on the HPSES did not differ to a statistically significant extent with respect to gender or the participants' racial identification.

Table 3. Correlations of HPSES Items with their Respective Subscales.

Item	Correlation With Subscale	Standard Error
Hurricane Response Possibilities		
1. It is possible for people in general to prepare and to secure their property ahead of time for a hurricane.	0.71	0.04
2. It is possible for people in general to develop a safety plan for how to deal with a hurricane.	0.46	0.05
3. It is possible for people in general to protect themselves against a hurricane.	0.72	0.04
4. It is possible for people in general to evacuate when necessary ahead of a hurricane.	0.41	0.05
Hurricane Personal Self-Efficacy		
5. I feel that I can prepare and secure my property ahead of time for a hurricane.	0.74	0.04
6. I have a safety plan for how to deal with a hurricane.	0.30	0.06
7. I can protect myself against a hurricane.	0.61	0.04
8. I can evacuate when necessary ahead of a hurricane.	0.50	0.05

Note: The correlations shown are standardized estimates of the relationship of the item with the subscale. Items #1 to #4 are correlated with the latent variable representing the Hurricane Response Possibilities subscale. Items #5 to #8 are correlated with the latent variable representing the Hurricane Personal Self-Efficacy Subscale. All correlations were statistically significant, $p < 0.0001$.

3.2. Predictors of Hurricane Personal Self-Efficacy

Social Cognitive Theory and Protection Motivation Theory both convey predictions about what may influence the degree of personal self-efficacy [32,33,42]. The knowledge or awareness that particular responses are possible for people in general presages a sense of personal self-efficacy. A person must learn and know that a particular response or set of responses is possible before he or she learns them personally [31–33]. From the perspective of SCT, people develop such awareness of the possibility of responses generally by observing others in person or via various media. Such observations provide a vicarious source of information about both the behaviors and the possible range of consequences that others might experience. These observations give a person a sense of what is possible for others in general to do or to accomplish. This also can contribute to a sense of personal self-efficacy beliefs as the person contemplates making the same responses him/herself [32, 33]. Similarly, SCT and PMT each predict that people learn from their experiences. Such enacted experiences, according to Bandura, can also contribute to personal self-efficacy beliefs, especially when people believe that their response to the experience was effective.

With this theoretical background in mind, it was expected that the variable of hurricane response possibility would predict the level of personal self-efficacy that people reported. It was also expected that prior experiences with hurricane evacuation or with damages that hurricanes produced would be predict personal self-efficacy. PMT also predicts that information people obtain from various sources and personal experiences affect personal self-efficacy along with the assessment of risk and the possible ways of responding to the risk that are part of the PMT psychological mediating processes [42,43].

Table 4. Descriptive Statistics and Pearson Correlation Coefficients for Variables Considered for the Structural Model of Hurricane Personal Self-Efficacy and its Contribution of Evacuation Likelihood (*n* = 331).

Variable	1	2	3	4	5	6	7	8	9	10
1. Seeking Weather Information (WxSQ Subscale)	–									
2. Sensing and Observing Weather (WxSQ Subscale)	0.44 *	–								
3. Previously evacuated due to a hurricane (yes/no)	0.13 *	0.02	–							
4. Self or family sustained hurricane damage previously (yes/no)	0.04	0.03	0.29 *	–						
5. Hurricane Response Possibility (Subscale)	0.08	0.15 *	0.10	0.09	–					
6. Hurricane Personal Self-Efficacy (Subscale)	0.12 *	0.16 *	0.21 *	0.21 *	0.56 *	–				
7. Weather Risk-Taking Difference Score (Rated Riskiness of Behavior—Perceived Benefit of Performing Behavior)	0.17 *	0.04	0.05	0.01	0.23 *	0.07	–			
8. Fear of Severe/Extreme Weather Effects (FOWS Subscale)	0.15 *	0.03	0.03	−0.01	0.02	−0.06	0.28 *	–		
9. Fear of Hurricane-Related Weather (FOWS Subscale)	0.09	−0.04	−0.05	−0.02	−0.04	−0.07	0.30 *	0.55 *	–	
10. Self-Reported Likelihood of Complying with Recommended Evacuation	0.13 *	0.02	0.02	0.04	0.10	0.14 *	0.38*	0.36 *	0.28 *	–
Mean	31.21	17.80	0.14	0.16	15.83	13.25	107.21	100.63	25.26	4.69
Std. Deviation	5.48	3.11	0.34	0.36	2.14	2.59	34.21	24.39	7.76	1.16

Note: * Statistical significance (*p*) < 0.01. The correlations reported here are the bivariate, zero-order Pearson coefficients. The numerical values may differ slightly from those reported in Figure 2, which were calculated in fitting the full path model to the variables.

The author used path analysis to model the contributions of the information subscale and the sensing and observing subscale of the WxSQ, indicators of past hurricane damage or evacuation experiences, and the HPSES response possibility subscale to hurricane personal self-efficacy. This analysis was performed as part of a larger path analysis to model the contributions of hurricane personal self-efficacy, fear, and risk appraisals to self-reported likelihood of evacuating. The fit statistics for the full path analysis will be reported in the next section.

The descriptive statistics and zero-order Pearson correlation coefficients for the variables considered for the path analysis appear in Table 4. The standardized values of the path coefficients appear in Figure 2 and were all statistically significant (*p* < 0.005). Hurricane response possibility was the strongest predictor of hurricane personal self-efficacy. Prior experiences with hurricane damages and with evacuating before a hurricane each made comparable and smaller contributions to levels of hurricane personal self-efficacy; people exhibited higher hurricane personal self-efficacy if they had experienced hurricane damages previously or if they had prior evacuation experience. Sensing and observing the weather directly, which is an aspect of weather salience, also made a small but statistically significant contribution to the path model. The general disposition of watching the weather and observing its changes enhanced the sense of personal self-efficacy in preparing for a hurricane.

Finally, the *seeking weather information* subscale of the WxSQ also exhibited a small correlation with hurricane personal self-efficacy. When the seeking weather information subscale was entered into the path model however, its relationship to the other variables decreased from the values shown in the initial correlation matrix (Table 4) and its path coefficient was not statistically significant. It appeared that the WxSQ *sensing and observing* subscale encompassed the aspects of information, knowledge or awareness that related to the other variables in the model. In addition, *the sensing and observing* scale demonstrated a greater degree of relationship with hurricane personal self-efficacy than did the *seeking weather information* subscale. For these reasons, the *seeking weather information* variable subsequently was removed from the path model. The adjusted R^2 value for this part of the path analysis was 0.36.

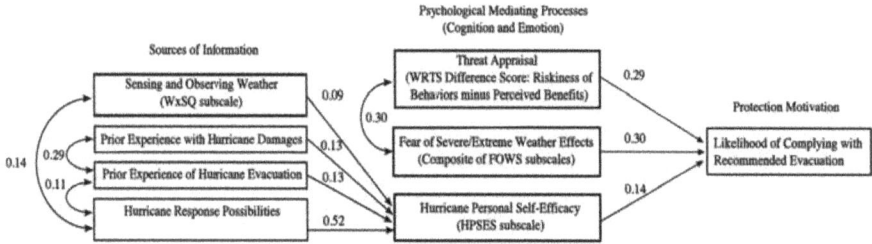

Figure 2. Path model operationalizing selected variables of Protection Motivation Theory to predict likelihood of evacuation. All path coefficients shown differed significantly from zero.

3.3. Hurricane Personal Self-Efficacy and the Likelihood of Evacuation

The PMT provides a framework for understanding how environmental inputs and the ensuing cognitive mediating processes affect the decisions people make to prepare or evacuate ahead of a hurricane (see Figures 1 and 2). In this section, the author expanded upon the path model of hurricane personal self-efficacy to assess its role in an enlarged model of the likelihood of evacuation. The PMT model brings together thinking and beliefs (cognitive processes) and emotional (or affective processes) in describing the psychological mediating processes that contribute to protection motivation. PMT specifies three components of psychological mediating processes: (1) threat appraisals posed by the phenomena under consideration and the potential maladaptive ways of responding to it; (2) fear; and (3) coping appraisals of performing adaptive responses. In this analysis, threat appraisals were operationalized by the WRTS difference score (i.e., rated riskiness—perceived benefits) of performing various behaviors during severe or extreme weather that could result in injury or death, as described above. The author operationalized fear via the FOWS Hurricane subscale and through the composite score encompassing various aspects of fear about the consequences of severe weather. Coping by performing an adaptive response was operationalized through the single variable of hurricane personal self-efficacy (i.e., beliefs that one could personally prepare for, evacuate, and remain safe during a hurricane).

Importantly, the present study did not attempt to operationalize or assess two other aspects that were part of the PMT model, namely the response efficacy and the response costs. Briefly, response efficacy pertains to the beliefs that a person has about the adequacy, effectiveness, or success of his/her responses. There were two reasons for not including response efficacy in this project, the first of which involved the scope of the author's work in designing the HPSES to assess only hurricane response possibilities and personal self-efficacy. Second, and relatedly, because only a portion of people in this study had reported a prior hurricane evacuation, gathering data about the efficacy of this response would not have been possible for those people who had not evacuated previously. Response costs pertain to the costs incurred to make the adaptive response in terms of things such as time, money, effort [42]. Again, because not everyone in the study had prior evacuation of hurricane experience, this variable was not operationalized.

The author used path analysis to evaluate the contributions of different model variables to the likelihood of complying with a recommended evacuation. The model is depicted in Figure 2, with information and personal disposition variables at the left, cognitive and emotional variables (psychological mediating processes) in the center, and the dependent variable, likelihood of evacuation, appearing on the right-hand side. All of the path coefficients depicted in Figure 2 were statistically significant. Some of the predictor variables also evidenced statistically significant zero-order inter-correlations and these, along with descriptive statistics, appear in Table 4.

Overall, the model fit the data well according to commonly-used indices. The chi-square test of model fit was not statistically significant, X^2 (df = 6) = 7.46, p = 0.28. This non-significant chi-square statistic suggested that there were no noteworthy discrepancies between the model and the data. The standardized root mean square residual was 0.02, indicating that the discrepancies between the model and the data were small in magnitude. Similarly, the root mean square error of approximation (RMSEA) was 0.03, indicating a good degree of model fit. The adjusted goodness of fit index was 0.95 and the Bentler-Bonett Non-normed fit index was 0.96. These indices of model fit (with values closer to 1.0 indicating better fit) suggested that the path model conformed to the data quite well. The adjusted R^2 for the full model was 0.26.

The prior experiences of hurricane damages or evacuation along with a disposition to sense and observe the weather exerted their influences on the likelihood of evacuating only through hurricane personal self-efficacy. In addition to its larger contribution to personal self-efficacy, hurricane response possibility exhibited a positive relationship with the threat appraisal of performing risky weather behaviors. That is, as the respondents' beliefs in the possibility of preparing for and evacuating ahead of a hurricane increased, so did the appraisals of threat posed by performing risky behaviors (i.e., the difference between perceived riskiness of the behaviors and the potential benefits of performing them). Personal self-efficacy was positively related to the likelihood of evacuating one's home when it was recommended; the value of the standardized path coefficient was 0.14 and was approximately one-half the magnitudes the paths of threat appraisal and of fear. The overall contribution of the variability in self-efficacy to the likelihood of evacuating was modest when considered in the context of the variability that the model accounted for in the reported likelihood to evacuate. That is, in substituting the mean values of the other predictor variables in the model, increasing hurricane self-efficacy scores from a minimum value of four (i.e., each of four HPSES items receive a rating of one) to a maximum possible value of 20 (four items each receive a rating of five), the likelihood of evacuation increased by 14.0%.

The authors of the PMT indicated that the threat appraisal is related to the degree of fear that people experience [41]. This relationship was observed in the results of this project. Statistically significant correlations of comparable magnitudes existed between threat appraisal and *fears of hurricane-related weather* (r = 0.30) and *fear of severe/extreme weather effects* (r = 0.28, see Table 4). This result suggested that changes in threat appraisal were positively associated with changes (increases) in both fears of hurricane-related weather (e.g., winds and storm surges) and of the effects of such severe weather (i.e., property damages, losses). *Fears of hurricane-related weather* and *fear of severe/extreme weather effects* were moderately correlated (r = 0.55). Although both fear variables were related positively to the likelihood of evacuation, the *fear of severe/extreme weather effects* exhibited a higher correlation (by approximately 0.10). In addition, when both fear variables were included in the path model, the contributions of the *fears of hurricane-related weather* to the likelihood of evacuation and to threat appraisals decreased in magnitude (from the initial correlation matrix in Table 4) and the path coefficients were no longer statistically significant. For this reason, the *fears of hurricane related weather* was removed from the model depicted in Figure 2. The *WRTS difference score* (threat appraisal, r = 0.29) and the *fear of severe/extreme weather effects* (r = 0.30) made comparable contributions to the prediction of the likelihood of evacuation.

4. Discussion and Conclusions

The data and analyses in this article were useful in the evaluation of a brief measure of hurricane response possibilities and personal self-efficacy. The author designed the HPSES to evaluate two aspects of hurricane preparedness: (1) Do people know that it is possible in general to develop a safety plan, to prepare property, to evacuate, and remain safe during a hurricane? and (2) Do respondents personally believe that they can develop their safety plans for a hurricane, prepare their property,

J. Mar. Sci. Eng. **2015**, *3*, 630–653

evacuate, and remain safe? The responses of 334 people to the measure revealed that it possessed acceptable internal reliability. The first four HPSES items related to hurricane response possibilities. The remaining four items assessed hurricane personal self-efficacy. The descriptive statistics for each HPSES subscale indicated that individual respondents varied in the degree to which they believed it was possible in general for people to prepare for a hurricane and in the extent to which they personally believed that they could prepare and evacuate ahead of a hurricane.

From a weather safety perspective, personal self-efficacy in hurricane preparedness is primary because it conveys a sense of confidence-to-act. A person with a higher level of hurricane personal self-efficacy feels enabled to prepare him- or herself and, perhaps, to assist others in with their preparations. The knowledge of response possibilities (i.e., knowing that response options exist generally) also is important because it precedes, both temporally and logically, the development of the specific responses that a person can make before a hurricane. This relationship was apparent in the prediction of personal self-efficacy by response possibilities. Knowing that particular behavioral responses are possible through observing others in person, via the broadcast media, or through other modes of information contributes to personal self-efficacy beliefs [31–33]. Similarly, the contribution of the sensing and observing subscale of the WxSQ to hurricane response possibilities and to personal self-efficacy also was an expected result. The mindful observation of the weather, along with information from other sources like broadcast and social media, can cue people that preparatory responses and evacuation may need to occur soon [62,63].

Because people generally learn from past experiences and the behaviors that they performed, it also was an expected result that the participants who had reported sustaining prior property damages from a hurricane, and thus who may have learned how to respond differently to subsequent hurricane threats, evidenced greater hurricane personal self-efficacy. Similarly, prior experiences with hurricane evacuation provide an important context for responding to subsequent hurricane threats. Once people experience the process of an initial evacuation, the learning which is accomplished at this time is available for use in future evacuation scenarios, especially if the person evaluated his or her behavior as effective or successful. This relationship was supported by the finding that prior hurricane evacuation experience positively predicted personal hurricane self-efficacy.

This project also evaluated the performance of the HPSES within a framework that was significantly informed by Protection Motivation Theory. As PMT would predict, information sources (prior experiences and environmental information) contributed to hurricane personal self-efficacy. In turn, personal self-efficacy, as it underlies making adaptive responses and coping in preparing for the storm, positively correlated with the likelihood of complying with an evacuation. The hurricane response possibilities subscale did not predict directly the likelihood of evacuating; instead, its influences were exerted through personal self-efficacy. This is an important result because it suggests that merely believing that preparation or evacuation is possible for people in general does not contribute directly to the likelihood of evacuation. General knowledge of response options was related positively to the threat appraisal as it was operationalized here as the difference between the riskiness of particular severe/extreme weather-related behaviors and the potential benefits of performing those high-risk behaviors. Thus general knowledge (i.e., hurricane response possibility) does play an important role that is distinct from personal self-efficacy.

Two additional results of this study also were consistent with what one would expect within the PMT framework. First, the positive correlation of threat appraisal with the likelihood of evacuation suggests a greater degree of motivation to protect oneself by evacuating as the perceived threats increase. Fear of the effects of severe and extreme weather exhibited a comparable degree of relationship with the likelihood of evacuation. PMT specifies that threat appraisal and fear would be related with each other and the results supported this aspect of PMT. Fear plays a noteworthy part in PMT and this theory is unique among other models in incorporating emotional experiences as these relate to protective actions. An important caveat to remember is that although the present study used a PMT-inspired framework for modeling the contributions of hurricane personal self-efficacy to the likelihood of evacuation, two constructs that are central to PMT, response efficacy and response costs were not assessed in this project. Thus, the present project does not constitute an evaluation of the PMT model, although it does illustrate its range of applicability and its usefulness in studies like this.

The establishment of the HPSES measurement properties and of its functionality in a broader PMT-inspired model of hurricane evacuation reveals the potential usefulness of the measure for research and applied purposes. In this regard, the present study has established HPSES measurement characteristics and illustrated the HPSES proof-of-concept in modeling the likelihood of hurricane evacuation. The measure's brevity makes it relatively easy to include along with other measures or to embed within a survey when exploring hurricane preparedness or evacuation. The brevity and straight-forward nature of the HPSES items also mean that local agencies (e.g., county emergency managers, coastal engineers or planners) could use the measure to assess the levels of personal readiness prior to the hurricane season.

The results of this research and potential prospective uses of the HPSES raise the important question of how to increase the level of hurricane response possibilities and personal self-efficacy so that people are better prepared for future storms. Although prior experience with hurricane damage and evacuation were associated with higher levels of personal self-efficacy, is there another route by which personal self-efficacy could be enhanced without actually having to experience a hurricane? Other researchers have raised similar questions about the ways that the experience of severe or extreme weather can make some people more cautious and proactive when anticipating future weather events [16,54,64].

Because hurricane self-efficacy pertains to beliefs in one's abilities to perform a behavior, efforts to increase efficacy should be at least two-tiered. First, communicating information related to hurricane safety, to evacuation routes, and the locations of inland shelters may enhance a person's knowledge of hurricane response possibilities [62,63]. From the results of this project, the knowledge or provision of information alone does not translate directly to the increased likelihood of making a response. The second tier should involve an effort that leads people to formulate and practice actual preparedness behaviors. Although coastal residents may be somewhat unlikely to perform these behaviors on their own, disseminating information and practicing responses in the context of work or community settings both provides the time and space for learning how to respond and also leverages the social learning aspect (i.e., observational learning of others) that is important for both becoming aware of response possibilities and in building personal self-efficacy beliefs [32,33]. Assembling supplies and safety kits for work settings, traveling along evacuation routes, and actually visiting the location of designated hurricane shelters may increase the likelihood of people making similar preparations at home when this becomes necessary. In addition, providing feedback about peoples' responses may build their response efficacy, especially in those cases where preparation behaviors were successful or effective.

The enhancement of hurricane personal self-efficacy also can occur within K-12 school settings. In this regard, the author has used the American Red Cross *Master of Disaster* (MoD) weather science and safety curriculum to build the knowledge and skills of students who complete the relevant lessons for hurricanes, floods, lightning, and tornadoes [65]. The MoD curriculum offers cutting-edge lessons that are designed to involve parents and family in practical activities like assembling supplies (food, potable water), safety kits, finding the locations of school and home on inundation maps, and evacuation planning. All of these things can build knowledge of possible response options and also increase personal self-efficacy in preparing and evacuating for hurricanes. When work, community, and school preparation efforts are coordinated with hurricane awareness weeks, this may maximize the possibility for practical learning to occur on a regular basis [66].

There are several implications of this project for coastal engineers and planners. First, for engineered structures such as sea walls or flood gates or for evacuation routes like bridges and highways, this project has demonstrated the importance of people knowing about the possible ways that they can respond ahead of a hurricane. For such engineered structures, this means that coastal residents can benefit from knowing about both the existence and performance capabilities of structures

J. Mar. Sci. Eng. **2015**, 3, 630–653

like sea walls and highways. For example, what height of storm surge can local sea walls withstand before flooding becomes likely? How many motor vehicles can cross a bridge or causeway in a given unit of time under heavy traffic conditions of an evacuation? Publicizing this information can help people to prepare. Knowing what response options are possible contributes to efficacy beliefs that a response (e.g., to use a particular evacuation route) can be made. Similarly, informing both coastal dwellers and emergency managers about areas that are likely to be inundated first from storm surge modeling efforts, is critical in alerting long-time residents and coastal visitors about how they need to respond and when. Third, and following from the discussion above, it is important for people in vulnerable coastal areas to practice hurricane preparation and evacuation ahead of time. Mock storm events (e.g., during hurricane preparedness weeks) can inform people about response options and help them to practice both using and benefitting from the engineered infrastructure.

This study possesses several limitations, one of which was the use of an undergraduate sample that did not dwell on the coast at the time during which the data was gathered. This raises the question about the extent to which the results from the present study may generalize to older adult residents living in areas that could be directly and severely impacted by hurricanes and the accompanying storm surge. Although this is a possibility, the author observed very comparable results between inland undergraduate samples and coastal residents in a study of the ways that hurricane damage perceptions may affect the likelihood of evacuation [17]. In addition, the participants in this study did have experiences with hurricanes: 13.6% of them had evacuated previously for a hurricane and 15.5% had sustained hurricane-related property damages. It is an important and necessary first step to use a suitable sample, such as undergraduates with hurricane experiences, for the purposes of developing the HPSES, exploring its psychometric properties, and assessing the extent to which it can be used to model the likelihood of hurricane evacuation. With the accomplishment of these initial efforts, the evaluation of the HPSES with a sample of coastal residents is a logical next step.

Another limitation of the study stemmed from the fact that the author gathered data for this project under conditions in which people were not experiencing any threat of a hurricane, storm surge, or the aftermath of these events. That is, the data in this article represent general and static dispositions to respond that may well differ from the kinds of responses people might make when faced with a particular hurricane situation in real-time. The data and analyses in this project are also cross-sectional with respect to time, rather than sequential. Similarly, the HPSES and other project variables modeled the self-reported likelihood of evacuation and not their actual behavioral responses. Although these are limitations, the present study does provide an informative baseline against which modeling done under more dynamic scenarios could be compared. In addition, evaluating measures like the HPSES in a more general project like this one can help to prepare it for deployment in near- or real-time hurricane situations.

Future research efforts with the HPSES among coastal residents could explore the combined contributions of observing neighbors' hurricane-related responses, information from social and broadcast media, and observing the weather to the levels of hurricane response possibilities and personal self-efficacy that people report. Similarly, to what extent does observing other people take weather-related risks as a hurricane and storm surge approach either inhibit or facilitate similar behaviors in the observer? As prior research has documented, numerous variables influence hurricane preparedness and evacuation [14–30]. The availability of an operational measure of hurricane response possibilities and personal self-efficacy now enables the quantification of these important concepts.

Acknowledgments: The author would like to thank Kristin Hunter and Azadeh Fatemi for their assistance in collecting the data for this project.

Author Contributions: The author designed the Hurricane Personal Self-Efficacy Scale, designed the research to evaluate the instrument, performed all statistical analyses, and wrote the manuscript.

Conflicts of Interest: The author declares no conflict of interest.

References

1. Rappaport, E.N. Fatalities in the United States from tropical cyclones. *Bull. Am. Meteorol. Soc.* **2014**, *95*, 341–345. [CrossRef]
2. Jonkman, S.N.; Maaskant, B.; Boyd, E.; Levitan, M.L. Loss of life caused by the flooding of New Orleans after Hurricane Katrina: Analysis of the relationship between flood characteristics and mortality. *Risk Anal.* **2009**, *29*, 676–698. [CrossRef] [PubMed]
3. Morss, R.E.; Hayden, M.H. Storm surge and "certain death": Interviews with Texas coastal residents following hurricane Ike. *Weather Clim. Soc.* **2010**, *2*, 174–189. [CrossRef]
4. Knabb, R.D.; Rhome, J.R.; Brown, D.P. Tropical Cyclone Report: Hurricane Katrina. NOAA National Weather Service, National Hurricane Center. Available online: http://www.nhc.noaa.gov/data/tcr/AL122005_Katrina.png (accessed on 20 March 2015).
5. Berg, R. Tropical Cyclone Report: Hurricane Ike. NOAA National Weather Service, National Hurricane Center. Available online: http://www.nhc.noaa.gov/data/tcr/AL092008_Ike.png (accessed on 20 March 2015).
6. Blake, E.S.; Kimberlain, T.B.; Berg, R.J.; Cangialosi, J.P.; Beven, J.L. Tropical Cyclone Report: Hurricane Sandy. NOAA National Weather Service, National Hurricane Center. Available online: http://www.nhc.noaa.gov/data/tcr/AL182012_Sandy.png (accessed on 20 March 2015).
7. Meyer, R.; Baker, J.; Broad, K.; Czajkowski, J.; Orlove, B. The Dynamics of Hurricane Risk Perception: Real-Time Evidence from the 2012 Atlantic Hurricane Season. *Bull. Am. Meteorol. Soc.* **2014**, *95*, 1389–1404. [CrossRef]
8. Morrow, B.H.; Lazo, J.K.; Rhome, J.; Feyen, J. Improving storm surge risk communication: Stakeholder perspectives. *Bull. Am. Meteorol. Soc.* **2015**, *96*, 35–48. [CrossRef]
9. Lazo, J.K.; Morrow, B.H. *Survey of coastal U.S. public's perspective on extra tropical—Tropical cyclone storm surge information*; National Center for Atmospheric Research: Boulder, CO, USA, 7 January 2013.
10. Saffir, H.S. Hurricane wind and storm surge. *Mil. Eng.* **1973**, *423*, 4–5.
11. Simpson, R.H. The hurricane disaster potential scale. *Weatherwise* **1974**, *27*, 169–186.
12. National Hurricane Center. Tropical cyclone storm surge probabilities. Available online: http://www.nhc.noaa.gov/surge/psurge.php (accessed on 20 March 2015).
13. Horn, D.P. Storm surge warning, mitigation, and adaptation. In *Coastal and Marine Hazards, Risks, and Disaster*; Ellis, J., Sherman, D.J., Shroder, J., Eds.; Elsevier: Waltham, MA, USA, 2015; pp. 153–180.
14. Blendon, R.J.; Benson, J.M.; Burh, T.; Weldon, K.J.; Herrmann, M.J. High-risk Area Hurricane Survey. Harvard School of Public Health Project on the Public and Biological Security. Available online: http://sphweb.sph.harvard.edu/news/press-releases/files/Hurricane_2008_Total_Release_Topline.doc (accessed on 20 February 2015).
15. Changnon, S.A.; Pielke, R.A., Jr.; Changnon, D.; Sylves, R.T.; Pulwarty, R. Human factors explain the increased losses from weather and climate extremes. *Bull. Am. Meteorol. Soc.* **2000**, *81*, 437–442. [CrossRef]
16. Horney, J.; Snider, C.; Malone, S.; Gammons, L.; Ramsey, S. Factors associated with hurricane preparedness: Results of a pre-hurricane assessment. *J. Disaster Res.* **2008**, *3*, 1–7.
17. Stewart, A.E. Gulf Coast residents underestimate hurricane destructive potential. *Weather Clim. Soc.* **2011**, *3*, 116–127. [CrossRef]
18. Baker, E.J. Hurricane evacuation behavior. *Int. J. Mass. Emerg. Disasters* **1991**, *9*, 287–310.
19. Dash, N.; Gladwin, H. Evacuation decision making and behavioral responses: Individual and household. *Nat. Hazards Rev.* **2007**, *8*, 69–77. [CrossRef]
20. Dow, K.; Cutter, S.L. Public orders and personal opinions: Household strategies for hurricane risk assessment. *Environ. Hazards* **2000**, *2*, 143–155. [CrossRef]
21. Dow, K.; Cutter, S.L. Crying wolf: Repeat responses to hurricane evacuation orders. *Coast. Manag.* **1998**, *26*, 237–252. [CrossRef]
22. Gladwin, H.; Peacock, W.G. Warning and evacuation: A night for hard houses. In *Hurricane Andrew: Ethnicity, Gender and the Sociology of Disasters*; Peacock, W.G., Morrow, B.H., Gladwin, H., Eds.; Routledge: New York, NY, USA, 1997; pp. 52–73.
23. Whitehead, J.C.; Edwards, B.; van Willigen, M.; Maiolo, J.R.; Wilson, K.; Smith, K.T. Heading for higher ground: Factors affecting real and hypothetical hurricane behavior. *Environ. Hazards* **2000**, *2*, 133–142. [CrossRef]

24. Lazo, J.K.; Waldman, D.M.; Morrow, B.H.; Thacher, J. Household evacuation decision making and the benefits of improved hurricane forecasting: Developing a framework for assessment. *Weather Forecast.* **2010**, *25*, 207–219.

25. Lindell, M.K.; Lu, J.; Prater, C.S. Household decision making and evacuation response to Hurricane Lili. *Nat. Hazards Rev.* **2005**, *6*, 171–179. [CrossRef]

26. Stein, R.; Buzcu-Guven, B.; Dueñas-Osorio, L.; Subramanian, D.; Kahle, D. How risk perceptions influence evacuations from hurricanes and compliance with government decisions. *Policy Stud. J.* **2013**, *41*, 319–342. [CrossRef]

27. Zhang, F.; Morss, R.E.; Sippel, J.A.; Beckman, T.K.; Clements, N.C.; Hampshire, N.L.; Harvey, J.N.; Hernandez, J.M.; Morgan, Z.C.; Mosier, R.M.; *et al.* An in-person survey investigating public perceptions of and responses to Hurricane Rita forecasts along the Texas coast. *Weather Forecast.* **2007**, *22*, 1177–1190. [CrossRef]

28. Elder, K.; Xirasagar, S.; Miller, N.; Bowen, S.A.; Glover, A.; Piper, C. African Americans' decisions not to evacuate New Orleans before Hurricane Katrina: A qualitative study. *Am. J. Public Health* **2007**, *97*, S124–S129. [CrossRef] [PubMed]

29. Kang, J.E.; Lindell, M.K.; Prater, C.S. Hurricane evacuation expectations and actual behavior in hurricane Lili. *J. Appl. Soc. Psychol.* **2007**, *37*, 887–903. [CrossRef]

30. Ricchetti-Masterson, K.; Horney, J. Social factors as modifiers of Hurricane Irene evacuation behavior in Beaufort County, NC. *PLoS Curr. Disasters* **2013**, *5*. [CrossRef] [PubMed]

31. Bandura, A. Self-efficacy: Toward a unifying theory of behavioral change. *Psychol. Rev.* **1977**, *84*, 191–215. [CrossRef] [PubMed]

32. Bandura, A. *Social Foundations of Thought and Action: A Social Cognitive Theory*; Prentice-Hall: Englewood Cliffs, NJ, USA, 1986; p. 617.

33. Bandura, A. Self-Efficacy: The Exercise of Control. *Br. J. Clin. Psychol.* **1998**, *37*, 470.

34. Carberry, A.R.; Lee, H.S.; Ohland, M.W. Measuring engineering design self-efficacy. *J. Eng. Educ.* **2010**, *99*, 71–79. [CrossRef]

35. Talley, E.; Goodwin, L.; Ruzic, R.; Fisler, A. Marine ecology as a framework for preparing the next generation of scientific leaders. *Mar. Ecol. Evol. Perspect.* **2011**, *32*, 268–277. [CrossRef]

36. Kulviwat, S.; Bruner, G.C.; Neelankavil, J.P. Self-efficacy as an antecedent of cognition and affect in technology acceptance. *J. Consum. Mark.* **2014**, *31*, 190–199. [CrossRef]

37. Riggs, I.; Enochs, L. Towards the development of an elementary teacher's science teaching efficacy belief instrument. *Sci. Educ.* **1990**, *74*, 625–637. [CrossRef]

38. Hyre, A.D.; Benight, C.C.; Tynes, L.L.; Rice, J.; DeSalvo, K.B.; Muntner, P. Psychometric properties of the hurricane coping self-efficacy measure following Hurricane Katrina. *J. Nerv. Ment. Dis.* **2008**, *196*, 562–567. [CrossRef] [PubMed]

39. Salmun, H.; Molod, A. The use of a statistical model of storm surge as a bias correction for dynamical surge models and its applicability along the U.S. east coast. *J. Mar. Sci. Eng.* **2015**, *3*, 73–86. [CrossRef]

40. Forbes, C.; Rhome, J.; Mattocks, C.; Taylor, A. Predicting the storm surge threat of Hurricane Sandy with the National Weather Service SLOSH Model. *J. Mar. Sci. Eng.* **2014**, *2*, 437–476. [CrossRef]

41. Davis, J.R.; Paramygin, V.A.; Vogiatzis, C.; Sheng, Y.P.; Pardalos, P.M.; Figueiredo, R.J. Strengthening the resiliency of a coastal transportation system through integrated simulation of storm surge, inundation, and nonrecurrent congestion in northeast Florida. *J. Mar. Sci. Eng.* **2014**, *2*, 287–305. [CrossRef]

42. Floyd, D.L.; Prentice-Dunn, S.; Rogers, R.W. A meta-analysis of research on Protection Motivation Theory. *J. Appl. Soc. Psychol.* **2000**, *30*, 407–429. [CrossRef]

43. Rogers, R.W.; Mewborn, C.R. Fear appeals and attitude change: Effects of a threat's noxiousness, probability of occurrence and efficacy of coping response. *J. Personal. Soc. Psychol.* **1976**, *34*, 54–61. [CrossRef]

44. Stewart, A.E. Psychological dimensions of adaptation to weather and climate. In *Biometeorology for Adaptation to Climate Variability and Change*; Ebi, K., Burton, I., McGregor, G., Eds.; Springer: London, UK, 2009; Volume 1, pp. 211–232.

45. Maddux, J.E.; Rogers, R.W. Protection motivation and self-efficacy: A revised theory of fear appeals and attitude change. *J. Exp. Soc. Psychol.* **1983**, *19*, 469–479. [CrossRef]

46. Rosenstock, I.M. Historical origins of the Health Belief Model. *Health Educ. Quart.* **1974**, *2*, 328–335.

47. Ajzen, I. The theory of planned behaviour. *Organ. Behav. Hum. Decis.* **1991**, *50*, 179–211. [CrossRef]

48. Lindell, M.K.; Perry, R.W. *Behavioral Foundations of Community Emergency Planning*; Hemisphere: Washington, DC, USA, 1992.

49. Lindell, M.K.; Perry, R.W. *Communicating Environmental Risk in Multiethnic Communities*; Sage: Thousand Oaks, CA, USA, 2004.

50. Grothmann, T.; Reusswig, F. People at risk for flooding: Why some residents take precautionary action while others do not. *Nat. Hazards* **2006**, *38*, 101–120. [CrossRef]

51. Milne, S.; Sheeran, P.; Orbell, S. Prediction and intervention in health-related behaviour: A meta-analytic review of Protection Motivation Theory. *J. Appl. Soc. Psychol.* **2000**, *30*, 106–143. [CrossRef]

52. Stewart, A.E.; Lazo, J.; Morss, R.; Demuth, J. The relationship of weather salience with the perceptions, uses, and values of weather information in a nationwide sample of the United States. *Weather Clim. Soc.* **2012**, *4*, 172–179. [CrossRef]

53. Simms, J.L.; Kusenbach, M.; Tobin, G.A. Equally unprepared: Assessing the hurricane vulnerability of undergraduate students. *Weather Clim. Soc.* **2013**, *5*, 233–243. [CrossRef]

54. Stewart, A.E. Minding weather: The measurement of weather salience. *Bull. Am. Meteorol. Soc* **2010**, *90*, 1833–1841. [CrossRef]

55. Cronbach, L.J. Coefficient alpha and the internal structure of tests. *Psychometrika* **1951**, *16*, 297–334. [CrossRef]

56. Cortina, J.M. What is Coefficient Alpha? An examination of theory and applications. *J. Appl. Psychol.* **1993**, *78*, 98–104. [CrossRef]

57. Kromrey, J.D.; Romano, J.; Hibbard, S. ALPHA_CI: A SAS® Macro for Computing Confidence Intervals for Coefficient Alpha. In Proceedings of the SAS Global Forum, Cary, NC, USA, 16–19 March 2008.

58. *SAS/STAT User's Guide, Version 9.1*; SAS Institute: Cary, NC, USA, 2004.

59. Panter, A.T.; Swygert, K.A.; Dahlstrom, W.G.; Tanaka, J.S. Factor analytic approaches to personality item-level data. *J. Personal. Assess.* **1997**, *68*, 561–589. [CrossRef] [PubMed]

60. Holgado-Tello, F.P.; Chacón-Moscoso, S.; Barbero-García, I.; Vila-Abad, E. Polychoric *vs.* Pearson correlations in exploratory and confirmatory factor analysis of ordinal variables. *Qual. Quant.* **2010**, *44*, 153–166. [CrossRef]

61. Loehlin, J.C. *Latent Variable Models: An Introduction to Factor, Path, and Structural Equation Analysis*; Lawrence Erlbaum: Mahwah, NJ, USA, 2004.

62. Demuth, J.L.; Morss, R.E.; Morrow, B.H.; Lazo, J.K. Creation and communication of hurricane risk information. *Bull. Am. Meteorol. Soc.* **2012**, *93*, 1133–1145. [CrossRef]

63. Sherman-Morris, K.; Antonelli, K.B.; Williams, C.C. Measuring the Effectiveness of the Graphical Communication of Hurricane Storm Surge Threat. *Weather Clim. Soc.* **2015**, *7*, 69–82. [CrossRef]

64. Norris, F.H.; Smith, T.; Kaniasty, K. Revisiting the experience-behavior hypothesis: The effects of Hurricane Hugo on hazard preparedness and other self-protective acts. *Basic Appl. Soc. Psychol.* **1999**, *27*, 37–47. [CrossRef]

65. Stewart, A.E.; Knox, J.A.; Schneider, P. Piloting and evaluating a workshop to teach Georgia teachers about weather science and safety. *J. Geosci. Ed.* **2015**. submitted.

66. National Weather Service. Weather preparedness events calendar. Available online: http://www.nws.noaa.gov/om/severeweather/severewxcal.shtml (accessed on 27 March 2015).

Journal of
*Marine Science
and Engineering*

MDPI

Article

Modeling Storm Surge and Inundation in Washington, DC, during Hurricane Isabel and the 1936 Potomac River Great Flood

Harry V. Wang [1,*], Jon Derek Loftis [1], David Forrest [1], Wade Smith [2] and Barry Stamey [2]

[1] Department of Physical Sciences, Virginia Institute of Marine Science, College of William and Mary
P.O. Box 1375, Gloucester Point, VA 23062, USA; jdloftis@vims.edu (J.D.L.); drf@vims.edu (D.F.)

[2] Noblis Inc., 3150 Fairview Park Drive South, Falls Church, VA 22042, USA;
oceanblue1492@verizon.net (W.S.); bbstamey.org@gmail.com (B.S.)

* Author to whom correspondence should be addressed; hvwang@vims.edu; Tel.: +1-804-684-7215.

Academic Editor: Rick Luettich

Received: 3 April 2015; Accepted: 1 July 2015; Published: 21 July 2015

Abstract: Washington, DC, the capital of the U.S., is located along the Upper Tidal Potomac River, where a reliable operational model is needed for making predictions of storm surge and river-induced flooding. We set up a finite volume model using a semi-implicit, Eulerian-Lagrangian scheme on a base grid (200 m) and a special feature of sub-grids (10 m), sourced with high-resolution LiDAR data and bathymetry surveys. The model domain starts at the fall line and extends 120 km downstream to Colonial Beach, VA. The model was used to simulate storm tides during the 2003 Hurricane Isabel. The water level measuring 3.1 m reached the upper tidal river in the vicinity of Washington during the peak of the storm, followed by second and third flood peaks two and four days later, resulting from river flooding coming downstream after heavy precipitation in the watershed. The modeled water level and timing were accurate in matching with the verified peak observations within 9 cm and 3 cm, and with R^2 equal to 0.93 and 0.98 at the Wisconsin Avenue and Washington gauges, respectively. A simulation was also conducted for reconstructing the historical 1936 Potomac River Great Flood that inundated downtown. It was identified that the flood water, with a velocity exceeding 2.7 m/s in the downstream of Roosevelt Island, pinched through the bank northwest of East Potomac Park near DC. The modeled maximum inundation extents revealed a crescent-shaped flooding area, which was consistent with the historical surveyed flood map of the event.

Keywords: Hurricane Isabel; 1936 Potomac River Great Flood; sub-grid modeling

1. Introduction

The subject of storm surge and inundation has long attracted physical oceanographer and coastal engineers, because these hazards can inflict tremendous damages and cause enormous impacts on human life and property. It thus becomes the most pressing issue to the coastal community as to how to minimize the impacts from storm surge and inundation during extreme weather conditions. The early studies focused on constructing the proper formulation on the structured grid for storm surge [1–4]. Recently, a breed of unstructured grid models coupled with the wind wave models became available and were applied in a relatively large domain [5–7]. For sub-grid modeling, the sub-grid scale parameterization was used for modeling inundation in a relatively small urban area (4 km^2) [8] and LiDAR (light detection and ranging) data was applied to study the effect of distributed roughness on flows in the flood plain [9]. Wang et al. (2014) [10] successfully used the semi-implicit formulation combined with sub-grids in simulating inundation in the New York City during Hurricane Sandy.

In the coastal areas of the U.S. eastern seaboard, many cities are located at the fall line near the headwaters of an estuary, where river flow combined with storm surges can present a flooding hazard.

During the 2003 Hurricane Isabel in Washington, DC, a storm surge of 8.8 feet (2.7 m) above mean sea level was recorded by the USGS (U.S. Geological Survey) gauge at Wisconsin Avenue and 10.1 feet (3.1 m) by a NOAA gauge on a pier in the Washington waterfront at the southwest portion of DC. Both observations, surpassed the previous records set by the 1933 Chesapeake Potomac Hurricane. After Isabel passed, the Potomac River crests reached DC two and four days later, as a result of the precipitation deluge in the Upper Potomac River Basin. In 1936, a flooding event affected much of the northeastern United States, ranging from the Ohio River Valley, New England south to the Potomac River Basin, as a result of a combination of heavy rain and snowmelt. At Washington, DC, the result was the 1936 Potomac River Great Flood. Damage was considerable along the Chesapeake and Ohio Canal, Harpers Ferry, WV, to Hancock, MD, with significant flooding in Washington, DC (USGS, 1937) [11]. The average daily flow on 18 March 1936 during the Potomac River Great Flood in Washington, DC was observed to be 12,100 m^3/s, which was 39-times the normal daily flow, and water levels of 8.5 m and 5.7 m were observed at the Chain Bridge and at the Key Bridge (3.2 km apart), respectively [11]. In the Upper Potomac River, the flow passes through the fall-line as a fluvial river, transitions into a tidal river and eventually becomes a major estuary downstream. In the process, the direction of currents, flow pattern, frictional resistance, the geomorphology and the sediment characteristics are all subject to change as the flow regimes change. Works have been attempted to make predictions in the region for the combined storm surge and riverine flooding, but with shortcomings or limited success. The USGS has developed a hydrodynamic model for simulating unsteady flow in a network of open channels and implemented the model for the tidal Potomac River [12]. The effect of freshwater inflow, tidal currents and meteorological conditions were tested, but have not been applied for storm conditions. Recently, Mashriqui et al. (2010) [13] initiated the CERIS (Coastal, Estuary, River Information Services) system to provide an integrated suite of water information for hazard mitigation, water resources and ecosystem management. The unsteady HEC-RAS (Hydrologic Engineering Centers River Analysis System) model was developed and tested for a 2003 Hurricane Isabel simulation, but the phase lagged by 4–6 h and the peak elevation under-predicted by 30–50 cm when compared with the observations measured at NOAA's Washington, DC, waterfront station. The National Weather Service's (NWS) Advance Hydrologic Prediction Service was responsible for the forecast of the river discharge into the Upper Tidal Potomac River, but with a disclaimer that their forecasts do not include the wind-induced storm surge. Lastly, EA Engineering, Inc. of Hunt Valley, MD (2001) [14] applied RMA2, RMA4 and SED2D, a suite of finite element hydrodynamic, transport, and sediment models developed by Army Corps of Engineers, in the upper reach of the Potomac River only for the dye-tracer and turbidity plume studies.

In the present paper, we focus on producing high resolution street-level scale inundation by using sub-grids from DEM (Digital Elevation Model) nested within the base grid cell to allow fine-scale topography features to be modeled without a heavy computational cost. The primary objective is to predict the water level and inundation caused by various storm conditions occurred in the Upper Tidal Potomac River near Washington, DC. They can be the combination of the storm tide coming upstream from the bay and river flooding originated in the Upper Potomac River basin. Section 2 provides an overview of the study area, including observation stations. Section 3 describes a sub-grid model incorporating LiDAR and high-resolution bathymetry data into a regular base grid and solved by a non-linear solver. Section 4 depicts the setup and modeling results of the 2003 Hurricane Isabel. Section 5 describes the modeling of 1936 Potomac River Great Flood during which the floodwater breached through Potomac bank and inundated downtown DC. Section 6 discusses outstanding issues and concludes the paper.

2. Study Area

Washington, DC, is the capital city of the United States and a major metropolitan area with a population of 5.8 million, including its surrounding suburban areas. It is located near the head of the Upper Tidal Potomac River (Figure 1, right panel) where the river flows across the fall line and meets the tide from the Chesapeake Bay. The Potomac River is the largest tributary of the Chesapeake Bay, whose length from the fall line to the mouth at Point Lookout is about 180 km. The Potomac River is like many other rivers along the Atlantic seaboard that has its hydraulic head in the Appalachian Mountains and flows eastward to the Atlantic Ocean. As it flows over the fall line, Potomac creates a spectacular landscape feature: the Great Falls. From the Great Falls to Theodore Roosevelt Island, the river goes through a series of rapids, narrow rock-girded channels twisting between cliffs, flow-topped bedrock and numerous islands composed of sand and gravel laid down by the river. The river transitions into the tidal portion of the river near the Chain Bridge, a few kilometers upstream from Roosevelt Island. The average flow is approximately 320 m^3/s; the flow may be less than 40 m^3/s in the summer and reaches 3800 m^3/s during flood periods. The tide in the Potomac is an integral part of the Chesapeake Bay system; originating from the Atlantic Ocean and propagating upstream along the main stem of the bay into the Potomac River. It can reach up to the Chain Bridge, where the tidal influence ends. The mean tidal range at Wisconsin Avenue is approximately 0.9 m. The tidal phase lags 11.5 h behind that at Hampton Roads at the mouth of the Chesapeake Bay. The model grid was constructed from the Little Falls (USGS Station 01646500; latitude 38°56′59.2″ longitude 77°07′39.5″), MD, at the fall line to Colonial Beach, VA, with a total length of about 120 km, as shown in Figure 2a. The domain covers about 2/3 of the tidal Potomac River area and contains a total of 18,259 base grids in square elements with a resolution of 200 m × 200 m and incorporates sub-grids (the sub-grids will be described in detail in Section 3). The bathymetry and topography associated with the model domain ranging from −10 m to 10 m (minus represents above ground) are displayed in Figure 2b. The observation stations used for the study include Little Falls, MD (USGS), Wisconsin Avenue (USGS), Washington, DC, waterfront (NOAA), Colonial Beach, VA (NOAA), and Lewisetta, VA (NOAA), whose locations are marked by solid symbols shown in the left panel of Figure 1. Among these stations, Washington, DC, is one of the longest-serving tide stations in the nation, which started operation 15 April 1931. For this study, the mean sea level is used as the datum.

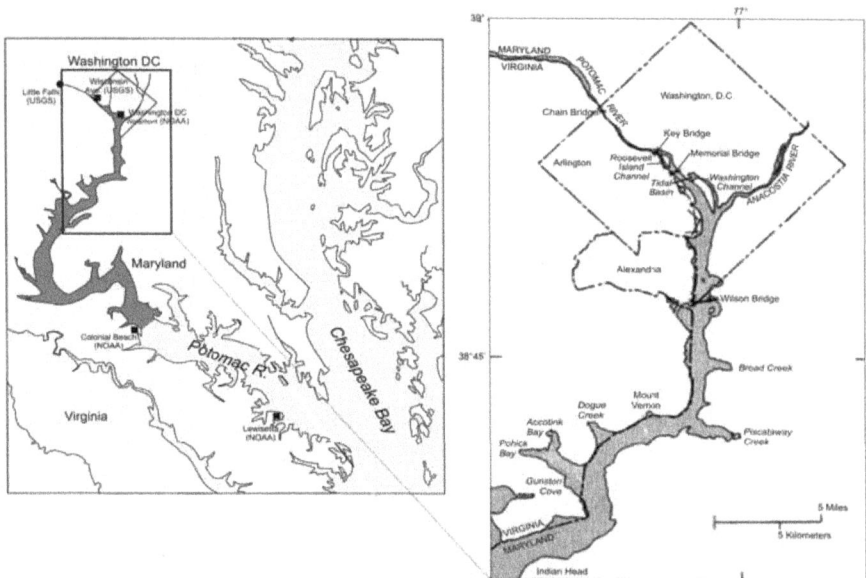

Figure 1. (Left) The Potomac River modeling domain (shaded) and observation stations used for the study; (right) the zoom-in map of Washington, DC, and the Upper Tidal Potomac River.

3. Storm Surge and Inundation Model Incorporating Sub-Grids in the Upper Tidal Potomac River

3.1. Model Description

This study makes use of a robust semi-implicit finite difference/finite volume model UnTRIM^2 (Unstructured Tidal Residual Intertidal Mudflat Model, power 2 version). The model is governed by the three-dimensional shallow-water equations with the Boussinesq approximation and is solved for free surface elevation, water velocities and salinity in a Cartesian coordinate system. The model was formulated with an efficient semi-implicit, Eulerian-Lagrangian scheme on unstructured orthogonal grids that includes both 3D barotropic and baroclinic processes pertaining to tide, wind and gravitationally-driven circulation on an f-plane [15–17]. The Potomac River is one of the major tributaries of the Chesapeake Bay, whose drainage area is 38,000 km^2, and the mean annual mean river discharge is 360 m^3/s. The Upper Tidal Potomac River is dominated by fresh water discharge, wind and tide. The tide from downstream can reach up to the Chain Bridge near DC, whereas the salt water intrusion normally only reaches up to the U.S. Route 301 Bridge near Colonial Beach. The length of the salt intrusion upstream from the lower Potomac River varies with the season depending on the magnitude of the freshwater discharge, but in general, the transition zone moves around Colonial Beach. The temperature-induced stratification is small because of the shallowness of the estuary. Thus, the region from the fall line to Colonial Beach, which was chosen as the model domain, can be reasonably considered as the vertically well-mixed tidal fresh zone with the downstream side ends in the transition zone. Because the interest of the present paper is focused on barotropic motions driven by the river, tide and wind-induced surge in the Upper Tidal Potomac River, using the vertically-averaged 2D version of the model (but including 2D temperature) for the domain chosen is thus appropriate.

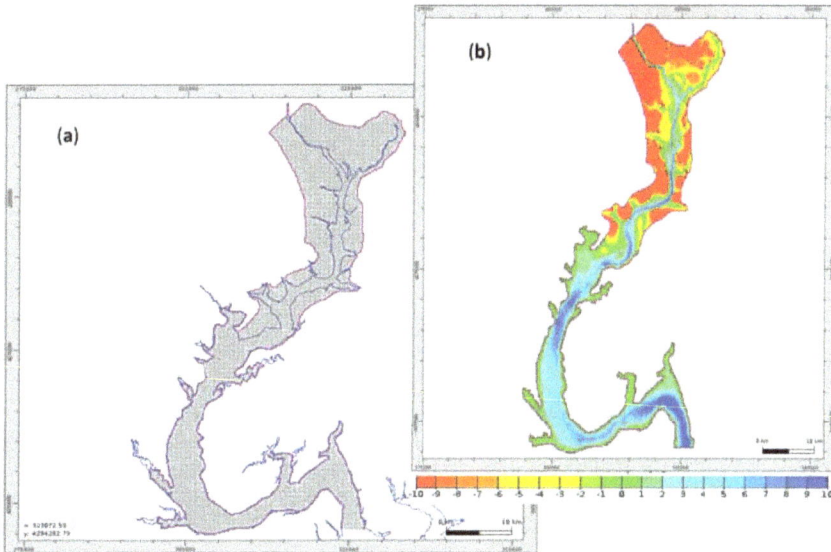

Figure 2. (a) The Potomac River grid domain including land area around Washington, DC; (b) the combined bathymetry data (30-m resolution) and LiDAR-derived topography (10-m resolution) used.

The sub-grids, in the form of raster DEM (a grid of squares) derived from LiDAR and high-resolution bathymetry, is nested within the base grid cell to allow the fine-scale topography features to be recognized. In the base grid and sub-grid framework, the core computation for solving the shallow water equations is performed on the base grid. Once the base grid finishes the calculations, the total flux on each edge of the base grid is then distributed to the individual sub-grid cells based on the analytic solution of the hydraulic conveyance approach. To illustrate the principle, it is sufficient to consider the following simplified 2D depth-averaged sub-grid momentum equation:

$$\frac{\partial u_j}{\partial t} + g\frac{\partial \varsigma}{\partial x} + cf\frac{||u_j||}{h} = 0 \quad \Rightarrow \quad \frac{\partial u_j}{\partial t} + g\frac{\partial \varsigma}{\partial x} + g\frac{||u_j||}{\Omega_j} = 0 \tag{1}$$

Where $\Omega_j = \sqrt{gh_j/cf}$ is hydraulic conveyance, u_j is sub-grid velocity, ς the surface elevation and cf is a dimension-less friction coefficient for which a formulation, such as Chezy's or Manning's, can be given.

For the validity of the equation, the inundation flow is considered to be frictionally dominated, so the advection term can be considered as the second order. Since the sub-grid formula is applied only one step at a time within each marching time step of the base grid, u_j is not a function of time. From this, it follows that the equation can be rewritten as $u_j = \Omega_j\sqrt{\varsigma_x}$. Assuming the pressure gradient is constant on each edge (but velocity, friction and depth are variable), it can be shown that the velocity of the individual sub-grid can be obtained by the base grid velocity times the ratio of the hydraulic conveyance of the sub-grid to that of base grid on each side of the grid edge, based on the following formula:

$$||u_j|| = \Omega_j \frac{||U||}{\Omega} \quad \text{with} \quad ||U|| = \frac{\sum_{j=1}^{J} h_j||u_j||}{\sum_{j=1}^{J} h_j} \quad \text{and} \quad \Omega = \frac{\sum_{j=1}^{J} h_j\Omega_j}{\sum_{j=1}^{J} h_j} \tag{2}$$

where $(||u_j||, \Omega_j)$ and $(||U||, \Omega)$ are the velocity and conveyance for the sub-grid and base grid, respectively, and J defines the total number of wet sub-grids within a base grid.

This means that the sub-grid velocity and its cross-section flux (the product of velocity and the cross-section area) at the edge of each of the sub-grid can be obtained analytically. Together with the bathymetry within each of the sub-grids, the water depth of each sub-grid and the status of its wetting (or drying) can be determined. Once the depth and the wetting and drying of the individual sub-grids are decided, the wetting, drying and/or partially-wetting-drying of the "base grid" can then be determined collectively by the distribution of the sub-grid population within that base grid. Here, it is important to recognize that the partially wetting-and-drying of the base grid, a desired feature for more accurately determining the inundation extent, which is unavailable by the traditional method, is now possible, attributed to the sub-grid approach. Another important aspect of the present approach is that sub-grid scale information does feed back to the base grid computation, one piece of which is the friction parameter. Now, the base grid friction is no longer dependent on the single gross-averaged depth on the base grid; rather, it is obtained as the collective contributions from each of the wet cells of the sub-grids according to the conveyance formulation above. One other important parameter, cross-sectional fluxes, essential to the base grid computations, are also based on the summation of the product of each sub-grid velocity times each sub-grid bathymetry, rather than the product of the averaged velocity and averaged cross-section area at the edge of the base grid. The accuracy of the cross-sectional calculation was further enhanced by using a nonlinear solver to determine the nonlinear relationship between water level and the cross-section area [18,19]. These combined approaches result in a more accurate determination of the cross-sectional flux, remove the requirement of using the minimum water depth (by the traditional models) and improve on the long wave propagation speed, which perhaps is crucial for determining the correct tidal phase. Computational-wise, all of this is done without having to resort to using a fine-scale computational mesh throughout the study area to compute the small-scale dynamic processes. The savings of computation time is thus quite significant.

3.2. Incorporation of LiDAR-Derived DEM into the Sub-Grid Model Domain

The horizontal computational domain comprises a set of non-overlapping convex three- or four-sided polygons. Each polygon side is designated as either a side of an adjacent polygon or as a boundary line in the model grid input file. The innovations in the UnTRIM model permit the use of a sub-grid mesh embedded within each base grid element with an inherent numerical scheme capable of partial wetting and drying [19]. Although the sub-grid can be implemented on either a triangular or rectangular grid, numerical accuracy is favorable when a uniform grid comprising quadrilaterals is used with high-resolution DEM (digital elevation model) sourced with LiDAR-derived data to best represent urban street-level flooding. The use of square grid structures has been demonstrated as a favorable method to preserve the city-block building structures with sufficient DEM resolution to resolve streets between buildings as an effective conduit for accurately modeling inundation [10,20]. For this reason, many of the grids developed using LiDAR-derived DEMs have been scaled to square grids congruent to the native pixel resolution of the topographic data contained in the DEM to avoid further interpolation. Using a square grid, the normal velocity on the faces of each polygon is calculated at the center point of the face, and the centers of two adjacent polygons are equally spaced from the shared face, which minimizes the associated discretization error in these computations. An unstructured, non-uniform grid can be utilized with a larger associated discretization error [21]. However, the benefits of mixing triangles and quadrilaterals to conform the grid shape to the bathymetric channels and shorelines are less significant with recent advancements, including the addition of sub-grids to the model, due to the partial wetting and drying scheme and the non-linear solver.

The setup and design of the model involves generating a base-grid mesh for which computations are geometrically calculated. The domain starts at Little Falls passing through the confluence of the Potomac and Anacostia Rivers and ends in Colonial Beach, VA. One more reason Colonial Beach was chosen is because it is a NOAA tidal station far downstream of DC. The Washington, DC, topography was a LiDAR-based DEM with 10-m resolution produced by Noblis with the NAD83 CORS 96 horizontal datum projected in UTM Zone 18N coordinates with a vertical datum of NAVD88 in meters. The topographic data were cast over a 200-m base grid with an embedded 20×20 10-m resolution sub-grid. This base grid resolution was chosen such that the main stem of the Potomac River channel would be multiple grid cells in width across the river for proper calculation of volume transport. The sub-grid scaling was chosen such that the topographic LiDAR-derived DEM would be at its native resolution (10 m) and not require further interpolation, which potentially could cause additional computational error due to stretching or distortion. Bathymetry point data ($\approx$30-m point spacing) were retrieved from five NOAA bathymetric surveys of the Potomac and Anacostia Rivers conducted in 1974 (NOAA Surveys: H09477, H09380, H09479, H09488, H09478). Using a shoreline polyline in ArcGIS 10.3, a power 2 inverse distance weighted interpolation was performed on the bathymetry data using the shoreline polyline as a barrier. The resulting interpolation product was then translated to NAD83 CORS 96 in UTM Zone 18N with a vertical datum of NAVD88 in meters. With the two datasets in the same projection and datum, they were merged such that the bathymetric data would overlap the LiDAR topographic data to resolve issues with bridges in the LiDAR DEM potentially blocking proper fluid movement into creeks and shallow water regions. Inherent uncertainty in the vertical data from the LiDAR (1-cm precision with $\pm$30-cm accuracy) and bathymetric datasets (10-cm precision with $\pm$0.5-m accuracy) contributes directly to uncertainty in inundation thickness, while uncertainty in the spatial extents is amplified by the slope of the DEM at the wetting and drying interface.

Spot checks of the combined DEM with known topography and bathymetry indicated reasonable agreement. The bathymetry data were subsequently verified with a report [14] submitted to U.S. Army Corps of Engineers Baltimore District, MD that provides transect data near Theodore Roosevelt Island and the Arlington Memorial Bridge (Figure 3). The resulting topography and bathymetry merged DEM was input to Janet v.2.2, an unstructured grid software by Smile Consulting Inc. (Hamburg, Germany), to provide elevations for the model base grid and sub-grid. Ultimately, the grid was constructed with a 200-m base grid with a 10-m resolution sub-grid for use in this study (Figure 4). The simulations for this modeling effort were performed on a Dell T3500 PC Workstation with Windows 7 Professional (64-bit edition) and an Intel Xeon Quad Core X5570 Processor (2.93GHz) (headquarter at Santa Clara,

CA, USA) with 24 GB RAM. The performance efficiency of the CPU during these modeling simulations was approximately 120-times faster than real time.

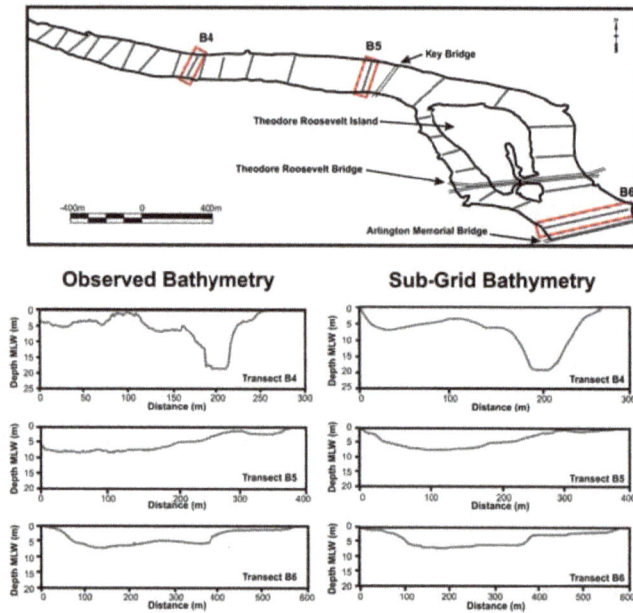

Observed Bathymetry
Sub-Grid Bathymetry

Figure 3. Three example of bathymetric transects in the Upper Potomac River used to verify bathymetry interpolation, with corresponding sounding data published in EA Engineering (2001) [14] (bottom left) and the model's sub-grid bathymetry (bottom right) in the vicinity of the Washington aqueduct.

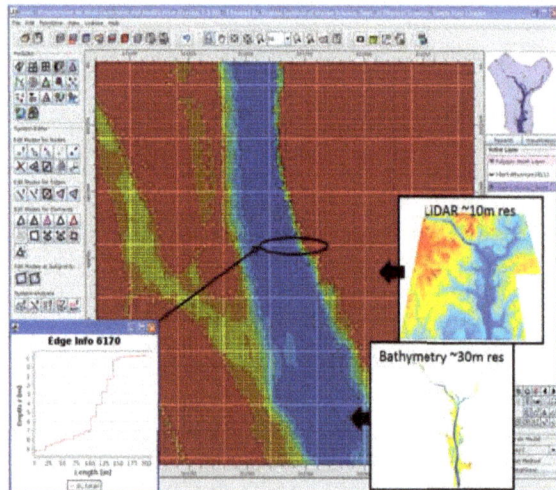

Figure 4. Detailed feature of the based grid *vs.* sub-grids in Washington, DC, near Roosevelt Island. The thicker white line shows the 200-m base grid with each grid cell containing a 20 × 20 of 10 m × 10 m sub-grid cells. The resolution is such that LiDAR data are in 10-m resolution and bathymetry in 30-m resolution. An example of the bathymetry cross-section is shown in the lower left corner.

4. Modeling the 2003 Hurricane Isabel Event

The 2003 Hurricane Isabel was the most devastating hurricane to ravage the Chesapeake Bay in recent history. It made landfall on the Outer Banks of North Carolina on 18 September 2003 and was reduced from a Category 5 to a Category 2 storm on the Saffir-Simpson hurricane intensity scale immediately prior to making landfall. Still, the hurricane storm surge that propagated up the Tidal Potomac River to the Washington, DC, area resulted in 160 homes, 60 condominiums flooded and an additional 2000 units of buildings reporting severe damage in Fairfax County and the City of Alexandria. In Washington, DC, the storm peak height was recorded at the Wisconsin Avenue gauge as 3.4 m and at the Washington, DC, gauge as 3.1 m. Heavy rainfall was reported in the Upper Potomac River Basin, peaking at 510 mm (20.2 inches) in Sherando, VA, which resulted in a peak discharge rate 4000–4500 m^3/s recorded at the USGS Little Falls, MD, station at the fall line.

4.1. Model Setup

Modeling the 2003 Hurricane Isabel event required two boundary conditions. One is the upstream river discharge boundary conditions specified at the Little Falls, MD, and the other is the water level open boundary condition specified at Colonial Beach, VA, 120 km downstream. The river discharge data provided were daily discharge, and the water level data provided at the downstream open boundary condition were every 6 min. The headwater of the Potomac River near the upstream boundary condition has spectacular landscape features. It cascades over a series of 20-foot (6 m) falls, falling a total of 76 feet (23 m) in elevation, with rapids, narrow rock-grid channels, twisted cliffs and flow-topped bedrock islands until reaching Chain Bridge. In addition, the Potomac narrows significantly as it passes through falls and the Mather Gorge, and the nearby shoreline can be inundated by floods caused by heavy rain or snow from the watershed upstream. Given these complex features, there is a question as to how to properly assign the upstream boundary condition. Written in terms of open channel non-uniform flows [22], the governing equation can be expressed as:

$$\frac{dH}{dx} = \frac{d}{dx}\left(z + y + \frac{u^2}{g}\right) = -S_f \tag{3}$$

$$\frac{d}{dx}\left(y + \frac{u^2}{g}\right) = -\frac{dz}{dx} - S_f \Rightarrow \frac{dE}{dx} = -\frac{dz}{dx} - S_f \tag{4}$$

where H is total energy; E is the specific energy; y is the vertical depth of flow above the channel bed; z is the height above the datum of the bed level; S_f is the friction slope.

According to the equations above, it can be deduced that, with a given bathymetry, the non-uniform flows are controlled by the external inputs of the water level and momentum, as well as the dissipation by the internal friction slope. Two types of boundary conditions were considered: (1) the horizontal flux boundary condition, for which the horizontal volume flux is specified as the product of horizontal velocity times the cross-section; and (2) the vertical volume flux given as a point source with which there is no horizontal momentum. Type 1 takes total discharge and represents it in the form of horizontal kinetic energy as the specific energy, whereas Type 2 takes total discharge and represents it in the form of the accumulation of vertical volume flux, which manifests as the elevated free surface equivalent to a form of potential energy in specific energy. For the Upper Tidal Potomac River, the water level and velocity downstream depend on how the specific energy is specified at the upstream boundary plus the internal dissipation by the friction slope. The internal dissipation induced by the Great Falls in the Potomac River is a complicated phenomenon, and the dynamics is not known, *a priori*. In our test, the Type 1 boundary condition was used at first, and it was found that the result during high flow was unsatisfactory (not shown). After several sensitivity tests, we found that dividing the total flow 50% for Type 1 combined with 50% for Type 2 performs the best, as shown in Figure 5, and so, this was used. It is conjectured that the reason that these divisions are needed may have to do with the presence of falls and rapids. After all, using the weighting between the momentum flux and point source is not the most rigorous approach we would like to have used; nevertheless, with the sensitivity test, it does suggest that with the presence of falls and rapids in a river, the conventional horizontal momentum inputs at the head of the river may not be the best boundary conditions to use.

The gauge at Colonial Beach, VA, was functioning during the period prior to the peak of the hurricane event, which destroyed the gauge platform. Thus, a portion of the time series of the water elevation was selected from another Potomac River station at Lewisetta, VA, about 60 km downstream from Colonial Beach (refer Figure 1). The water level of the two stations between Colonial Beach, VA, and Lewisetta, VA, is well correlated, and the tidal phase required a phase shift of 45 min. The friction coefficient, Manning's "n", was used to calculate the bottom shear stress, and the "n" value was obtained by comparing with the independent astronomical tide prediction dataset provided by the NOAA Tides and Currents website for 15 stations. Tests were conducted, including the comparison of modeled mean tidal range and the time of high water to the observations. With the sub-grid feature and nonlinear solver used, it was found that standard Manning's $n = 0.025$ works adequately for the domain from Colonial Beach, VA, up to the Washington, DC, station. However, from the Washington, DC, station to the upper reaches near the headwaters from Little Falls, MD, required a coefficient of $n = 0.040$ to produce the best results. This is expected because the headwaters are in a fluvial river environment, which can exert extra shear stress on the passing flow. The wind and atmospheric pressure data were retrieved from NOAA observation data at Washington, DC (Station No. 8594900). These data were interpolated to 5-min intervals of the model time step for the same time period and prepared as a uniform input throughout the domain for Hurricane Isabel. The Garratt (1977) wind drag formula [23] was used to calculate the wind stress with a cap on drag coefficient at U_{10} (at 10 m height) for wind speed greater than 40 m/s [24]. Model simulations began on 00:00 GMT 1 September 2003 and ended at 00:00 GMT 1 October 2003 with a five-day spin up period.

4.2. Results

The modeled water level compared with observations at Wisconsin Avenue and at Washington, DC, is shown in Figure 5a,b. The statistical measures of the water level comparisons in terms of R^2 (R-squared value), RMS (root mean square) and peak difference were 0.94, 14.3 cm and 9.2 cm for Wisconsin Avenue and 0.98, 7.3 cm and 2.4 cm for Washington, DC, respectively, as shown in Table 1. For a hurricane event with the peak water level reaching 3 m, the prediction skill of the current model is quite reasonable. In further analysis of the individual uncertainties over the comparisons, the largest uncertainties, based on NOAA Co-OPS's user manual, were associated with the seasonal effect of the tidal river, local wind and weather patterns and thermal expansion. The errors can also be associated with the datum selected (1–5 cm) and the measurement technique (1–2 cm). In our effort in simulating storm tide of the 2003 Hurricane Isabel, the observed wind, pressure, river discharge and temperature fields were prescribed to the model; thus, seasonal effects were not be a major issue for the uncertainty. There were still base errors, which were embedded in the datum selection and the measurement itself, which amounts to about 2–7 cm. Our water level prediction at the Washington, DC, station was close to this lower limit.

Table 1. Statistical comparison of modeled time series *vs.* observations during Hurricane Isabel.

Statistic	Wisconsin Ave.	Washington, DC
R^2	0.94	0.98
RMS	14.3 cm	7.3 cm
MAE	11.4 cm	4.8 cm
Peak Difference	9.2 cm	2.4 cm

From Figure 5a,b, it was noted that the hurricane-induced storm surge peak, which was characterized by a single peak at a height close to 3 m, arrived first on Julian Day 262.5 immediately after the hurricane made landfall. This first peak was followed by the gentler second and third flood peaks on Julian Days 264.5 and 266.5, respectively. Examination of water level variation and the Little Falls river discharge curves together showed that the second and third peaks were associated with the river floods of two peak flows: 4500 m^3/s and 4000 m^3/s. These two peak flows did not arrive until two and four days after the Hurricane passed, an indication of the delay of the watershed in collecting the precipitation dumped by the hurricane. To test the hypothesis quantitatively that the second and third peak were indeed the river discharge induced, a sensitivity tests was conducted with a scenario in which the no flux boundary condition was assigned at the head of the river. The scenario was dubbed "without" river discharge. The model results, under "without" river discharge, under-predicted the observed water level (along with the "with" river discharge model results) during the high flow period by about 1 m for the second peak and about 0.5 m for the third peak at Wisconsin Avenue station, as shown in Figure 6a. The under-prediction of water level during high flow periods was obvious at the Wisconsin Avenue station; however, it was not as clear at the Washington, DC, station.

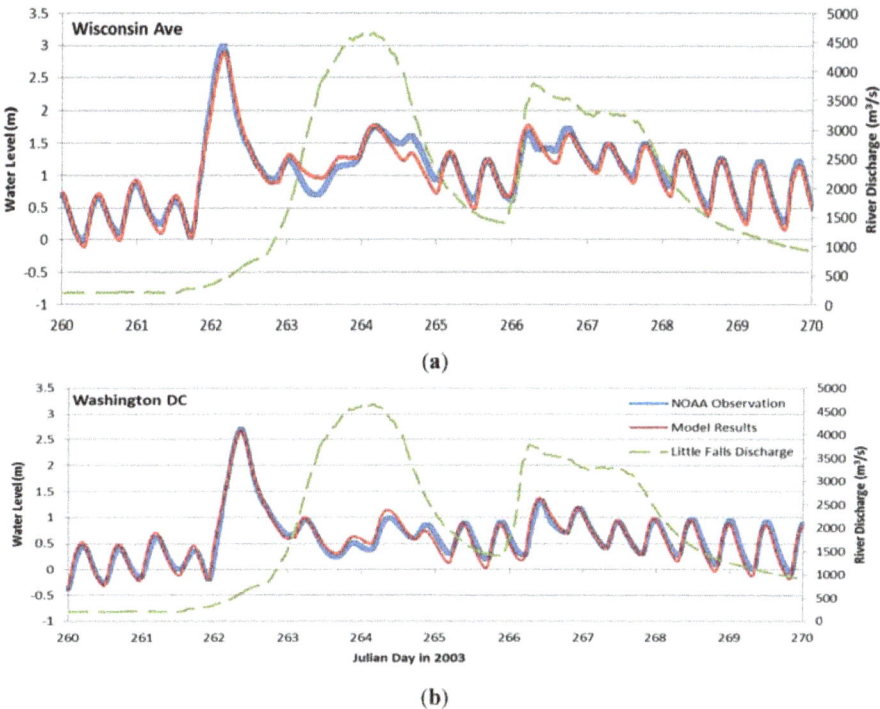

Figure 5. The 2003 Hurricane Isabel model simulation results for (a) Wisconsin Avenue and (b) for Washington, DC, compared with the gauge measurement. Model results are shown in red and observation in blue; also included is the Potomac River discharge at Little Falls, MD, shown in green (with the scale on the right-hand side scale).

With a close examination of the results presented in Figures 5b and 6b at the Washington, DC, station, it was found that the "without" river discharge scenario resulted in a decreasing water level by only 0.2 m and 0.1 m for the second and third peak, respectively, at the Washington, DC, station when the "with" and "without" river discharge scenarios were compared. This was approximately one fifth of the water elevation under-prediction occurring at Wisconsin Avenue. Our explanation regarding the fact that the same high river floods have less influence on the water level at that Washington, DC, station than that of the station at Wisconsin Avenue was attributed to the following two factors: (1)

significant widening of the Potomac River channel downstream of Roosevelt Island; the width at the confluence of Potomac and Anacostia is 3.5-times that near Wisconsin Avenue; and (2) expansion of the flood plains during floods. The Potomac River downstream of Roosevelt Island, including the Tidal Basin, Washington Channel, East Potomac Park, Virginia Park and Reagan Airfield, is a vast area of low lying lands with channels and land intertwined. The region acts like a major floodplain during the floods and can assimilate a large volume of river discharge by expanding the inundation region in several directions horizontally. The extent of the inundated area during the 1936 Potomac Great Flood can be found later in Section 5.2, which showed the vast expansion of flood zone downstream of the Roosevelt Island.

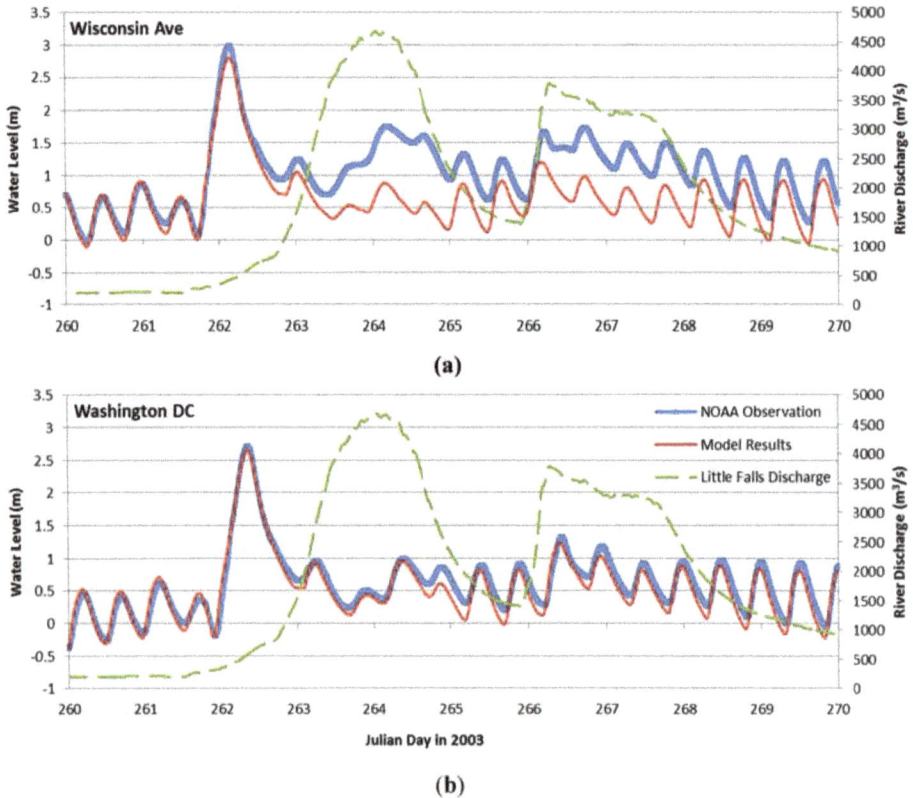

Figure 6. The 2003 Hurricane Isabel model simulation results for (a) Wisconsin Avenue and (b) for Washington, DC, similar to Figure 5, except using zero upstream river discharge at Little Falls, MD. Although not used as an upstream boundary condition, the Potomac River discharge (in green) at Little Falls, MD, was retained for reference.

A spatial inundation map for the flooding associated with Hurricane Isabel was also generated at the City of Alexandria, where a substantial area of the docks and business district along the waterfront of the Potomac River experienced extensive flooding, as shown in Figure 7. There is no official gauge measurement available in the region for a rigorous verification, but a picture taken by a citizen observer at the King Street and Union Street intersection right after the storm, as shown in the upper left corner, provided an interesting validation. It is visible and can be identified in the picture that the water marks reach 10.2 feet at the wall (marked by the red line). After examining the wrack line from several pictures taken at the same location, but at different times, we came to the conclusion that the water level was between 4 and 6 feet (1.2 m–1.8 m) above the ground during the peak height of the flooding, which was consistent with the model-produced peak inundation of 4.9–6.6 feet (1.5–2.0 m) at the site. Similar results were reported by Stamey et al. (2007) [25].

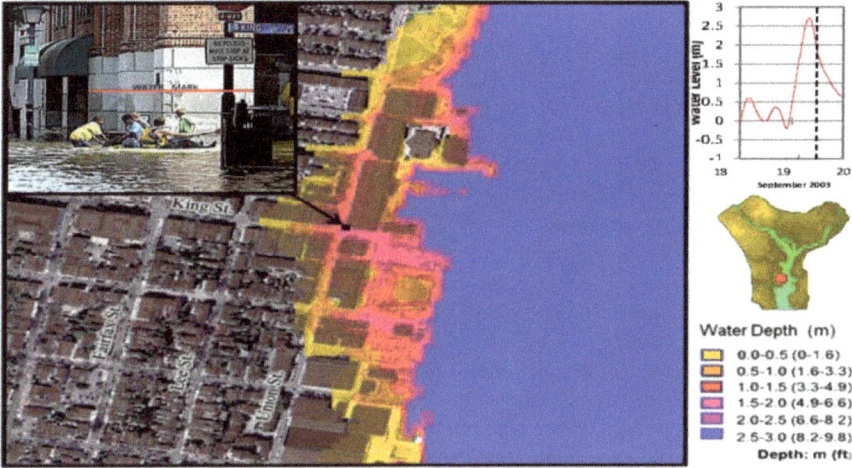

Figure 7. The modeled 2003 Hurricane Isabel-induced inundation in the City of Alexandria, VA. The photograph depicts a high water mark at the intersection of King Street and Union Street showing that the floodwater reached 10.2 feet, approximately 5.5 feet (1.7 m) above the ground level, which is consistent with the modeled inundation at the location, which is between 4.9 and 6.6 feet (1.5–2.0 m).

5. Inundation Simulation for the Potomac River Great Flood of 1936

In March 1936, a combination of warmer-than-normal temperatures and torrential rain after a cold and snowy winter resulted in rapid melting of snow and rainfall runoff in much of the northeastern US, triggering the historic 1936 Great Flood. The result was significant flooding in much of the northeastern region of United States, ranging from the Ohio River Valley, New England and south to Washington, DC in the Potomac River Basin, prompting to the passage of Flood Control Act of 1936 by the Congress [26]. The damage was considerable along the Potomac River, ranging from Harpers Ferry, WV, to Hancock, MD, with significant flooding at Washington, DC. The water level height recorded at four miles (6.4 km) above Chain Bridge was 8.75 m, while at the NOAA station in Washington, DC was 5.70 m [27]. The peak river discharge during the Great Flood in Washington, DC, was observed to be 14,500 m^3/s, which is 39-times of the normal daily flow.

5.1. Model Setup

To drive the inundation model, hourly discharge data obtained from the USGS station at Little Falls, MD, were used. Since only the hourly tidal water level at the Washington, DC, gauge station was available at the time of the flood in the Tidal Potomac River, the downstream open boundary condition was established by using the time series of water level measured at the Washington, DC, gauge station and extrapolating downstream to Colonia Beach, VA, by adjusting the tidal phase by 360 min advance in time. The Great Falls of the Potomac River mark the geological boundary whereby the elevation of the water rapidly changes, and thus, the boundary condition should account for both the kinetic and potential energy. The friction parameters used are obtained from the calibration over an independent astronomical tide dataset. Both the boundary condition and the selection of friction parameter were described in Section 4.1. The simulation period was 00:00 GMT March 01 to 00:00 GMT 31 March 1936.

The 20 days simulation including a five-day spin up took about 4 h of real time in a seven processor PC to finish.

5.2. Results

For the illustration of the 1936 Potomac Great Flood results, it should be noted that, only the Washington, DC station had measurement data at the time. The station locations selected for time series comparison are two: the present day USGS Wisconsin Avenue station and the long-term Washington, DC waterfront station. The modeled water level from 10 March through 25 March of 1936 were shown for 15 days' simulation in Figure 8a,b. In order to demonstrate the sub-grid modeling capability, two types of model results were presented—"With" the sub-gird (shown in red) and "without" the sub-grid (shown in gray). For "without sub-grid", the grid used is a 200 m × 200 m base grid, and the topography is the average of the bare ground over the grid size. The green line superposed on the model results was the observed river discharge obtained from USGS Little Falls, MD, with a unit of m^3/s using the scale on the right-hand side.

From the time series, one can observe two major temporal variabilities of the water level: one is a low frequency variation and the other the tidal frequency. For the low frequency component, it was seen that the variation is quite consistent with that of the river discharge (marked by the green line), an indication that those are river-induced water level variations. On the tidal frequency component, it was revealed that the "with" the sub-grid simulation did very well in terms of both amplitude and phase, but there is a problem using the "without" sub-grid approach associated with the tidal phase. The statistics of the time series comparison were given in Table 2. The R^2, RMS and peak difference were 0.98, 5.8 cm and 2.9 cm for the "with" sub-grid approach but 0.77, 41 cm and 23.8 cm for the "without" sub-grid approach at the Washington gauge station. The errors of using the "without" sub-grid approach were almost 8-times larger, and the R-squared drops below 0.8. The mismatch of the phase was well-documented in the USGS and NOAA's prior efforts. The fact that the "without" sub-grid approach encountered similar problems in producing the incorrect tidal phase, but can be overcome, highlighted the power of the high-resolution sub-grid approach and the nonlinear solver it uses. In terms of uncertainties in the model-data comparison, we feel that the sub-grid approach has reduced the large errors imbedded in the "without sub-grid" approach to the point that it reached the inherent error associated with the datum selection and equipment measurement itself at about 2–7 cm, as shown in Table 2. The comparison of water level and river discharge time series revealed that the peak water level can reach Washington, DC, with very little delay from the time when the flood peak passes the fall line. Having over several million people living in the metropolitan area, this means that Washington, DC, will have very little time to prepare and evacuate for a flash river flooding without a proper early warning system.

Table 2. Statistical comparison of modeled time series results with and without a 10-m sub-grid at Washington, DC, during the 1936 Potomac River Flood.

Statistic	"With" Sub-Grid	"Without" Sub-Grid
R^2	0.98	0.77
RMS	5.8 cm	41.0 cm
MAE	3.7 cm	36.0 cm
Peak Difference	2.9 cm	23.8 cm

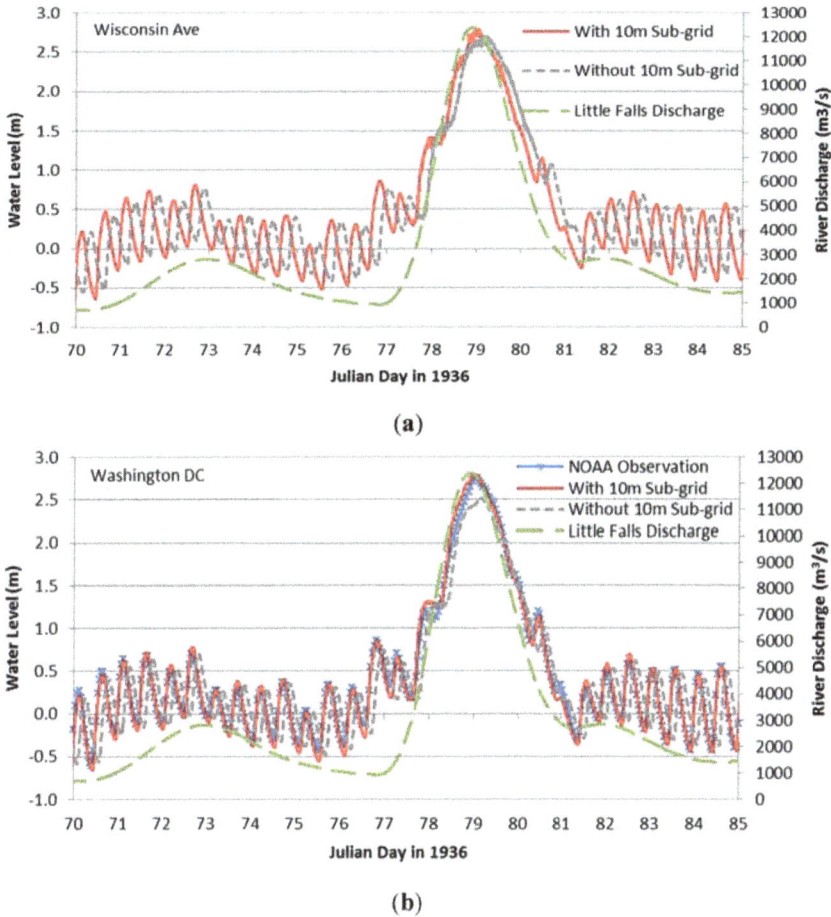

Figure 8. Time series plots comparing modeled results for (**a**) Wisconsin Avenue and (**b**) for Washington, DC "with" 10 m × 10 m sub-grids (red) and "without" sub-grids (gray dashed line) during the 1936 Potomac River Great Flood. The comparison was made at the Wisconsin Avenue (top) and Washington, DC, stations (bottom). The observation record was available only at the Washington, DC, station; river discharge from Little Falls, MD, is superposed (green) for reference.

A snapshot of the velocity/elevation distribution from the animation is shown in Figure 9. The velocity vector superimposed with the water level at the peak of the flooding highlights that the river bank north of East Potomac Park near DC was flooded over by the large (>2.7 m/s) water velocity deflected from Roosevelt Island. As a result, the water flooded eastward and southward to form the crescent shape of the inundated area through downtown DC. The spatial extent of the flooded area in downtown DC was verified from the historic records collected by the U.S. Army Corps of Engineers and archived by the National Capital Planning Commission (2008) [28], as shown in the right panel of Figure 10. The inundation simulation also showed that the flooded area was widespread in the Washington, DC, metropolitan area, including East Potomac Park and Golf Course, Washington Harbor, the Washington Navy Yard in southeast DC and areas across the river in the southern bank, as shown in the left panel of Figure 10. Table 3 shows the area of the inundation (in square km) in various

J. Mar. Sci. Eng. **2015**, *3*, 607–629

locations around Washington, DC, obtained from the model results. What the model has provided is essentially a reconstruction of a detailed historical flooding map of the 1936 Potomac River Great flood.

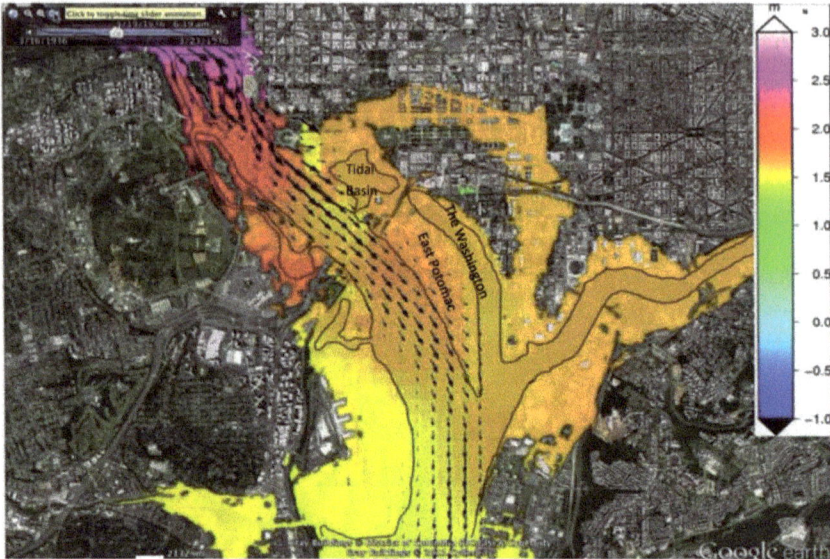

Figure 9. Visualization of the velocity vectors and water level (background color) from sub-grid model simulation results for the Washington, DC, metropolitan area during the 1936 Potomac River Great Flood. The shoreline is shown superposed in black. It is revealed that at the height of the flooding, the river bank north of East Potomac Park near DC was pinched by large (>2.7 m/s) velocities deflected from Roosevelt Island and subsequently flooded the downtown area.

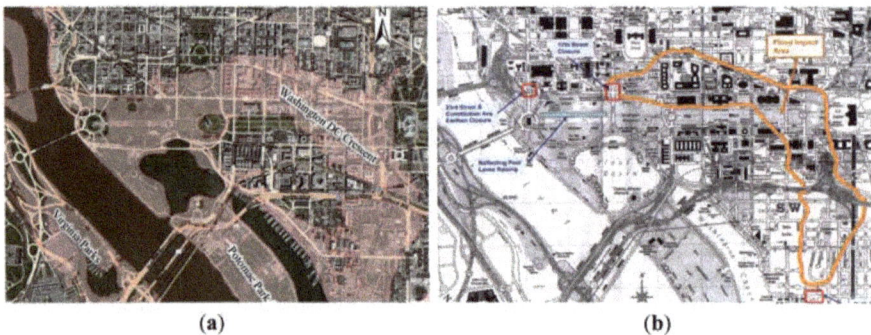

(a)   (b)

Figure 10. Modeled maximum inundation extent for the Greater Washington, DC (**a**), and surveyed downtown DC flood area (**b**) during the 1936 Potomac River Great Flood.

The responses of the Upper Tidal Potomac River to the 1936 Potomac River Great Flood and the 2003 Hurricane Isabel were different. The time series plots in Figure 8 during the 1936 flood showed no tidal signal during the four-day peak of the flood, an indication that the tide was overwhelmed by the river discharge, such that it disappeared completely from the Washington, DC, station. This is the result of river discharge 3–4-times greater than that during Hurricane Isabel in 2003. The hurricane-induced storm surge, on the other hand, originated in the Chesapeake Bay and propagated up river as a surge wave toward the fall line, with Washington, DC, in its path. Due to the long distance it propagated, the strength and the speed could be dissipated by the shallow embayment and narrow channels, and thus, some warnings can be obtained from the downstream stations. More importantly, it lasted only for a few hours and quickly receded; thus, the flood gate and coastal levees could probably hold up without the worry of breaching and backwash by the rainwater from precipitation. Comparatively, Washington, DC, is much more vulnerable to a river flash flood carried from the Upper Potomac River Basins across the fall line into the Tidal Potomac River. It can be directly hit by the enormous momentum and the sediments that the flood carries in a short travel distance from the fall line within one-half hour and can last for several days. One key element by which the riverine flood differs from storm surge is in that it has a significant magnitude of velocity with persistent uni-direction flows going downstream that can continuously scour the bank for days and potentially breach vulnerable spots of the shoreline. When this happens, the enormous water volume will be diverted onto the land through the pinched point, as occurred in the 1936 Potomac Great Flood, which flooded downtown Washington, DC.

Table 3. Model simulated inundation region in different parts of the DC area during the 1936 Potomac Great Flood. The individual (top) and total square km and miles (bottom) are listed.

Modeled Flood Area	m^2	km^2	mi^2
Potomac Park & Golf Course	3,118,210.81	3.12	1.20
Washington DC Crescent *	2,466,778.05	2.47	0.95
Washington Harbor	1,167,493.83	1.17	0.45
DC Naval Yard	633,843.92	0.63	0.24
Reagan Airfield	1,819,267.96	1.82	0.70
Virginia Parks	1,778,806.55	1.78	0.69
Anacostia-Bolling Base & Park	1,632,464.54	1.63	0.63
Total	12,616,865.65	12.62	4.87

* DC Crescent Modeled Flood Area	m^2	km^2	mi^2
Upper Crescent	1,815,294.67	1.82	0.70
Lower Crescent	651,483.375	0.65	0.25
Total	2,466,778.05	2.47	0.95

Note: * A potential flood zone area identified in downtown Washington DC from 17th street and Constitution avenue east to the capital and south toward Fort McNair, which is protected by the National Mall levee [28].

6. Discussion and Conclusions

The forecast for Potomac River water level exceeding flooding stage and its potential impacts on metropolitan Washington, DC, require multi-discipline cooperation across meteorology, hydrology, hydraulics and coastal hydrodynamics. A service gap exists between the missions of NOAA NOS (National Ocean Service), USGS (U.S. Department of the Interior) and NWS (National Weather Service), resulting in the lack of an operational forecast system for predicting real-time water level in the Washington, DC, area. The NWS Middle Atlantic River Forecast Center presently operates an Advance Hydrologic Prediction Service (AHPS), which issues probabilistic forecasts of river discharge three days in advance at the fall line. Traditionally, the processes being considered are precipitation, snow melting, ground water flow, aquifer discharge, evaporation and flood wave routing. Efforts have been made to incorporate tide and wind, but with shortcomings and limited success. Given sub-grid modeling's breakthrough, it is recognized that there are three elements that play a pivotal role in making progress: sufficiently large modeling domain downstream, the sub-grid approach incorporating LiDAR and high-resolution bathymetry data and the application of the nonlinear solver. Based on our benchmarking using a high-end PC, the sub-grid inundation model can finish within one-half hour for making a three-day forecast, which should be sufficient to meet AHPS's operational criteria. Our results suggest integrating the sub-grid model technology with the real-time

J. Mar. Sci. Eng. **2015**, *3*, 607–629

river discharge forecast for predicting inundation under both storm surge and riverine floods for the Washington, DC, metropolitan area is feasible.

The sub-grid modeling differs significantly from the popular "bathtub model" used by many geographic information system analysts in that the water level of the present model is not homogeneous everywhere in the domain, and there is a pressure gradient force derived from the governing equation to drive the flow. The flow velocity field could be very important, especially when the situation could affect the stability of the shoreline and local structures. It is also not a steady-state model; the time varying nature of the water level, such as affected by tides and winds, is calculated each time step rather than treated as a static variable. On the historical 1936 Potomac River Great Flood, although the simulation of the inundation using recently acquired LiDAR data was a success and demonstrated the potential of the sub-grid technology, notable caution is needed in interpreting the model results, since the urban landscape of 1936 may have been quite different from that of the 2000s.

The model as it stands now has the capability to simulate the effect of a structure in the sub-grid scale as long as the structure can be aligned with the side of the grid. However, one of the limitations of the model applied for this application is that the grid has to be orthogonal (with respect to the circumcenter), and many structures do not always align with the grid side. Furthermore, the finite volume scheme used presently to construct the sub-grid approach is limited to second order accuracy. It will be desirable to incorporate sub-grid features in the higher-order scheme, such as the finite element scheme, to further enhance the model accuracy for the shallow water region. Lastly, neither the effect of wind wave nor the effect of flood gate and coastal control structures operations were considered in this study, which is beyond the scope of this effort.

In summary, the hydrodynamic modeling of major storm surge and inundation events, such as the 2003 Hurricane Isabel and the 1936 Potomac River Great Flood, provided us the opportunity to experiment with state-of-the-art sub-grid modeling techniques that incorporate high-resolution LiDAR topography and bathymetry data for making efficient and accurate simulations for the storm surge and inundation. The study is critical to improving the understanding of the processes that lead to major flooding preventative measures that can potentially mitigate the loss of property and human life for the coastal community. Although the model could simulate the storm surge for Hurricane Isabel remarkably well, further testing could be performed with different storm surge cases, such as the 2009 November Nor'easter and the 2011 Hurricane Irene in Chesapeake Bay, and for additional study sites to further validate the capability of the sub-grid model approach. Additionally, the model may warrant further testing with the coupled inclusion of precipitation, infiltration processes [29,30] and a general framework of the interaction between the surface and sub-surface flow to improve model simulation in more complicated land use and flooding caused by rainfall. When coupled with a Google Earth interface for a true street-level inundation simulation, this can be a powerful tool serving as a flood early warning system for emergency managers, city administrators, policy-makers, scientists and the general public alike. Since learning from the lessons of the historical events is extremely useful and informative, the velocity and water level animation and the spatial distribution of the maximum extent of flooding of the 1936 Potomac River Great Flood were produced and shared by one of the co-authors, Dave Forrest. The same results have been presented by Smith (2012) [31] at the Metropolitan Washington Council of Governments climate seminar in the past. The two animation files are available in .mov format at VIMS Physical Sciences web link (2015) [27].

Acknowledgments: The authors would like to acknowledge NOAA Tides and Currents' public domain site from which the historical water level elevation and meteorological observations were accessed and used in driving the model and to verify and validate the model presented in this study. The authors also would like to give thanks to Noblis, Inc., for support on modeling grid construction and providing the visualization capability in the Chesapeake Bay Inundation Project.

Author Contributions: The sub-grid modeling effort is a product of the intellectual environment of H.V. Wang and J.D. Loftis, and each has contributed in various capacities to the research efforts and execution of these numerical experiments. D. Forrest, W. Smith, and B. Stamey contributed to the data collection, visualization, animation and analysis of the model results.

Conflicts of Interest: The authors declare no conflict of interest.

Abbreviations

USACE	US Army Corps of Engineers, Department of Defense
NOAA Co-OPS	National Oceanic and Atmospheric Administration, Center for Operational Oceanographic Products and Services
NAVD88	North America vertical datum of 1988
NAD83 CORS96	North America datum of 1983, readjusted based on the Continuous Operating Reference Stations (CORS)

References

1. Heaps, N.S. On the numerical solution of the three-dimensional hydrodynamic equations for tides and storm surge. *Mem. Soc. R. Soc. Sci. Liège Coll Huit* **1972**, *2*, 143–180.
2. Reid, R.O.; Bodin, R.O. Numerical model for storm surges in Galveston Bay. *J. Waterw. Harbor Div.* **1968**, *94*, 33–57.
3. Davis, A.M. Three-dimensional modelling of surge. In *Flood due to High Winds and Tides*; Peregrine, D.H., Ed.; Academic Press: London, UK, 1981; pp. 45–74.
4. Jelesnianski, C.P.; Chen, J.; Shaffer, W.A. *SLOSH: Sea, Lake, and Overland Surges from Hurricanes*; National Weather Service: Silver Spring, MD, USA, 1992.
5. Kerr, P.C.; Donahue, A.S.; Westerink, J.J.; Luettich, R.A., Jr.; Zheng, L.Y.; Weisberg, R.H.; Huang, Y.; Wang, H.V.; Teng, Y.; Forrest, D.R.; et al. US IOOS coastal and ocean modeling testbed: Inter-model evaluation of tides, waves, and hurricane surge in the Gulf of Mexico. *J. Geophys. Res.* **2013**, *118*, 5129–5172.
6. Chen, C.; Beardsley, R.C.; Luettich, R.A., Jr.; Westerink, J.J.; Wang, H.; Perrie, W.; Xu, Q.; Donahue, A.S.; Qi, J.; Lin, H.; et al. Extratropical storm inundation testbed: Intermodel comparisons in Situate, Massachusetts. *J. Geophys. Res. Oceans* **2013**, *118*, 5054–5073.
7. Roland, Aron; Zhang, Y.; Wang, H.V.; Meng, Y.; Teng, Y.; Maderichd, V.; Brovchenkod, I.; Dutour-Sikirice, M.; Zankea, U. A fully coupled 3D wave-current interaction model on unstructured grids. *JGR Oceans* **2012**, *117*, C00J33. [CrossRef]
8. Neelz, S.; Pender, G. Sub-gird scale parameterisation of 2D hydrodynamic models of inundation in the urban area. *Acta Geophys.* **2007**, *55*, 65–72. [CrossRef]
9. Casas, A.S.; Lane, N.; Yu, D.; Benito, G. A method for parameterizing roughness and topographic sub-grid scale effects in hydraulic modelling from LiDAR data. *Hydrol. Earth Syst. Sci. Discuss.* **2010**, *7*, 2261–2299.
10. Wang, H.V.; Loftis, J.D.; Liu, Z.; Forrest, D.; Zhang, J. The storm surge and sub-grid inundation modeling in New York City during Hurricane Sandy. *J. Mar. Sci. Eng.* **2014**, *2*, 226–246.
11. United States Geological Survey (USGS). Part3. Potomac, James, and Upper Ohio Rivers (Water-Supply Paper 800). In *The floods of March 1936*; United States Geological Survey: Denver, CO, USA, 1937.
12. Schaffranek, R. *A Flow Simulation Model of the Tidal Potomac River—A Water-Quality Study of the Tidal Potomac River and Estuary*; United States Geological Survey: Denver, CO, USA, 1987; p. 2234-D.
13. Mashriqui, H.S.; Halgren, J.S; Reed, S.M. Toward Modeling of river-estuary-ocean interactions to enhance operational river forecasting in the NOAA National Weather Service. In Proceedings of the 2nd Joint Federal Interagency Conference, Las Vegas, NV, USA, 27 June–1 July 2010.
14. EA Engineering, Science, and Technology, Inc. *Water Quality Studies in the Vicinity of the Washington Aqueduct*; Baltimore District, US Army Corps of Engineers, Washington Aqueduct Division: Washington, DC, USA, October 2001.
15. Casulli, V. A semi-implicit finite difference method for non-hydrostatic, free-surface flows. *Int. J. Numer. Methods Fluids* **1999**, *30*, 425–440. [CrossRef]
16. Casulli, V.; Walters, R.A. An unstructured grid, three-dimensional model based on the shallow water equations. *Int. J. Numer. Methods Fluids* **2000**, *32*, 331–348. [CrossRef]
17. Casulli, V.; Zanolli, P. High resolution methods for multidimensional advection-diffusion problems in free-surface hydrodynamics. *Ocean Model.* **2005**, *10*, 137–151. [CrossRef]
18. Casulli, V. A high-resolution wetting and drying algorithm for free-surface hydrodynamics. *Int. J. Numer. Methods Fluid Dyn.* **2009**, *60*, 391–408. [CrossRef]

19. Casulli, V.; Stelling, G. Semi-implicit sub-grid modeling of three-dimensional free-surface flows. *Int. J. Numer. Methods Fluid Dyn.* **2011**, *67*, 441–449. [CrossRef]
20. Loftis, J.D.; Wang, H.V.; Hamilton, S.E.; Forrest, D.R. Combination of LiDAR Elevations, Bathymetric Data, and Urban Infrastructure in a Sub-Grid Model for Predicting Inundation in New York City during Hurricane Sandy. *Comput. Environ. Urban Syst.* **2015**. under review.
21. Casulli, V.; Zanolli, P. A three-dimensional semi-implicit algorithm for environmental flows on unstructured grids. In Proceedings of the Conference on Methods for Fluid Dynamics, University of Oxford, Oxford, UK; 1998; pp. 57–70.
22. Hederson, F.M. *Open Channel Flow. 1996*; The Macmillan Company: New York, NY, USA; p. 522.
23. Garratt, J.R. Review of drag coefficients over oceans and continents. *Mon. Weather Rev.* **1977**, *105*, 915–929. [CrossRef]
24. Powell, M.D.; Vickery, P.J.; Reinhold, T.A. Reduced drag coefficient for high wind speeds in tropical cyclones. *Nature* **2003**, *422*, 279–283. [CrossRef] [PubMed]
25. Stamey, B.; Wang, H.V.; Koterba, M. Predicting the Next Storm Surge Flood. *Sea Technol.* **2007**, *48*, 10–15.
26. Arnold, J.L. *The Evolution of the Flood Control Act of 1936*; United States Army Corps of Engineers: Fort Belvoir, VR, USA, 1988.
27. VIMS Physical Science. 1936 Potomac River flood simulation. Available online: http://web.vims.edu/physical/3DECM/DC19360301/ (accessd on 11 July 2015).
28. National Capital Planning Commission (NCPC). Flooding and Stormwater in Washington, DC, 2008. Available online: http://www.ncpc.gov/DocumentDepot/Publications/FloodReport2008.png (accessed on 17 January 2008).
29. Loftis, J.D.; Wang, H.V.; DeYoung, R.J.; Ball, W.B. Integrating LiDAR Data into a High-Resolution Topo-bathymetric DEM for Use with Sub-Grid Inundation Modeling at NASA Langley Research Center. *J. Coast. Res.* **2015**, in press.
30. Loftis, J.D. Development of a Large-Scale Storm Surge and High-Resolution Sub-Grid Inundation Model for Coastal Flooding Applications: A Case Study during Hurricane Sandy. Ph.D. Thesis, College of William & Mary, Williamsburg, VA, USA, 2014.
31. Smith, W. Climate Change Symposium, Washington Metropolitan Council of Governments, May 21, 2012. Available online: http://www.mwcog.org/environment/climate/adaptation/Presentations/5-%20Smith.png (accessed on 22 May 2012).

Journal of
Marine Science and Engineering

MDPI

Article

Observations and Predictions of Wave Runup, Extreme Water Levels, and Medium-Term Dune Erosion during Storm Conditions

Serge Suanez [1,*], Romain Cancouët [1], France Floc'h [2], Emmanuel Blaise [1], Fabrice Ardhuin [3], Jean-François Filipot [4], Jean-Marie Cariolet [5] and Christophe Delacourt [2]

[1] LETG-Brest-Géomer UMR 6554 CNRS, Institut Universitaire Européen de la Mer, Rue Dumont d'Urville, Plouzané 29280, France; romain.cancouet@univ-brest.fr (R.C.); emmanuel.blaise@univ-brest.fr (E.B.)
[2] Laboratoire Domaines Océaniques (LDO) UMR 6558 CNRS, Institut Universitaire Européen de la Mer, Rue Dumont d'Urville, Plouzané 29280, France; france.floch@univ-brest.fr (F.F.); christophe.delacourt@univ-brest.fr (C.D.)
[3] Laboratoire de Physique des Océans (LPO) UMR 6523 CNRS-Ifremer-IRD, Institut Universitaire Européen de la Mer, Rue Dumont d'Urville, Plouzané 29280, France; Fabrice.Ardhuin@ifremer.fr
[4] France Energies Marines, 15 rue Johannes Kepler, Site du Vernis, Technopôle Brest-Iroise, Brest 29200, France; jean.francois.filipot@france-energies-marines.org
[5] Lab'Urba—EIVP, Université Paris Est, 80 rue Rébeval, Paris 75019, France; jean-marie.cariolet@eivp-paris.fr
* Author to whom correspondence should be addressed; serge.suanez@univ-brest.fr; Tel.: +33-02-98-498-610; Fax: +33-02-98-498-703.

Academic Editor: Rick Luettich
Received: 13 June 2015; Accepted: 13 July 2015; Published: 24 July 2015

Abstract: Monitoring of dune erosion and accretion on the high-energy macrotidal Vougot beach in North Brittany (France) over the past decade (2004–2014) has revealed significant morphological changes. Dune toe erosion/accretion records have been compared with extreme water level measurements, defined as the sum of (*i*) astronomic tide; (*ii*) storm surge; and (*iii*) vertical wave runup. Runup parameterization was conducted using swash limits, beach profiles, and hydrodynamic (H_{m0}, $T_{m0,-1}$, and high tide water level—HTWL) data sets obtained from high frequency field surveys. The aim was to quantify *in-situ* environmental conditions and dimensional swash parameters for the best calibration of Battjes [1] runup formula. In addition, an empirical equation based on observed tidal water level and offshore wave height was produced to estimate extreme water levels over the whole period of dune morphological change monitoring. A good correlation between this empirical equation ($1.01H_{m0}\xi_o$) and field runup measurements (R_{max}) was obtained (R^2 85%). The goodness of fit given by the RMSE was about 0.29 m. A good relationship was noticed between dune erosion and high water levels when the water levels exceeded the dune foot elevation. In contrast, when extreme water levels were below the height of the toe of the dune sediment budget increased, inducing foredune recovery. These erosion and accretion phases may be related to the North Atlantic Oscillation Index.

Keywords: macrotidal beach; runup; storm; dune; erosion; extreme water level; NAO

1. Introduction

Extreme events such as storms or hurricanes play a major role in dune erosion [2–7]. In these conditions, the foredune is severely scarped due to flooding processes that exacerbate wave attack on the dune foot [6,8–17]. Based on this principle, Sallenger et al. [18] and Ruggiero et al. [19] proposed different models designed to assess the foredune's sensitivity to erosion generated by the impact of storm waves. This methodological approach was used for assessing the vulnerability of barrier islands to hurricanes along the eastern coast of the USA [7,18] and analyzing decadal-scale variations in dune erosion and accretion rates on the Sefton coast in northwest England [20]. These models examine the relationship between the extreme water level elevation and relevant beach morphology corresponding to the height of dune foot. Extreme water level is defined as the sum of (*i*) astronomic tides; (*ii*) storm surges; and (*iii*) vertical wave runup, including both setup and swash. Sallenger [18] has defined four storm-impact regimes (*swash, collision, overwash*, and *inundation*) related to increased water levels from storms that shift the runup and location of wave attack higher on the profile, making berms or foredunes more vulnerable to erosion and overtopping. In this storm-impact scaling model, the borders between the impact regimes represent thresholds across which the magnitudes and processes of dune erosion are substantially different. Ruggiero's [19] model appears simpler; it simply examines predicted extreme water elevations with measured elevations of the junctions between the beach face and the toe of foredunes or sea cliffs. The aim is to evaluate the frequency with which water can reach the property, providing an evaluation of the susceptibility to potential erosion.

If storm surge (wind and pressure surge) can be deduced from the observed tide using tide gauge measurements, estimation of wave runup is a more complicated issue because of the complex processes driving the swash zone [21]. It corresponds to the time-fluctuating vertical position of the swash limit on the upper part of the beach, and was first studied in relation to engineering structures such as dykes [22] or rock-rubble structures [23]. It is defined as the difference between discrete water elevation maxima and still water level corresponding generally to observed tide level [21,24,25]. The complexity of processes that govern the swash zone are related to incident band wave energy transferred to both higher and lower frequencies through the surf zone [26]. Therefore, wave runup is largely dependent on environmental conditions such as the local beach slope (synthesized through dissipative to reflective context generally given by Iribarren number [27]) and the infragravity-to-incident offshore wave energy which dominates the inner-surf zone [24,28–30]. A simple formula was first proposed by [23] using significant wave height (H_s) and slope (S):

$$R = H_s S \tag{1}$$

Battjes [1,27] has shown that runup was better related to a morphodynamic component defined by a dimensionless surf similarity parameter called the Iribarren number, expressed by the following equation:

$$\frac{R}{H_s} = C\xi_o \tag{2}$$

where C is a constant, and ξ_o is the Iribarren number given by [17]:

$$\xi_o = \frac{\tan\beta}{(H_o/L_o)^{1/2}} \tag{3}$$

where *tanβ* is the beach slope, H_o corresponds to significant offshore wave (equivalent to H_s in deep water), and L_o is deep water wavelength.

Following this approach, a statistical analysis of wave run-up ($R_{2\%}$) was proposed by Holman from field data collected on a natural intermediate-to-reflective beach (beach slope *tanβ* from 0.07 to 0.2) [25]. He found a clear relationship between the 2% exceedence value of runup normalized by H_s and ξ_o, and fit this equation to field data collected at Duck, NC (USA) using the intermediate depth (18 m) H_{m0} and T_{pic} (where H_{m0} is wave height estimates based on spectral moments, and T_{pic} is the period associated with the largest wave energy known as the peak period). Based on the laboratory tests, Mase [31] developed a predictive equation using deep water wave parameters for irregular wave runup on uniform impermeable slopes (*tanβ* from 0.03 to 0.2). He found that the runup was approximately twice as large as values measured in the field by Holman [25], and explained this discrepancy by the effect of beach profile geometry. The runup spectrum measured on natural sandy beaches on the coast of New South Wales (Australia) indicated proportionality between the best-fit of runup elevation distribution and the beach slope for a steeper beach (*tanβ* $\geq$ 0.10). For the flatter beaches (*tanβ* $\leq$

0.10), the slope became largely unimportant and the vertical scale of the runup distribution was scaled directly with $(H_0L_0)^{0.5}$ [32]. Ruessink et al. [29] came to the same conclusion by examining runup under highly dissipative conditions (beach slope from 0.01 to 0.03) at Terschelling (The Netherlands). They found that the significant infragravity swash height (R_{ig}) was about 30% of the offshore wave height H_0, and that the slope in the linear H_0 dependance of R_{ig} amounted to only 0.18, considerably smaller than the value of 0.7 observed on steeper beaches by Guza and Thornton [24]. More recently, a synthesis of empirical parameterization of extreme $R_{2\%}$ runup, based on several natural beach and laboratory experiments, indicated that in an infragravity-dominated dissipative context, the magnitude of swash elevation was dependent only on offshore wave height and wavelength [21]. In an intermediate and reflective context with complex foreshore morphology, beach slope was on the contrary much more important in practical applications of the runup parameterization. Therefore, the authors have elaborated different runup equations according to the beach morphodynamic context. For a dissipated state ($\xi_o < 0.3$), Formula (4) is used, while for an intermediate state ($0.3 < \xi_o < 1.25$) it is recommended to use Formula (5). Formula (6) is used for a reflective state ($\xi_o > 1.25$):

$$R_{2\%} = 0.043 \, (H_0L_0)^{1/2} \qquad (4)$$

$$R_{2\%} = 1.1 \left(0.35\beta_f(H_0L_0)^{1/2} + \frac{\left[H_0L_0(0.563\beta_f^2 + 0.004)\right]_{1/2}}{2} \right) \qquad (5)$$

$$R_{2\%} = 0.73\beta_f \, (H_0L_0)^{1/2} \qquad (6)$$

where $R_{2\%}$ corresponds to the height reached by 2% of the highest runups, β_f is the slope calculated by the whole length of the upper part of the beach, and H_0 and L_0 are deepwater wave height and wavelength, respectively.

In a recent study, a methodological approach for calculating runup from the analysis of morphodynamic conditions on macrotidal sandy beach in Vougot (Brittany, France) was published [33]. The goal of this work was to improve simple parameterization for a maximum runup elevation based on the earlier empirical formula produced by Battjes [1]. The method was based on field measurements of wrack lines related to the highest high-tide swash runup elevation and the analysis of morphological and hydrodynamic conditions. This allowed us to calibrate runup formula effectiveness on a macrotidal sandy beach and to determine the best slope parameters to estimate runup in this coastal environment that has a tidal range of about 7 m. The results suggest that on the macrotidal sandy beach, the slope of the active section of the upper beach should be used to obtain the most relevant estimation of observed runup elevations (Figure 1). The work presented in this paper extends the analysis of runup on the same study site (Vougot beach in north western Brittany) in order to estimate extreme water levels. Based on the Sallenger [18] and/or Ruggiero [19] models, the aim is to evaluate the frequency with which these extreme water levels have reached the toe of the dunes, providing an evaluation of the susceptibility to potential erosion. First, a new parametrization of the runup equation was accomplished following the same methodological approach as Cariolet and Suanez [33]. This analysis was based on a new data set obtained between June 2012 and June 2013 and includes the one used in the previous study [33]. Secondly, calibration of a general empirical formula based on tide and offshore wave measurements was achieved in order to predict extreme water levels over the last decade (2004–2014). Thirdly, the relationship between the elevation of extreme water levels and relevant beach morphology (in this case the toe of the dune) was analyzed from 2004 onwards, this being the period during which the survey of dune morphological changes started. The aim was to identify and explain the dune system's phases of erosion and recovery related to long term meteo-oceanic condition variations. Emphasis was put on storms events causing erosion and retreat of dune fronts.

2. Geomorphological and Hydrodynamic Setting

The study area is the Vougot beach located on the North coast of Finistère in Brittany (France) (Figure 2). The general morphological setting comprises large rocky outcrops representing the submerged part of the Léon plateau. Contact between the coastal platform and the continental part of the plateau consists of a partly tectonic scarp 30 to 50 m high. In the Vougot beach area, the scarp is disconnected from the sea by the existence of a dune which was formed during the Holocene [34]. This dune, anchored on the Zorn abandoned cliff, stretches over about 2 km in a southwest to northeast direction (Figure 2b). It culminates at an altitude of 13 m (NGF) (i.e., above sea level—asl); the altimetric reference NGF refers to French datum. In our case this reference is situated 3.5 m above the lowest astronomic tide level (LAT). It represents a massive dune complex 250 to 400 m wide. Over the last decades, the dune of Vougot beach has experienced erosion. A historical shoreline change analysis based on a series of aerial photographs and field measurements from 1952 to 2014 shows that the retreat of the dune principally affected the eastern part of Vougot beach. Erosion was caused by the construction of the Enez Croas Hent jetty in 1974 (Figure 2b), which completely modified the hydrodynamics and interrupted the westward sand drift, inducing an increase in sediment loss for the Vougot beach/dune system [35]. Calculation of erosion rates over the 1978–2000 period (following the building of the jetty in 1974) showed that the maximum retreat of the dune reached −0.6 m/year; and this rate has increased from −0.6 m/year to −1.5 m/year over the last decade (from 2000 to 2009) due to the impact of a major storm on 10 March 2008 [35,36]. However, from spring 2008 to summer 2013, almost five years of dune recovery occurred. It was characterized by dune progradation reaching +12 m on the zones with the most accretion [37]. Finally, during the winter of 2013–2014, a cluster of about 12 storm events hit the coast of Brittany with an exceptional frequency [38]. Dune erosion of Vougot beach during this period (between December 2013 and March 2014) reached almost −15 m on the most retreated part. Therefore, the maximum retreat of the dune between 2008 and 2014 was about −0.7 m/year.

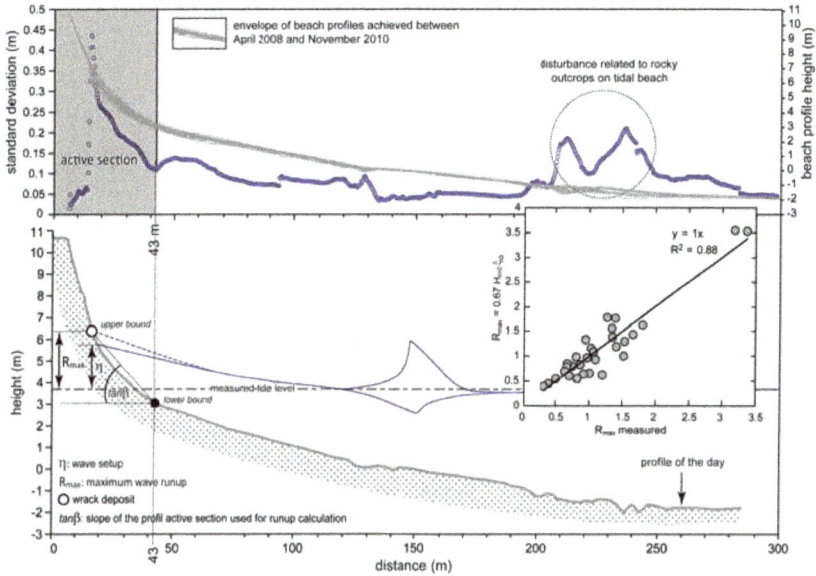

Figure 1. Method used by Cariolet and Suanez [33] to calculate beach slope for the runup calculation. The lower bound corresponds to the limit of the profile section where changes of elevation are the most significant. This section concerns the upper part of the profile, and it is called "active section". This limit of 43 m has been defined by calculating the standard deviation of height changes of the beach profiles (gray lines) measured between April 2008 and November 2010 (see Table 1). The upper bound corresponds to field measurements of the swash height given by the water mark limit or wrack line deposit.

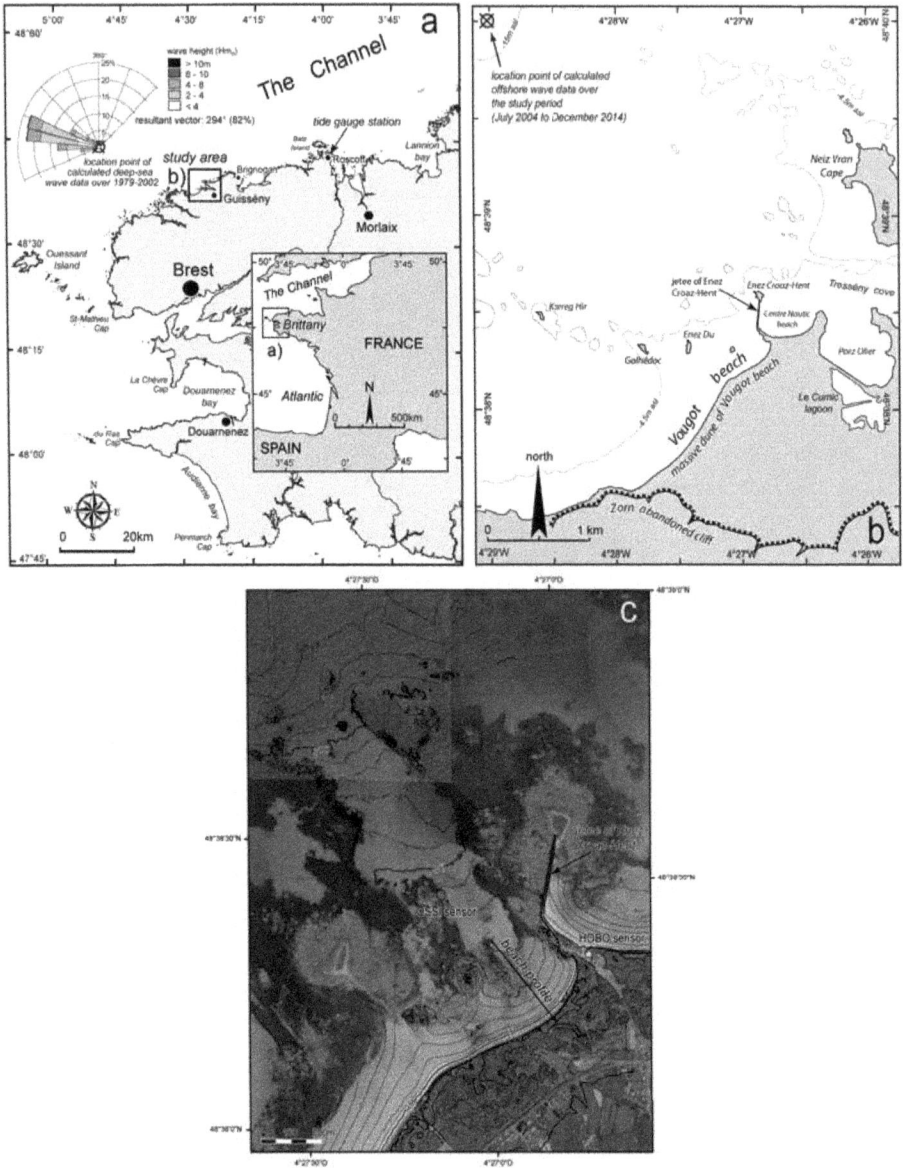

Figure 2. Location map. Regional setting and Roscoff tidal gauge station (**a**); local setting and location point where offshore wave data (H_{m0} and $T_{m0,-1}$) were calculated using (WW3) modeling (**b**); and aerial photography of Vougot beach showing the beach/dune profile location and both wave/water level and atmospheric pressure sensors (**c**).

Offshore incident waves obtained over the period 1979–2002 show that they come mostly from the west–northwest direction (242°) (Figure 2a). The most frequent wave height (H_{m0}) is between 1.5 and 3 m with an average height reaching 2.2 m, and the most frequent period (T_{pic}) is between 9 and 11 s, with an average period of 10.6 s. The maximum wave height and period related to storm events reached respectively 14 m and 20 s. Because of a tidal range reaching about 7 m between MHWS and MLWS, the Vougot beach is characterized as a macrotidal environment. The beach profile of the section studied is characterized by different morphodynamic environments according to the composite slope and concave beach (Figure 3). The lower part of the tidal beach, between MHWN and MLWN, is mainly associated with low Iribarren parameter values of ≤0.3 and a very gentle slope, *tanβ*, reaching 0.034 to 0.014. These morphodynamic conditions correspond to a dissipative environment. In contrast, the upper beach, between MHWN and HAT, is characterized by intermediate conditions with Iribarren values >0.72 and beach slope *tanβ* reaching 0.18 between HAT level and the foot of the dune. Therefore, depending on the tide's water level, waves break at high tide on different morphodynamic environments. Under neap tide conditions, wave-breaking processes are related to rather dissipative conditions, while under spring tides, intermediate to moderately reflective conditions (Iribarren parameters up to 1.6) prevail. This environmental context is important because the behavior of runup under dissipative conditions is different than during reflective and intermediate conditions [21,39].

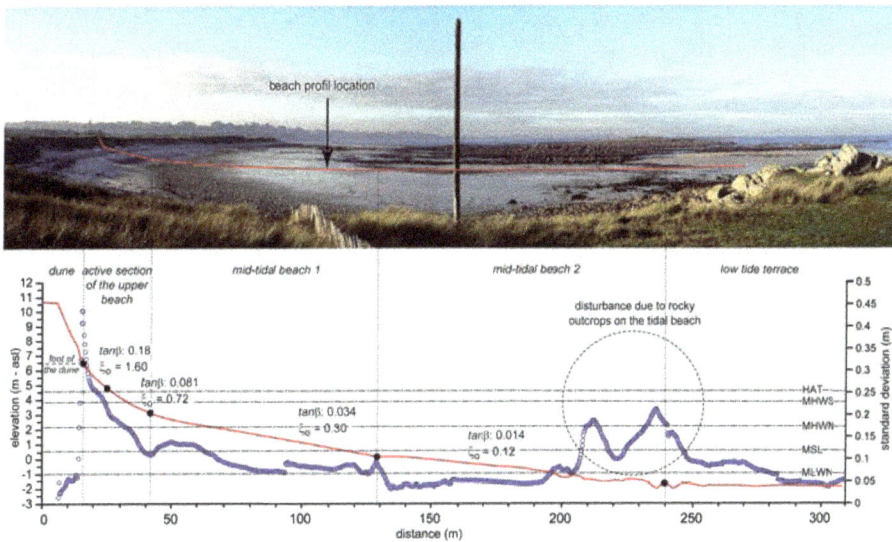

Figure 3. (**Red line**): Mean cross-shore profile of the surveyed Vougot beach section. (**Blue spot**): The standard deviation of profile elevation change rates (see Figure 1). Morphodynamic conditions (dissipative to reflective conditions) have been analyzed along the concave beach profile using Iribarren value.

3. Methods

3.1. Monitoring of Dune Morphological Changes

The monthly monitoring of dune morphological changes started in July 2004. It consisted of beach/dune profile measurements carried out along the cross-shore transect presented in Figure 2c, using a Trimble 5700/5800 Differential GPS. Data points described by three coordinate values (x, y, z) were collected in Real Time Kinematic (RTK) mode. Measurements were calibrated using the geodesic marker from the French datum and the geodesic network provided by the IGN located about 2 km from the study area. Several control points set up in the field were used to assess the accuracy of the survey reaching ±4–5 cm (X and Y) and ±1–2 cm (Z). These values were used to calculate the margin of error associated with the dune sediment budget.

3.2. Survey of Beach Profile and Maximum Swash Elevation (Runup) R_{max}

Between July 2012 and June 2013, 59 measurements of beach profile and maximum swash elevation were carried out using the same method as the one followed in the previous study by Cariolet and Suanez [33]. Maximum swash elevation was determined by the wrack deposit and/or the limit of the water mark identified by a tonal change from dark wet foreshore sand to light dry sand on the upper beach (Figure 4a). We assume that this limit corresponded to the highest level reached by the runup during the previous high tide. Therefore, it corresponds to R_{max} (maximum runup) instead of the generally used random variable $R_{2\%}$ that corresponds to vertical runup distance exceeded by two percent of wave runups. In addition to the swash elevation measurement, the beach/dune profile was also measured in order to recover the morphological parameters needed to analyze runup processes. These measurements were acquired along the same transect and according to the same DGPS method as described previously. This data set was added to the 31 surveys conducted as part of the study of Cariolet and Suanez [33]. In total, a set of 90 morphological and runup measurements was used in this study (Figure 4b and Table 1).

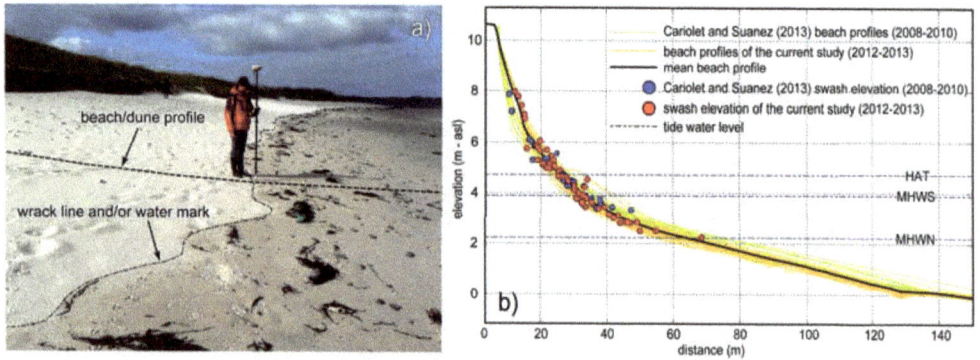

Figure 4. The limit between dry and wet sand (water mark) at the level of high tide deposit (wrack line) shows the level reached by the swash processes (**a**); DGPS measurement of beach/dune profiles and the maximum runup elevation reached during the previous high tide (**b**).

Table 1. Overview of environmental conditions and dimensional swash parameters where H_{m0} (m), $T_{m0,-1}$ (s), and L_0 (m) correspond to WW3 offshore wave, R_{max} (m) is runup field measurements, and $HTWL$ (m) is high tide water level.

Date	H_{m0} (m)	$T_{m0,-1}$ (s)	L_0 (m)	Slope ($tan\beta$)	ξ_0	R_{max} (m)	$HTWL$ (m)
08 April 2008	0.6	8.1	101	0.118	1.502	0.95	4.36
29 August 2008	1.0	8.3	107	0.061	0.623	0.30	3.10
29 September 2008	0.8	7.5	88	0.093	1.004	0.74	3.80
12 January 2009	3.6	12.9	261	0.096	0.825	3.24	4.01
13 February 2009	1.8	11.0	188	0.107	1.111	1.46	4.14
29 April 2009	1.7	8.7	119	0.071	0.591	0.93	3.37
17 December 2009	0.8	10.0	156	0.080	1.100	0.75	3.56
22 December 2009	1.0	8.0	101	0.067	0.680	0.67	3.22
23 December 2009	1.0	7.9	98	0.056	0.559	0.64	2.97
30 December 2009	1.6	10.5	171	0.073	0.749	1.11	3.40
04 January 2010	1.6	8.3	107	0.110	0.906	1.06	4.19
07 January 2010	2.3	6.6	69	0.058	0.319	0.69	3.09
13 January 2010	2.8	11.6	210	0.054	0.462	1.36	3.03
14 January 2010	2.6	11.8	217	0.069	0.635	1.43	3.36

Table 1. *Cont.*

Date	H_{m0} (m)	$T_{m0,-1}$ (s)	L_0 (m)	Slope ($tanβ$)	$ξ_0$	R_{max} (m)	$HTWL$ (m)
16 January 2010	2.3	11.9	223	0.091	0.903	1.52	3.83
21 January 2010	2.8	12.9	261	0.054	0.525	1.40	3.04
28 January 2010	2.0	7.5	88	0.049	0.324	0.89	2.86
01 February 2010	1.5	6.6	68	0.123	0.837	1.57	4.53
03 February 2010	1.8	7.6	90	0.111	0.781	1.63	4.23
05 February 2010	2.9	11.5	207	0.068	0.570	1.14	3.35
05 February 2010	4.5	13.5	285	0.049	0.393	1.75	2.95
06 February 2010	4.1	12.0	226	0.039	0.293	1.02	2.41
26 February 2010	2.1	6.0	56	0.065	0.336	1.01	3.24
28 February 2010	1.5	5.6	50	0.124	0.710	1.50	4.56
03 March 2010	1.1	5.7	50	0.127	0.846	1.26	4.62
29 March 2010	1.2	7.6	89	0.113	0.965	1.08	4.28
31 March 2010	4.5	9.2	132	0.114	0.616	3.47	4.52
10 June 2010	1.3	6.2	61	0.051	0.347	0.47	2.87
13 July 2010	1.1	8.6	116	0.097	0.991	0.64	3.89
12 October 2010	1.8	6.3	62	0.082	0.482	0.63	3.61
08 November 2010	2.6	7.3	82	0.115	0.648	1.67	4.39
05 July 2012	1.7	9.2	132	0.097	0.847	1.50	3.95
02 October 2012	2.6	10.7	180	0.086	0.712	1.97	3.73
17 October 2012	3.6	12.5	246	0.123	1.018	3.36	4.63
02 November 2012	2.8	8.9	123	0.080	0.531	1.58	3.60
06 November 2012	1.5	6.8	72	0.033	0.225	0.57	2.24
12 November 2012	1.9	9.7	146	0.078	0.686	1.73	3.52
19 November 2012	1.6	9.3	134	0.076	0.697	1.44	3.46
23 November 2012	3.3	11.5	205	0.039	0.306	1.00	2.45
26 November 2012	2.4	8.5	112	0.065	0.440	1.06	3.25
30 November 2012	1.0	9.6	144	0.077	0.903	0.10	3.49
03 December 2012	3.0	10.4	170	0.058	0.434	1.51	3.13
06 December 2012	1.7	10.0	157	0.036	0.337	1.27	2.39
11 December 2012	1.1	7.6	89	0.063	0.575	0.68	3.15
13 December 2012	0.8	10.8	183	0.107	1.584	1.21	4.13
14 December 2012	1.8	11.3	198	0.125	1.304	2.48	4.64
17 December 2012	4.8	12.2	233	0.097	0.678	3.38	4.14
07 January 2013	1.8	10.8	181	0.034	0.337	0.58	2.25
08 January 2013	1.8	10.8	184	0.038	0.384	0.55	2.50
09 January 2013	1.8	12.1	227	0.053	0.596	0.47	2.92
16 January 2013	1.1	6.7	71	0.098	0.774	1.05	3.91
23 January 2013	4.0	13.1	266	0.038	0.312	1.54	2.33
24 January 2013	2.4	10.9	187	0.041	0.362	1.07	2.63
25 January 2013	2.1	11.2	195	0.053	0.502	0.85	2.95
27 January 2013	3.1	10.1	160	0.089	0.635	2.11	3.81
28 January 2013	5.1	13.9	301	0.074	0.567	3.25	3.68
29 January 2013	5.0	14.3	318	0.089	0.708	3.91	3.98
04 February 2013	2.7	9.4	139	0.038	0.270	1.01	2.43
05 February 2013	5.7	11.9	220	0.039	0.241	1.35	2.20
06 February 2013	6.1	12.4	242	0.040	0.251	1.50	2.29
07 February 2013	3.3	9.3	136	0.040	0.259	1.09	2.53
14 February 2013	3.1	10.2	163	0.094	0.678	1.83	3.93
19 February 2013	1.8	12.5	242	0.031	0.361	0.65	1.60
21 February 2013	1.5	7.4	87	0.032	0.240	0.99	1.87
22 February 2013	1.9	8.5	112	0.034	0.266	0.54	2.30
04 March 2013	0.9	5.6	49	0.057	0.422	0.12	2.99
05 March 2013	0.5	4.8	36	0.042	0.359	0.22	2.63
10 March 2013	1.5	12.7	250	0.095	1.211	1.43	3.89
14 March 2013	1.2	5.5	48	0.104	0.656	0.98	4.04
28 March 2013	0.9	7.2	82	0.109	1.038	0.98	4.17

Table 1. *Cont.*

Date	H_{m0} (m)	$T_{m0,-1}$ (s)	L_0 (m)	Slope ($tan\beta$)	ξ_0	R_{max} (m)	HTWL (m)
29 March 2013	0.9	6.8	73	0.116	1.058	0.86	4.32
08 April 2013	1.9	11.1	191	0.081	0.814	1.24	3.57
09 April 2013	2.0	8.4	111	0.099	0.734	1.19	3.98
07 May 2013	1.7	12.4	241	0.062	0.747	0.95	3.15
09 May 2013	2.4	9.0	126	0.076	0.546	1.51	3.50
23 May 2013	1.5	6.8	73	0.063	0.438	0.84	3.17
23 May 2013	2.2	6.4	65	0.072	0.396	1.09	3.40
24 May 2013	1.8	6.2	60	0.077	0.440	1.17	3.49
25 May 2013	2.1	6.3	63	0.086	0.466	1.32	3.71
27 May 2013	0.6	7.6	91	0.104	1.246	1.04	4.04
12 June 2013	1.6	8.2	106	0.060	0.499	1.02	3.10
13 June 2013	2.8	9.3	135	0.049	0.344	0.99	2.92
14 June 2013	1.4	8.6	114	0.045	0.405	1.08	2.73
18 June 2013	0.9	9.1	128	0.034	0.408	0.10	2.39
19 June 2013	1.9	7.8	94	0.035	0.249	0.14	2.36
20 June 2013	1.9	7.9	97	0.043	0.305	0.45	2.69
21 June 2013	1.6	9.4	138	0.057	0.530	0.91	3.02
23 June 2013	3.8	9.9	154	0.091	0.578	1.89	3.91
24 June 2013	2.7	8.9	124	0.089	0.605	1.89	3.79
25 June 2013	1.1	7.8	95	0.096	0.911	0.78	3.89

3.3. Hydrodynamic Condition Measurements

Wave analysis is based on two data sets acquired between June 2012 and June 2013. The first one corresponds to records taken on the intertidal zone using the OSSI-010-003C pressure sensor (accuracy ±1.5 cm specification) (Ocean Sensor Systems, Inc®, Coral Springs, FL, USA), which was deployed along the morphological profile mentioned above, at −2.5 m asl which corresponds about to the low water spring tide level (Figure 2c). A recording frequency of 5 Hz was chosen to reproduce as accurately as possible the wave spectrum. The sensor was calibrated before and after each deployment by comparing the pressure measured at the low tide level (when the sensor is out of the water and thus measures atmospheric pressure) with the atmospheric pressure recorded *in situ*. The atmospheric pressure was measured using the HOBO U20 Water Level Logger sensor (Onset Computer Corporation®, Bourne, MA, USA) which was positioned on the outside wall of the nautical center (Figure 2c). The second set of wave data concerns simulations acquired from the WAVEWATCH III model (WW3), which reproduced the offshore wave conditions at the calculation point 4°29′24″ W, 48°40′12″ N at a water depth of 18.3 m [40,41].

Wave parameters such as wave height (H_{m0}) and period ($T_{m0,-1}$) were extracted from both data sets for the time periods corresponding to the high tide level (Figure 5). Results showed that the monitoring period was marked by a high variability of hydrodynamic conditions. Between the end of November 2012 and mid-February 2013, ten episodes marked by high offshore waves (>4 m) were recorded, including the storm of 6 February, which was characterized by significant heights of >6 m. One can also note the two episodes of 14 May and 23 June, where the swells were often above 4 m. A validation of the offshore wave data set obtained using WW3 modeling was achieved by comparing these data to those measured in the tidal zone by a wave gauge sensor. The correlation shows a good relationship between both sets of data, especially for the wave height, with, however, less correlation regarding the periods (Figure 5).

Analysis of the tides is also based on records taken in the tidal zone using gauge sensor OSSI-010-003C (Figure 2c). The observed water level was computed taking into account atmospheric pressure measured by the HOBO U20 Water Level Logger sensor (Onset Computer Corporation®,

Bourne, MA, USA) set up on the study site (Figure 2c). It was then possible to calculate the pressure exerted by the water column and thus to calculate the height of the latter with the following expression:

$$H_{(water\ level)} = (P_{sensor} - P_{atmosphere})/\rho \cdot g \qquad (7)$$

where H is the height of the water column (in m), P_{sensor} is the pressure measured by the sensor (in Pa), $P_{atmosphere}$ is the atmospheric pressure (in Pa), ρ is the density of water (=1025 kg/m^3), and g is the acceleration of gravity (=9.81 m/s^2).

Water levels were smoothed to a moving average of 10 min to filter out deformations of the water surface related to wave action, and water levels corresponding to both daily high tides were extracted. A similar calculation was done using data recorded at a permanent tide gauge station near Roscoff located at about 30 km east of the study site (Figure 2). Both the time series from Guissény and Roscoff were used to estimate the differences in high tide water level between the two sites (Figure 6). As Figure 6b shows, more than 500 high tide level records were used for the statistical analysis, showing a very good correlation between both the Roscoff and Guissény sites. The mean deviation is 18 cm, with variations between 25 cm for spring tides and 5 cm for neap tides (Figure 6c).

Observed tide levels show that during the survey period some episodes characterized by high spring tide levels, with a tide coefficient close to 100 or higher, occurred (Table 2). When these events were combined with a storm, the measurement of runup elevation was considerably higher because of the storm surge effect. This is mainly the case for the following seven episodes: 17 October and 17 December of 2012, 29 January, 11 February, 11 March, 28 May, and 23 June of 2013.

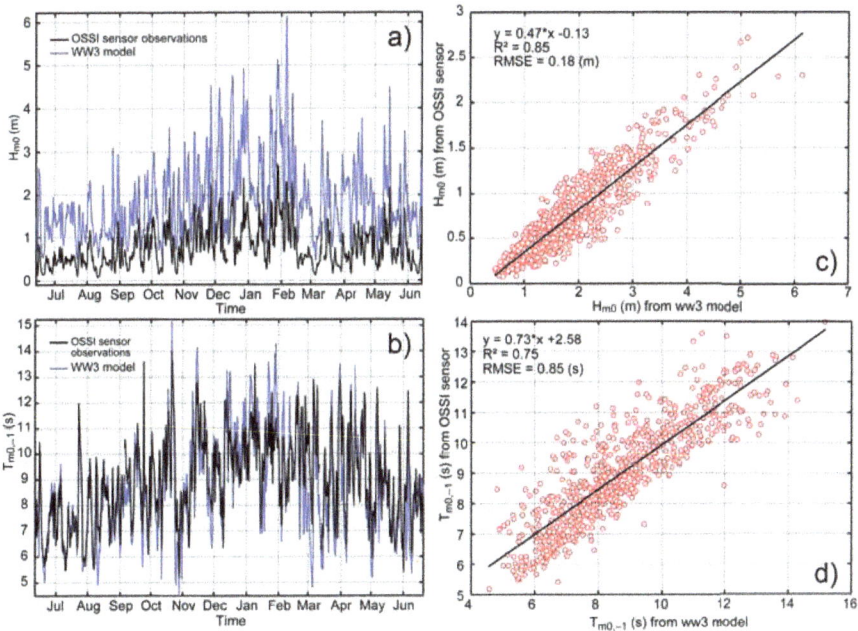

Figure 5. Offshore (WW3) and shallow (OSSI) wave heights (**a**) and periods (**b**) obtained between July 2012 and June 2013. Correlations between offshore and shallow wave heights (**c**) and periods (**d**).

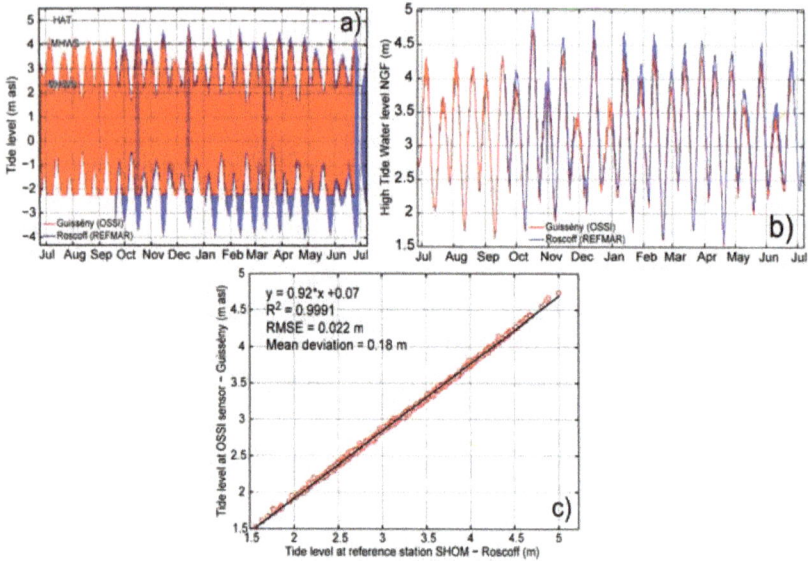

Figure 6. Comparison between tide level (**a**) and daily high tide water levels (**b**) recorded at Vougot beach, using the OSSI-010-003C sensor, and at the permanent Roscoff tide gauge station. (**c**) Correlation between high tide water levels recorded at Vougot beach and at the permanent Roscoff tide gauge station.

Table 2. Inventory of high spring tide events characterized by a tide coefficient ≥ 100. In France, the magnitude of the tide from its average value is indicated by a coefficient expressed in hundredths, which lies between 20 and 120. A coefficient of 100 is associated with a maximum astronomical tidal range in Brest, calculated by the *Service Hydrographique et Océanographique de la Marine* (SHOM). It is defined as follows: $C = (H - N_o)/U$, where, H: high tide water level, N_o: mean water level at Brest: 4.13 m, U: height unit specific to the locality at Brest: 3.05 m. Tidal coefficients higher than 70 correspond to spring tides, below 70 they correspond to neap tides. A tidal coefficient of 95 corresponds to mean spring tide level, 45 corresponds to mean neap tide level.

Date and High Tide Time	Tide Coefficient	Predicted Tide Level (m)	Observed Tide Level (m)	Surge (m)
17/09/2012—(17:25)	104	4.3	4.34	0.04
18/09/2012—(18:15)	106	4.29	4.27	−0.02
19/09/2012—(06:20)	103	4.13	4.07	−0.06
16/10/2012—(10:30)	107	4.27		
17/10/2012—(05:20)	109	4.36	4.73	0.37
18/10/2012—(06:05)	105	4.29	4.55	0.26
14/11/2012—(04:15)	104	4.26	4.22	−0.04
15/11/2012—(05:00)	107	4.40	4.35	−0.05
16/11/2012—(05:55)	104	4.36	4.39	0.03
14/12/2012—(04:50)	104	4.32	4.66	0.34
15/12/2012—(05:40)	104	4.38	4.58	0.20
12/01/2013—(04:30)	102	4.23	4.38	0.15
13/01/2013—(05:20)	106	4.39	4.42	0.03
14/01/2013—(06:11)	104	4.37	4.31	−0.04
11/02/2013—(05:11)	106	4.33	4.42	0.09
12/02/2013—05:52)	106	4.35	4.32	−0.03
12/03/2013—(04:50)	102	4.14		

Table 2. *Cont.*

Date and High Tide Time	Tide Coefficient	Predicted Tide Level (m)	Observed Tide Level (m)	Surge (m)
13/03/2013—(05:27)	103	4.16	4.26	0.1
28/03/2013—(17:20)	103	4.06	4.22	0.16
29/03/2013—(05:40)	105	4.16	4.32	0.16
26/04/2013—(16:55)	103	4.12	4.05	−0.07
27/04/2013—(17:35)	106	4.18	4.15	−0.03
26/05/2013—(17:20)	104	4.2	4.22	0.02
27/05/2013—(18:15)	104	4.16	4.31	0.15
24/06/2013—(17:15)	102	4.22	4.16	−0.06
25/06/2013—(18:00)	105	4.27		
26/06/2013—(18:49)	103	4.15		

4. Results

4.1. Calibration of Battjes (1971) Runup Formula

Fit analysis between observed runup values and morphodynamic variables was achieved following the same methodological approach as the one used by Cariolet and Suanez [33]. Morphodynamic parameters such as $H_{m0}\xi_0$ have been used to characterize runup processes in as far as it was demonstrated that these variables were best correlated with runup when using the slope of the active section. A new correlation between observed runup and $H_{m0}\xi_0$ was calculated including the data set used by Cariolet and Suanez [33] (Figure 7). The relation can be expressed as:

$$R_{max} = 0.68 H_{m0}\xi_0 \tag{8}$$

It gives the same result as the previous Cariolet and Suanez [33] study with a constant equal to 0.68 (95% confidence intervals [0.65; 0.71]).

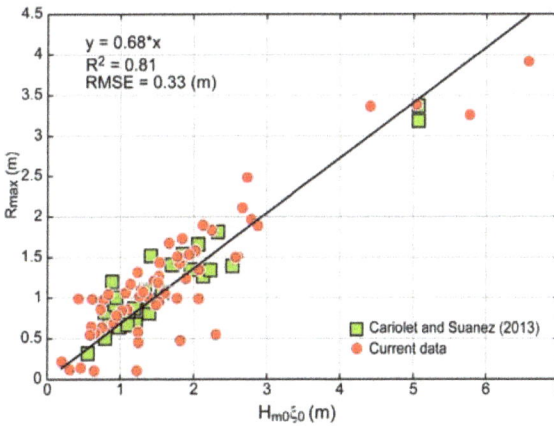

Figure 7. Correlation between observed runup (R_{max}) and $H_{m0}\xi_0$. Equation $R_{max} = 0.68 H_{m0}\xi_0$ is obtained; it was $R_{max} = 0.67 H_{m0}\xi_0$ from Cariolet and Suanez [33] previous study.

4.2. Elaboration of a General Empirical Equation

This part of the work focused on the parameterization of a general empirical equation that is no longer dependent on morphodynamic parameters obtained from high-frequency field measurements such as *(i)* the daily beach profile and *(ii)* the position of swash elevation along this profile. The approach is therefore to quantify the runup using hydrodynamic parameters such as offshore wave and water level, which are continuously recorded by wave and tide gauge stations. Morphological parameters such as the beach slope, $tan\beta$, are meanwhile deduced from the mean beach profile assuming that the measurement of a daily beach profile is no longer taken, as we said earlier. However, the mean beach profile must be calculated from a series of measurements already available. In this case, the mean beach profile was calculated using all profile measurements recorded between June 2012 and June 2013.

Considering the previous method exposed in Section 4.1, the main problem encountered when predicting wave runup on a beach with composite-slopes (or concave shape) is how to define the upper and lower bounds of the beach profile section for which the slope is calculated when they are no longer measured on the field. Following the approach of [21,42], the slope has been calculated using the observed high tide water level (*HTWL*) and a fraction of the offshore wave height (H_{m0}) from which the horizontal beach slope section (*HBSS*) was defined (Figure 8). Different correlation tests have shown that $1/4H_{m0}$ gives the best result in this case. Therefore, the upper and lower bounds of the beach slope profile width is calculated as follows

$$Bound_{\text{up and low}} = HTWL \pm 1/4H_{m0} \qquad (9)$$

Swash runup is in this case best parameterized with a best-fit R^2 (0.85) and RMSE (0.29 m). The coefficient of the regression line is 1.01 with 95% confidence intervals [0.97; 1.05]. In this case, the relationship can be expressed as (Figure 8).

$$R_{max} = 1.01 H_{m0}\xi_0 \qquad (10)$$

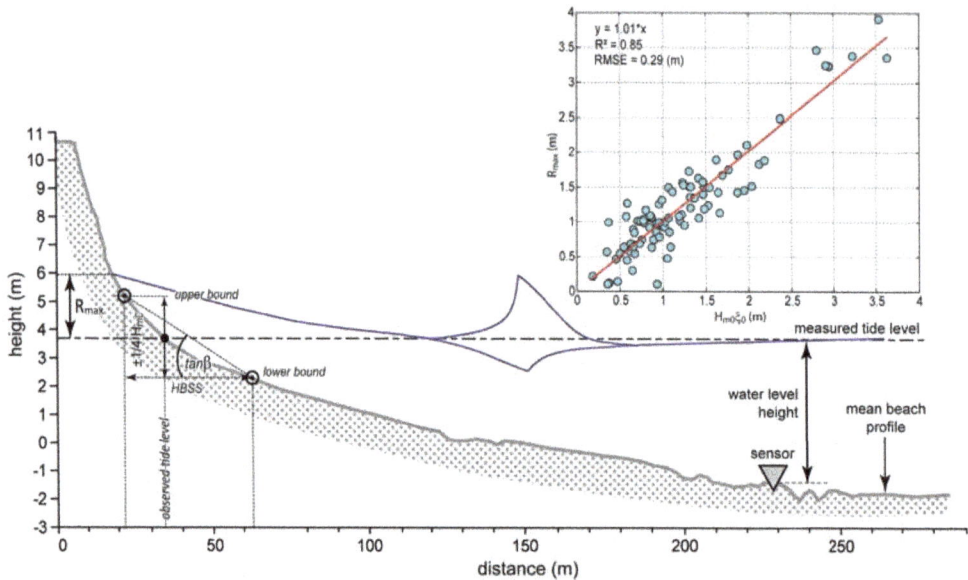

Figure 8. Method used to calculate beach slope for the runup calculation. It is based on measured tide level from which the beach section is defined. In contrast to the previous method, note that this approach is using the mean beach profile instead of daily beach profile.

The use of wave height for the calculation of *HBSS* gives a physical meaningful approach that is applicable from low wave energy conditions to storm wave events. The width of this beach section ranges from 3 m (H_{m0} = 0.6 m and *HTWL* = 4.4 m asl) to 80 m (H_{m0} = 6.1 m and *HTWL* = 2.3 m asl), with an average value of 17 m and standard deviation of 14 m (the mean of H_{m0} for the dataset is 2.2 m, with standard deviation 1.2 m), depending on the position of the *HTWL* on the beach profile. Figure 9 shows an overview of the measured runup (R_{max}) dependencies of estimated runup using both Equations (8) and (10), and Equations (4)–(6) of Stockdon et al. [21]. According to the 95% confidence intervals (see Figure 9), the three correlations show that the three equations give very similar results. Nevertheless, Equation (10) best fits the observed runup.

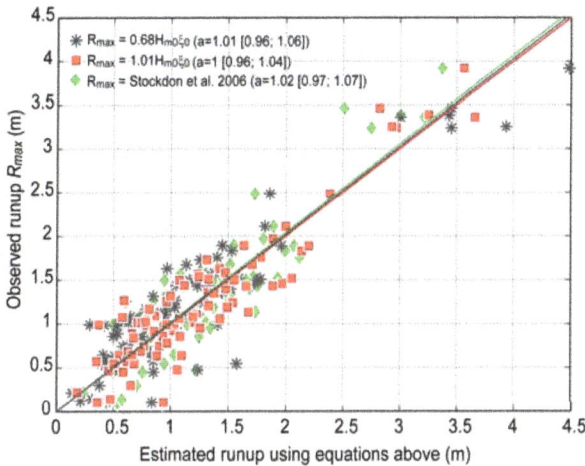

Figure 9. Overview of the measured runup (R_{max}) dependencies of $CH_{m0}\xi_0$ for both Equations (8) and (10), and Equations (4)–(6) of Stockdon et al. [21].

4.3. Long-Term Dune Changes Related to Storm Event Erosion and Recover

This part of the study focused on the relationship between the evolution of the dune sediment budget (in terms of accretion and erosion) and extreme water levels since July 2004. Following the "Property Erosion Model" method proposed by Ruggiero et al. [19] and/or the storm-impact scaling model proposed by Sallenger [18], the aim was to assess the sensitivity of the dune to extreme water levels by considering that erosion is experienced when the dune foot elevation is below extreme water level. Therefore, morphodynamic analysis was performed to identify the erosion stages related to extreme events, combining storm surge and high spring tide level, and on the other hand, the recover periods associated to calm wave conditions and/or low neap tide level.

Extreme water level was estimated for each daily high tide level by summing the swash runup elevation calculated from the Equation (10), and the measured tide level at Roscoff calibrated to the Vougot beach site. The altitude of the foot of the dune was obtained from monthly beach/dune profile measurements. However, the great morphological changes of the upper beach/dune section over the last ten years made it very difficult to identify this morphological proxy (Figure 10a). When we analyze in more detail the data set, three main phases related to strong dune erosion were identified. These three erosion phases have induced a landward displacement of the foot of the dune. As shown in Figure 10b, the foot of the dune was situated at 15.5 m from the head profile mark at the beginning of the survey. It retreated over more than 3 m during the big storm of 10 March 2008 [36,37], and retreated again over 3 m during the storm of 1 February 2014 [38]. These three reference distances (15.5 m, 12 m, and 9.5 m) were used for the calculation of the dune foot height.

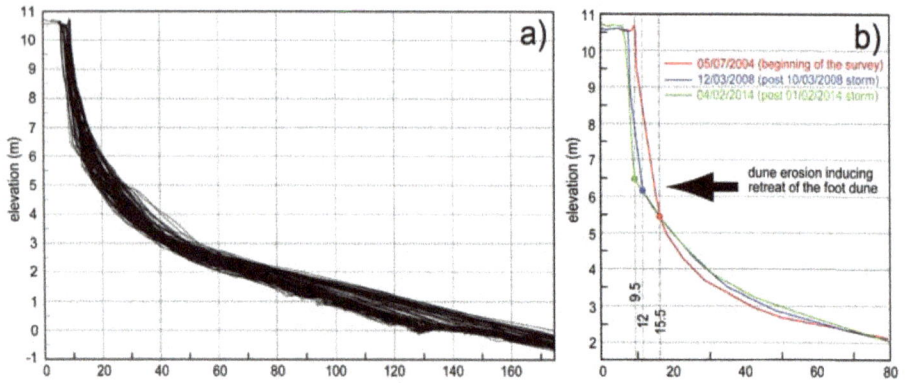

Figure 10. Envelop of beach/dune profiles measured from July 2004 to December 2014 (**a**); The front of the dune retreated during the two main storm events, which took place during the past 10 years (i.e., 10 March 2008 and 1 February 2014) inducing landward displacement of the foot of the dune (**b**).

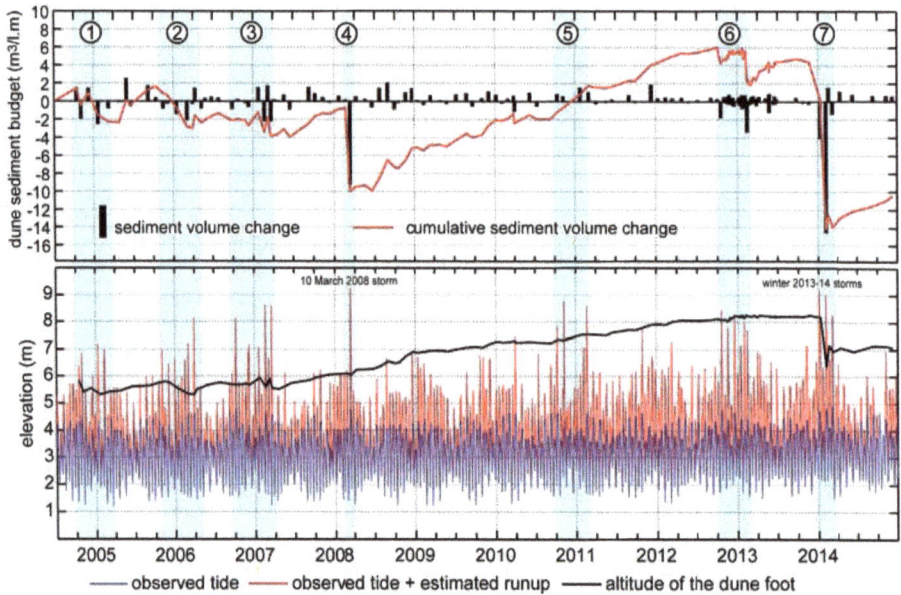

Figure 11. Evolution of the dune sedimentary budget related to extreme water levels over the period from July 2004 to December 2014.

The results show a good relationship between the negative sediment budget of the dune and phases during which extreme water levels exceed the height of the foot of the dune. On the contrary, when extreme water levels are below the height of the foot of the dune, the sediment budget increases (Figure 11). Seven phases of high extreme water levels are identified: from 28 October 2004 to 12 February 2005 (maximum water level: 7.19 m 12 January 2005), from 03 November 2005 to 31 March 2006 (maximum water level: 8.12 m 31 March 2006), from 08 October 2006 to 20 March 2007 (maximum water level: 8.61 m 20 February 2007), the 10 March 2008 storm event (9.23 m), from 09 October 2010 to 20 February 2011 (maximum water level: 8.79 m 09 November 2010), from 17 October 2012 to 11 February 2013 (maximum water level: 8.43 m 17 October 2012), and from 01 January 2014 to 03 March 2014 (maximum water level: 9.18 m 04 January 2014) (Figure 11). For six of them, extreme water levels are well-related to an erosion phase of the dune, with the exception of the period from 09 October 2010 to 20 February 2011 (Figure 12a–i). During this last period, three episodes characterized by extreme water levels higher than the foot of the dune were recorded without erosion of the dune (09 October 2010: 7.83 m; 09 November 2010: 8.79 m; 20 February 2011: 8.56 m). However, we notice that these three extreme events have occurred during a long phase of dune recovery, which started in spring 2008 (post-storm of 10 March 2008) and ended during the autumn of 2012 (Figure 12—From (j) to (l)). During these four years, the sediment budget of the dune increased considerably, inducing an elevation of the foot of the dune up to 2 m. Therefore, extreme water levels never hit the foot of the dune during this entire period except during the short stage mentioned above. The two major dune erosion stages which were recorded are related to the big storm event of 10 March 2008 [36,37], and to a cluster of storms occurring during the winter of 2013–2014 [38].

Figure 12. Photos illustrating dune erosion which occurred during the high extreme water levels phases inventoried in Figure 11: (**a**) post-12 January 2005 extreme water level; (**b**) a day before 31 March 2006 extreme water level; (**c**) post-20 February 2007 extreme water level; (**d**) 10 March 2008 storm event; (**e**) post-10 March 2008 storm event; (**f**) post-17 October 2012 extreme water level; (**g**) post-04 January 2014 extreme water level; (**h**) post-02 February 2014 extreme water level; and (**i**) 03 March 2014 storm event. Photos illustrating the dune recover phase post-storm of 10 March 2008 to September 2012 (**j**–**l**).

J. Mar. Sci. Eng. **2015**, 3, 674–698

5. Discussion

Following the previous study [33], this experiment has more deeply examined new parameterization of the runup formula [1] for predicting total swash elevation in the extreme water level calculation. The focus was put on parameters that provide a first-order description of the beach morphodynamic environment, such as deep wave height (H_0), period (T), and beach steepness $tan\beta$, which are expressed in terms of the non-dimensional surf parameter (Iribarren number ξ_0) [27]. Thus, measurements of local wave height have confirmed the validity of the use of the deep-water wave height (H_0 at 18 m water depth) obtained by modeling. Similarly, *in-situ* measurements of water levels have improved the estimation of extreme water levels at the coast and determined the tidal range of the tide gauge shifts between Roscoff tide gauge station and the site of Guissény. The mean deviation is 18 cm, with variations between 25 cm for spring tides and 5 cm for neap tides. It is close to the 13 cm mean deviation calculated in the previous study that was based on a shorter data set and a less accurate method [33]. Concerning the beach steepness, the new set of data used in this study has confirmed the complexity of defining the best beach slope for use in the runup formula when the beach exhibits composite-slope and/or a concave profile. As demonstrated by Cariolet and Suanez [33], the slope of the active section of the upper beach gives a good fit in comparison to the field measurement (R^2: 81%; RMSE of 33 cm). Nevertheless, a best-fit was obtained when beach slope was calculated using observed water level (R^2: 85%; RMSE of 29 cm). This approach, based on sea level changes due to tides and/or storm surges, allows for better consideration of beach slope variations in the context of a concave beach profile. As already indicated by Mayer and Kriebel [43] the use of fixed bounds (upper or lower bounds) or an averaged planar slope for the calculation of beach steepness is therefore inappropriate when beaches exhibit complex morphology with a composite-slope, especially in a macrotidal environmental context. If we take into consideration the steep slope face of the upper concave beach ($0.08 > tan\beta > 1.8$), this experiment also confirms the findings of Nielsen and Hanslow [32], attesting that the best-fit distribution is proportional to the surf similarity parameter (ξ_0) on intermediate to reflective beaches in agreement with Hunt's formula for runup of regular waves on steep slopes. However, statistical tests have indicated that the reflective-specific Equation (5) of Stockdon et al. [21] was also best fitted to runup field measurements (R_{max}), and therefore both Equations (8) and (10).

Long-term erosion of the dune related to extreme water levels shows different pluri-decadal phases. From 2004 to 2006, the dune sediment budget indicated normal functioning characterized by erosion and high water levels during winter and accretion associated to low water levels during summer. However, dune sediment budget slightly decreased during these two first years. From the winter of 2006–2007 to the storm of 10 March 2008, the dune experienced a phase of significant sediment budget decrease related to several high extreme water level events. As mentioned earlier, this stage was followed by a long phase of dune recovery that ended during the winter of 2012–2013. The increase of the dune sediment budget was explained by supply from post-storm sediment transport between the upper intertidal beach and the lower intertidal beach—The nearshore to shoreface zone [37]. This sand supply took place during low extreme water levels associated with cold winters. The last phase was again characterized by a significant loss of dune sediment budget due to the erosion effects of the stormy winters of 2012–2013 and 2013–2014.

The inter-annual variability in dune erosion and accretion may be related to the winter North Atlantic Oscillation (NAO) index. Figure 13 presents NAO index fluctuation for the whole survey period (2004–2014). It shows three different phases that could be related to dune morphological changes. From 2000 to 2008, a positive index (denoted NAO+) is observed (Figure 13). Generally, this is associated to strong southwesterly winds that bring warm air deep into Europe. This results in mild and wet winters characterized by active storms that hit Western Europe at the latitude of England and Brittany. The winter of 1989–1990 is a good example of this weather pattern [44,45], as well as the winter of 2013–2014, during which a cluster of a dozen storms hit the Brittany coast [38,46]. In contrast, the second phase, from 2008 to 2012, is characterized by a negative index (NAO−). In this context, western European areas suffer cold dry winters and storm tracks are shifted towards the south of Europe (North Spain and Mediterranean areas). These meteorological conditions are favorable for the regeneration of dune systems because the dry weather is generally associated with effective aeolian transit to the dune. At the same time, the absence of major winter storms plays an important role in the low erosion of the dunes during these periods. The third and last phase that began in the winter of 2012–2013 is characterized by a positive NAO index. It accompanied warm and stormy winters, especially the winter of 2013–2014 [38,46]. Between December 2013 and March 2014, a cluster of about 12 storm events hit the coast of Brittany with an exceptional frequency. It was in February that these storm events were the most frequent and particularly virulent. The significant wave heights measured off Finistère reached, respectively, 12.3 and 12.4 m during the Petra and Ulla storms on 5 and 14 February. However, analysis of hydrodynamic conditions showed that only three episodes promoted extreme morphogenetic conditions because they were combined with high spring tide level. The first one occurred from 1–4 January, it was followed by events during 1–3 February, and 2 and 3 March. As indicated on Figure 12, these three events generated high extreme water levels and strong dune erosion. The maximum retreat of the front of the dune during this period reached more than −16 m [38].

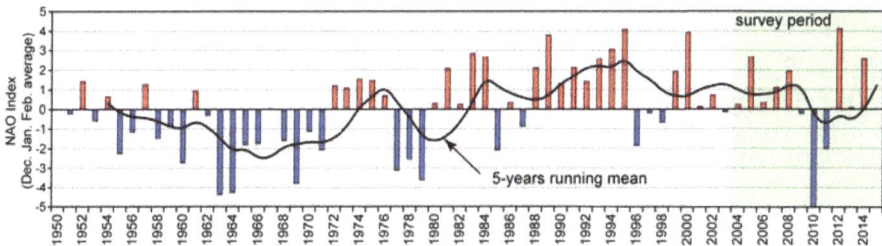

Figure 13. North Atlantic Oscillation Index (NAO) from 1950 to 2015 [47].

Consistent with this assumption, many studies have suggested that the North Atlantic Oscillation (NAO) may control the occurrence of storm events in the Atlantic, and thus potentially influence coastal morphological changes. The role of the NAO in coastal morphological dynamics has been suggested by Masselink et al. [48] to explain medium-term outer sand bar dynamics in the southwest of England (Perranporth). It was also suggested by Thomas et al., after analyzing beach rotation at South Sands Tenby in West Wales [49,50] and O'Connor et al. [51] concerning long-term shoreline and ebb channel evolution in northwest Ireland. The same conclusion was put forward by Vespremeanu-Stroe et al. [52] who showed that shoreline changes at decadal time scales were also driven by the NAO which controls the storminess on the Danube delta coast. Nevertheless, analysis of long-term dune morphological changes on the Sefton coast (west England) indicated only a modest relationship between dune erosion/accretion rates and the North Atlantic Oscillation index [20]. The authors suggested that these dune erosion/accretion phases are also related to the long-term beach sediment budget that governs essential changes in the morphology of the nearshore and offshore zones. Similarly, Montreuil and Bullard indicated that the winter North Atlantic Oscillation phase was not a good indicator of storminess on the east coast of England but may be a useful proxy for quiescence [53]. For this specific coastal area, the authors found the Jenkinson daily weather type classification to be a better proxy for the occurrence of strong onshore storm winds.

6. Conclusions

The runup process is still relatively too complex to parameterize in a macrotidal environment where beach profiles exhibit a composite-slope. This morphology is quite often found in North Brittany along the Channel coast where the tidal range is considerable. This study revealed a number of points related to the runup processes:

- The methodological approach of measuring the maximum swash elevation using wrack deposit and/or the limit of the water mark in the field is relatively easy to implement and requires much less post-treatment compared to classic video measurements. However, this method is extremely time consuming and does not allow for collection of a large dataset, which notably limits the statistical analysis.

- This experiment confirms that the beach slope scaled with ξ_0 plays a key role in the parameterization of the runup equation when the morphodynamic context of the beach is shifting from intermediate to reflective according to high–neap or spring–tide water level. In this context, beach slope may be much more important in runup elevation distribution than a wave component such as H_0 or L_0.

- In comparison to the previous study of Cariolet and Suanez, the use of observed water level changes due to astronomical tide and/or storm surges for the calculation of beach slope gives better results (RMSE decreasing from 0.33 to 0.29 m for Equations (8) and (10), respectively). This is explained by the fact that both the upper and lower bounds defining the beach section on which the slope is calculated are shifting according to the sea level changes. Therefore, the slope values obtained are much more fair and accurate, especially when the beach profile is concave and tidal range is large ($\approx$7 m), as is the case in this study.

- Taking into account the environmental conditions and dimensional swash parameters of the Vougot beach, the Stockdon's Equations (4)–(6) [21] may also be used with the appropriate beach slope value β_f.

- Dune retreat, and hence volume of sand eroded, depends on extreme water level (and therefore the frequency and intensity of each runup event) when its height is greater than that the toe of the dune.

- A good relationship seems to be revealed between erosion phases of the dune due to high extreme water levels and NAO+. In contrast, NAO$-$ is associated to phases of dune recovery during cold and non-stormy winters.

Acknowledgments: This work was supported by the French *"Agence Nationale de la Recherche"* through the *"Laboratoire d'Excellence"* LabexMER (ANR-10-LABX-19) program, and co-funded by a grant from the French government through the *"Investissements d'Avenir"* program and ANR COCORISCO by means of the *"Changements Environnementaux Planétaires & Sociétés"* (CEP&S) 2010" (ANR-10-CEPL-0001) research program. It was also supported by the French *"Institut National des Sciences de l'Univers"* (INSU) under-program, SNO-DYNALIT. Tide data came from the REFMAR database (refmar.shom.fr), and were provided by the *"Service Hydrographique et Océanographique de la Marine"* (SHOM), who we thank.

Author Contributions: Serge Suanez designed the experiments, interpreted results, prepared the figures and wrote most the text; Romain Cancouët and France Floc'h helped interpret results and write text; Romain Cancouët helped in developing figures; Serge Suanez, Emanuel Blaise and Jean-Marie Cariolet achieved the monitoring of dune morphological changes, and survey of beach profile and maximum swash elevation (Runup) R_{max}; Romain Cancouët and Serge Suanez achieved the hydrodynamic condition measurements (*in-situ* wave and water level measurements); Fabrice Ardhuin designed the WW3 model grid and ran the offshore wave data; Romain Cancouët, Jean-François Filipot, France Floc'h and Fabrice Ardhuin achieved hydrodynamic analysis (waves and water levels); Christophe Delacourt helped fund the study.

Conflicts of Interest: The authors declare no conflict of interest.

Abbreviations

asl	above sea level
IGN	Institut Géographique National
MHWS	Mean High Water Spring
HAT	Highest Astronomical Tide
MHWN	Mean High Water Neap

J. Mar. Sci. Eng. **2015**, 3, 674–698

NAO   North Atlantic Oscillation
NGF   Nivellement Général Français
RMSE   Root Mean Square Error
SHOM   Service Hydrographique et Océanographique de la Marine
WW3   WAVEWATCH III model

References

1.   Battjes, J.A. Run-up distributions of waves breaking on slopes. *J. Waterw. Harb. Coast. Eng. Div. ASCE* **1971**, *97*, 91–114.
2.   Edelman, T.I. Dune erosion during storm conditions. In Proceedings of the 11th Conference on Coastal Engineering, London, UK, September 1968; pp. 719–722.
3.   Van der Meulen, T.; Gourlay, M.R. Beach and dune erosion tests. In Proceedings of the 11th Conference on Coastal Engineering, London, UK, September 1968; pp. 701–707.
4.   Edelman, T. Dune erosion during storm conditions. In Proceedings of the 13th Conference on Coastal Engineering, Vancouver, BC, Canada, 10–14 July 1972; pp. 1305–1311.
5.   Van de Graaff, J. Dune erosion during a storm surge. *Coast. Eng.* **1977**, *1*, 99–134. [CrossRef]
6.   Van de Graaff, J. Probabilistic design of dunes; an example from the Netherlands. *Coast. Eng.* **1986**, *9*, 479–500. [CrossRef]
7.   Stockdon, H.F.; Sallenger, A.H.; Holman, R.A.; Howd, P.A. A simple model for the spatially-variable coastal response to hurricanes. *Mar. Geol.* **2007**, *238*, 1–20. [CrossRef]
8.   Vellinga, P. Beach and dune erosion during storm surges. *Coast. Eng.* **1982**, *6*, 361–387. [CrossRef]
9.   Fisher, J.S.; Overton, M.F. Numerical model for dune erosion due to wave uprush. In Proceedings of the 19th Coastal Engineering Conference, Houston, TX, USA, 3–7 September 1984; pp. 1553–1558.
10. Kriebel, D.L.; Dean, R.G. Numerical simulation of time-dependent beach and dune erosion. *Coast. Eng.* **1985**, *9*, 221–245. [CrossRef]
11. Kriebel, D.L. Verification study of a dune erosion model. *Shore Beach* **1986**, *54*, 13–21.
12. Overton, M.; Fisher, J.; Young, M. Laboratory Investigation of Dune Erosion. *J. Waterw. Port. Coast. Ocean Eng.* **1988**, *114*, 367–373. [CrossRef]
13. Carter, R.W.G.; Stone, G.W. Mechanisms associated with the erosion of sand dune cliffs, Magilligan, Northern Ireland. *Earth Surf. Process. Landf.* **1989**, *14*, 1–10. [CrossRef]
14. Pye, K.; Neal, A. Coastal dune erosion at Formby Point, north Merseyside, England: Causes and Mechanisms. *Mar. Geol.* **1994**, *119*, 39–56. [CrossRef]
15. Larson, M.; Erikson, L.; Hanson, H. An analytical model to predict dune erosion due to wave impact. *Coast. Eng.* **2004**, *51*, 675–696. [CrossRef]
16. Erikson, L.H.; Larson, M.; Hanson, H. Laboratory investigation of beach scarp and dune recession due to notching and subsequent failure. *Mar. Geol.* **2007**, *245*, 1–19. [CrossRef]
17. Claudino-Sales, V.; Wang, P.; Horwitz, M.H. Factors controlling the survival of coastal dunes during multiple hurricane impacts in 2004 and 2005: Santa Rosa barrier island, Florida. *Geomorphology* **2008**, *95*, 295–315. [CrossRef]
18. Sallenger, A.H. Storm impact scale for barrier islands. *J. Coast. Res.* **2000**, *16*, 890–895.
19. Ruggiero, P.; Komar, P.D.; McDougal, W.G.; Marra, J.J.; Beach, R.A. Wave runup, extreme water levels and erosion of properties backing beaches. *J. Coast. Res.* **2001**, *17*, 407–419.
20. Pye, K.; Blott, S.J. Decadal-scale variation in dune erosion and accretion rates: An investigation of the significance of changing storm tide frequency and magnitude on the Sefton coast, UK. *Geomorphology* **2008**, *102*, 652–666. [CrossRef]
21. Stockdon, H.F.; Holman, R.A.; Howd, P.A.; Sallenger, A.H. Empirical parameterization of setup, swash, and runup. *Coast. Eng.* **2006**, *53*, 573–588. [CrossRef]
22. Wassing, F. *Model Investigation on Wave Run-Up Carried Out in the Netherland during the Past Twenty Years*; Amer. Soc. Civil Engrs: Gainesville, FL, USA, 1957; Volume 6, pp. 700–714.
23. Hunt, I.A. Design of seawalls and breakwaters. *J. Waterw. Harb. Div.* **1959**, *85*, 123–152.
24. Guza, R.T.; Thornton, E.B. Swash oscillations on a natural beach. *J. Geophys. Res. Ocean.* **1982**, *87*, 483–491. [CrossRef]

25. Holman, R.A. Extreme value statistics for wave run-up on a natural beach. *Coast. Eng.* **1986**, *9*, 527–544. [CrossRef]

26. Longuet-Higgins, M.S.; Stewart, R.W. Radiation stress and mass transport in gravity waves, with application to. *J. Fluid Mech.* **1962**, *13*, 481–504. [CrossRef]

27. Battjes, J.A. Surf similarity. In Proceedings of the 14th Conference on Coastal Engineering, Copenhagen, Denmark, 24–28 June 1974; pp. 466–480.

28. Holman, R.A.; Sallenger, A.H. Setup and swash on a natural beach. *J. Geophys. Res. Ocean* **1985**, *90*, 945–953. [CrossRef]

29. Ruessink, B.G.; Kleinhans, M.G.; van den Beukel, P.G.L. Observations of swash under highly dissipative conditions. *J. Geophys. Res.* **1998**, *103*, 3111–3118. [CrossRef]

30. Ruggiero, P.; Holman, R.A.; Beach, R.A. Wave run-up on a high-energy dissipative beach. *J. Geophys. Res. Ocean.* **2004**, *109*. [CrossRef]

31. Mase, H. Random Wave Runup Height on Gentle Slope. *J. Waterw. Port Coast. Ocean Eng.* **1989**, *115*, 649–661. [CrossRef]

32. Nielsen, P.; Hanslow, D.J. Wave Runup Distributions on Natural Beaches. *J. Coast. Res.* **1991**, *7*, 1139–1152.

33. Cariolet, J.-M.; Suanez, S. Runup estimations on a macrotidal sandy beach. *Coast. Eng.* **2013**, *74*, 11–18. [CrossRef]

34. Guilcher, A.; Hallégouët, B. Coastal Dunes in Brittany and Their Management. *J. Coast. Res.* **1991**, *7*, 517–533.

35. Suanez, S.; Cariolet, J.-M.; Fichaut, B. Monitoring of recent morphological changes of the dune of Vougot beach (Brittany, France) using differential GPS. *Shore Beach* **2010**, *78*, 37–47.

36. Suanez, S.; Cariolet, J.-M. L'action des tempêtes sur l'érosion des dunes: Les enseignements de la tempête du 10 mars 2008. *Norois* **2010**, *215*, 77–99. [CrossRef]

37. Suanez, S.; Cariolet, J.-M.; Cancouët, R.; Ardhuin, F.; Delacourt, C. Dune recovery after storm erosion on a high-energy beach: Vougot Beach, Brittany (France). *Geomorphology* **2012**, *139–140*, 16–33. [CrossRef]

38. Blaise, E.; Suanez, S.; Stéphan, P.; Fichaut, B.; David, L.; Cuq, V.; Autret, R.; Houron, J.; Rouan, M.; Floc'h, F.; *et al.* Bilan des tempêtes de l'hiver 2013–2014 sur la dynamique de recul du trait de côte en Bretagne. *Géomorphol. Relief Process. Environ.* **2015**, in press.

39. Senechal, N.; Coco, G.; Bryan, K.R.; Holman, R.A. Wave runup during extreme storm conditions. *J. Geophys. Res. Ocean* **2011**, *116*. [CrossRef]

40. Rascle, N.; Ardhuin, F. A global wave parameter database for geophysical applications. Part 2: Model validation with improved source term parameterization. *Ocean Model.* **2013**, *70*, 174–188. [CrossRef]

41. Roland, A.; Ardhuin, F. On the developments of spectral wave models: Numerics and parameterizations for the coastal ocean. *Ocean Dyn.* **2014**, *64*, 833–846. [CrossRef]

42. De Bakker, A.T.M.; Tissier, M.F.S.; Ruessink, B.G. Shoreline dissipation of infragravity waves. *Cont. Shelf Res.* **2014**, *72*, 73–82. [CrossRef]

43. Mayer, R.H.; Kriebel, D.L. Wave runup on composite-slope and concave beaches. In Proceedings of the 24th Coastal Engineering Conference, ASCE, Kobe, Japan, 23–28 October 1994; pp. 2325–2339.

44. McCallum, E.; Norris, W.J.T. The storms of January and February 1990. *Meteorol. Mag.* **1990**, *119*, 201–210.

45. Betts, N.L.; Orford, J.D.; White, D.; Graham, C.J. Storminess and surges in the South-Western Approaches of the eastern North Atlantic: The synoptic climatology of recent extreme coastal storms. *Mar. Geol.* **2004**, *210*, 227–246. [CrossRef]

46. Castelle, B.; Marieu, V.; Bujan, S.; Splinter, K.D.; Robinet, A.; Sénéchal, N.; Ferreira, S. Impact of the winter 2013–2014 series of severe Western Europe storms on a double-barred sandy coast: Beach and dune erosion and megacusp embayments. *Geomorphology* **2015**, *238*, 135–148. [CrossRef]

47. NOAA's Climate Prediction Center Home Page. Available online: http://www.cpc.ncep.noaa.gov (accessed on 7 February 2015).

48. Masselink, G.; Austin, M.; Scott, T.; Poate, T.; Russell, P. Role of wave forcing, storms and NAO in outer bar dynamics on a high-energy, macro-tidal beach. *Geomorphology* **2014**, *226*, 76–93. [CrossRef]

49. Thomas, T.; Phillips, M.R.; Williams, A.T. Mesoscale evolution of a headland bay: Beach rotation processes. *Geomorphology* **2010**, *123*, 129–141. [CrossRef]

50. Thomas, T.; Phillips, M.R.; Williams, A.T.; Jenkins, R.E. Medium timescale beach rotation; gale climate and offshore island influences. *Geomorphology* **2011**, *135*, 97–107. [CrossRef]

J. Mar. Sci. Eng. **2015**, *3*, 674–698

51. O'Connor, M.C.; Cooper, J.A.G.; Jackson, D.W.T. Decadal Behavior of Tidal Inlet-Associated Beach Systems, Northwest Ireland, in Relation to Climate Forcing. *J. Sediment. Res.* **2011**, *81*, 38–51. [CrossRef]
52. Vespremeanu-Stroe, A.; Constantinescu, S.; Tatui, F.; Giosan, L. Multi-decadal Evolution and North Atlantic Oscillation Influences on the Dynamics of the Danube Delta Shoreline. *J. Coast. Res.* **2007**, *2007*, 157–162.
53. Montreuil, A.-L.; Bullard, J.E. A 150-year record of coastline dynamics within a sediment cell: Eastern England. *Geomorphology* **2012**, *179*, 168–185. [CrossRef]

Journal of
Marine Science and Engineering

MDPI

Article

Wind and Wave Setup Contributions to Extreme Sea Levels at a Tropical High Island: A Stochastic Cyclone Simulation Study for Apia, Samoa

Ron Karl Hoeke *, Kathleen L. McInnes and Julian G. O'Grady

CSIRO Oceans and Atmosphere Flagship, PMB#1 Aspendale, VIC 3195, Australia;
kathy.mcinnes@csiro.au (K.L.M.); julian.o'grady@csiro.au (J.G.O.)

* Author to whom correspondence should be addressed; ron.hoeke@csiro.au; Tel.: +61-3-9239-4400 (ext. 123); Fax: +61-3-9239-4444.

Academic Editor: Rick Luettich
Received: 29 June 2015; Accepted: 14 September 2015; Published: 22 September 2015

Abstract: Wind-wave contributions to tropical cyclone (TC)-induced extreme sea levels are known to be significant in areas with narrow littoral zones, particularly at oceanic islands. Despite this, little information exists in many of these locations to assess the likelihood of inundation, the relative contribution of wind and wave setup to this inundation, and how it may change with sea level rise (SLR), particularly at scales relevant to coastal infrastructure. In this study, we explore TC-induced extreme sea levels at spatial scales on the order of tens of meters at Apia, the capitol of Samoa, a nation in the tropical South Pacific with typical high-island fringing reef morphology. Ensembles of stochastically generated TCs (based on historical information) are combined with numerical simulations of wind waves, storm-surge, and wave setup to develop high-resolution statistical information on extreme sea levels and local contributions of wind setup and wave setup. The results indicate that storm track and local morphological details lead to local differences in extreme sea levels on the order of 1 m at spatial scales of less than 1 km. Wave setup is the overall largest contributor at most locations; however, wind setup may exceed wave setup in some sheltered bays. When an arbitrary SLR scenario (+1 m) is introduced, overall extreme sea levels are found to modestly decrease relative to SLR, but wave energy near the shoreline greatly increases, consistent with a number of other recent studies. These differences have implications for coastal adaptation strategies.

Keywords: storm surge; coral reefs; waves; sea level; islands; climate; tropical cyclone

1. Introduction

Tropical cyclones pose a significant hazard for many small island nations and have been estimated to account for 76% of natural disasters in the Pacific [1]. The generation of extreme coastal sea levels from storm surges and tides (referred to here as storm tides) and wind-waves are particularly hazardous for coastal communities, many of which are expanding through population pressure and tourism [2]. Understanding the likelihood of extreme sea levels under present and future climate conditions is, therefore, important for resilient coastal habitation and development. However, sea level observations, from which the likelihood of extreme sea levels can be assessed, are often of insufficient number and length to enable robust assessments, particularly given the relative infrequency of tropical cyclone (TC) occurrences at any given coastal location [3].

Extreme sea levels arise from a combination of factors including astronomical tides, storm surges, and wave breaking processes that lead to wave setup and run-up. Storm surge is caused by the inverse barometer effect (IBE) together with surface wind stress acting over coastal seas (which produces wind setup, see for example Pugh [4]). Wave setup is the increase in mean water levels due to the wind-wave dissipation and wave run-up is the maximum extent of the instantaneous wave uprush at the coast. These contributions have a morphological dependence: storm surge tends to be the dominant contributing factor on wide-shelved continental coastlines, and wave setup is often assumed to contribute less than 10% to the total water level (e.g., [5,6]). Wave setup and run-up, on the other hand, have been shown to be an important contributor to extreme sea levels, particularly along steep-shelved coastlines and narrow fringing reefs that characterize many small islands and atolls, e.g., [6–9]. In addition, whereas the cyclone-induced storm surge tends to be concentrated in the region of maximum onshore winds close to the cyclone center, wind-waves propagate with little loss of energy over the deep ocean, and so can increase the scale and duration over which damaging coastal impacts occur during a TC event [10].

McInnes et al. [11] applied a stochastic and numerical modelling approach to assess the frequency and magnitudes of storm tides around the Samoa archipelago under TC conditions. However, that study did not consider wind-wave processes and was regional in scale: extreme sea-level patterns below the scale of a kilometer were not resolved. An additional study, reported in Hoeke et al. [12], used a high-resolution storm surge model, coupled with a spectral (phase-averaged) wave model as well as a one-dimensional phase-resolved wave model, and inputs from McInnes et al. [11] to produce TC-induced inundation hazard information for the redevelopment of the Samoan parliament building, located in Apia (the capitol city of Samoa). In this study, we expand on these two earlier works by performing a series of additional simulations with the purpose of investigating the scale and spatial variation of the relative contribution of wind setup and wave setup to TC-induced extreme sea levels, as well as how these dynamics may change with SLR. This is accomplished by forcing the aforementioned high-resolution coupled wave and storm surge model of the Apia coastline with two sets (ensembles of 30 members each) of stochastically generated TCs. The two sets of cyclones produce storm tides near the statistical 50-year and 100-year recurrence interval (return period) levels at Apia, respectively, according to the findings of McInnes, et al. [11]. Three forcing "regimes" are considered: "wind", "wave", and "all". These forcing regimes allow the separate contribution of wind and wave setup, and how they interact, to be examined; the ability to separate forcing mechanisms is an advantage of numerical modelling exploited by past studies [13]. Two background sea-level scenarios are also considered: a "baseline" scenario which sets sea level to the mean local level between 1980 and 1999 relative to recently surveyed topography and bathymetry, and a "SLR" scenario where 1 m is added to the baseline scenario. The SLR scenario is not intended as a quantitative projection of local sea level rise, and no time horizon is assigned to it, although it is within the upper range of regional sea level projections for 2081–2100 [14]. The SLR scenario is instead a proxy to examine potential changes in wind and wave setup with an increased background sea level.

The coastal morphology of Apia (and of most of the Samoan Archipelago) is that of a basaltic island surrounded by fringing reefs, a common morphology of tropical and sub-tropical oceanic islands worldwide and may be considered representative of morphological analogues in the Indian and Atlantic, as well as the Pacific.

The remainder of the paper is organized as follows. The next section gives an overview of the study site, the methodological approach and the implementation of models that are used. Model results are presented in Section 3, followed by a discussion and conclusions in Sections 4 and 5, respectively.

2. Experimental Section

2.1. Study Site and Context

The larger islands of the Samoan Archipelago are volcanic high islands, surrounded by a relatively narrow littoral strip of complex fringing reefs, frequently incised with channels and embayments, often associated with terrestrial streams. Apia, on the northern shore of the island of Upolu (Figure 1), is the capitol city of the Independent State of Samoa and site of most of the nation's infrastructure.

Tides in Samoa are mixed semi-diurnal with mean daily range of 0.86 m and a typical spring tidal range of 1.3 m. While the northern coast of Upolu is frequently impacted by both local trade wind seas and long-period swell associated with northern hemisphere mid-latitude storms, analysis of a 34-year wave hindcast [15] indicates that all historical wave events with greater than a five-year recurrence interval were associated with the passage of named tropical cyclones [12]. Thus, extreme wind-wave events impacting Apia are most likely generated by TCs.

Several tropical cyclones have affected Samoa in recent decades, including TC Ofa in 1990, TC Val in 1991, TC Tui in 1998, TC Heta in 2004 and most recently TC Evan in 2012. In February 1990, TC Ofa killed seven people and caused damages estimated at US$130 million [16]. TC Val in December of the following year was more destructive due to the slow movement of the cyclone, causing 15 deaths, the loss of 95% of homes in Samoa and approximately US$200 million damage [17,18]. The tide gauge at Apia was not installed until 1993 [19], so only anecdotal information is available on extreme water levels arising from the earlier cyclones. Much of the destruction during TC Ofa was caused by waves and high water levels [18]. Deep water wave heights during TC Val were reported to reach 7.5 m while maximum water levels were up to 1.6 m and resulted in the Apia Observatory (on the Mulinu'u Peninsula, Figure 1) being submerged to 0.5 m [20,21]. These reports, though providing only qualitative comparisons to the modelling outputs for the most part, have been used to assess and validate the models and results presented in this study.

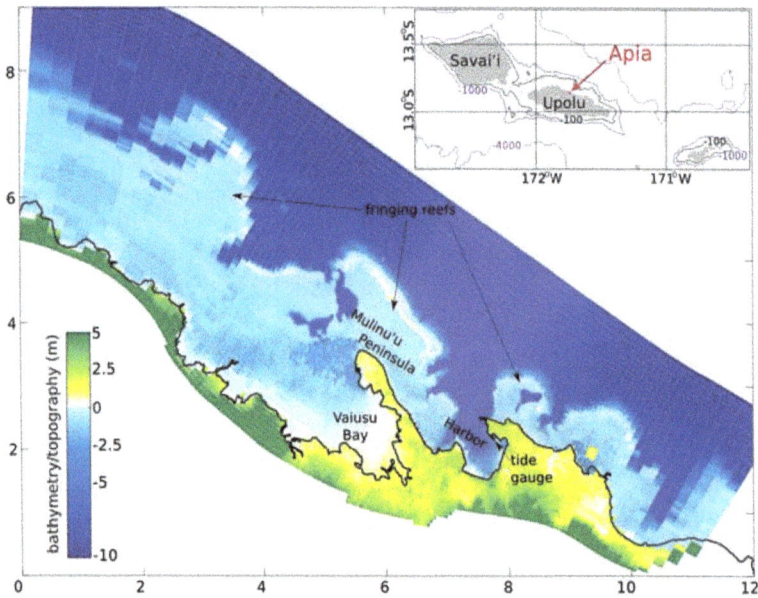

Figure 1. Location map of Apia in the Samoan Archipelago (inset, upper left) and the Apia Model's curvilinear grid and bathymetry/topography (main panel) with key geographical features labeled. Note grid resolution varies from approximately 100 m near the boundaries to approximately 10 m near the center. The numbers on the x- and y-axes indicate distance in kilometers.

2.2. Ensemble Methodology

The TC ensembles used in this study are drawn from the stochastic cyclone and storm tide modelling study of McInnes et al. [11]. The methodology for generating and selecting cyclones, and subsequently forcing the (high-resolution) Apia model simulations, is illustrated schematically in Figure 2. Steps 1 to 3 summarise the approach used in McInnes et al. [11], in which synthetically-generated populations of tropical cyclone histories were developed for Samoa. Steps 1 to 3 of [11] are described briefly here for completeness. Step 1 involved analysing historical cyclones in an 8° radius of Samoa; the large radius being to ensure sufficient events to develop empirical probability density functions (PDFs) for cyclone attributes of track, translation, speed, and intensity. The PDFs were then sampled to generate synthetic storm tracks. In step 2 these were used in conjunction with the analytical cyclone ("vortex") model described in Holland [22] to develop wind and pressure gridded fields for each synthetic storm track for a 1-km resolution archipelago-scale hydrodynamic model that simulated the water levels due to the resultant storm surge. An example of Holland vortex winds and the extent of the 1-km resolution archipelago-scale hydrodynamic model are given in Figure 3a. Astronomical tides were also included by randomly assigning to each storm track a commencement time during the cyclone season within a tidal epoch and simulating the associated tide variations to ensure random phasing of tides with each synthetic cyclone event. Analysis of the synthetically generated sea level maxima in step three involves assigning an average frequency of occurrence according to the observed cyclone frequency within the region covered by the model grid to develop storm tide recurrence intervals (RIs). The 50 and 100-year storm tide RIs were assessed to be 0.81 m and 0.92 m respectively [11].

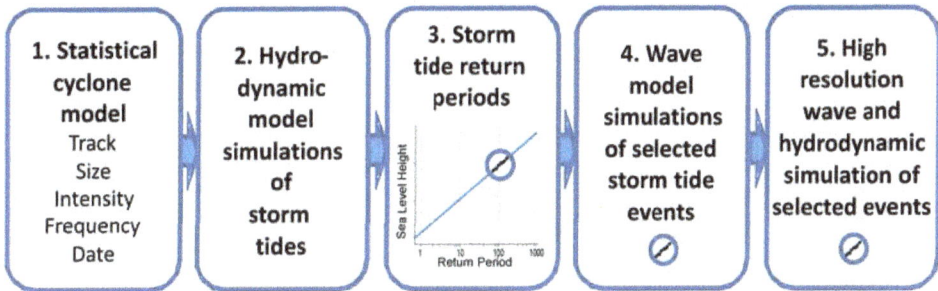

Figure 2. Schematic illustrates the method adopted in this study. Steps 1 to 3 relate to the study of McInnes et al. [11] in which synthetic cyclone and hydrodynamic models are used to estimate storm tide return periods. Steps four and five are undertaken in this study to evaluate the additional contribution from wave setup.

Given the large deep-water fetches and long period waves typical of Pacific TCs [2], the "wave" and "all" forcing regimes of the Apia model require wave modelling of a large ocean region to produce realistic wave boundary conditions (step four of Figure 2). The archipelago spectral wave model implemented to do this (see next section), as well as the wave module of the Apia model itself, are an order of magnitude more computationally expensive than the depth integrated storm surge model at similar scales; therefore, simulating wave fields for the thousands of TC events generated in step one would be prohibitive for most practical applications. In this study the problem has been made tractable by selecting, from the total population described by McInnes [11], the 30 cyclones with maximum storm tide values nearest to the 50 and 100-year RI values at Apia, respectively. A total of 30 synthetic TC ensemble members was chosen because this (qualitatively) appeared to capture most probable cyclone approaches to the Apia shoreline for both the 50 and 100 year RI ensembles; these two RIs were selected because they are commonly of interest for engineering and planning applications. This "reduced ensemble" approach allows the range of variability in wave setup to be explored for a given storm tide level. As will be shown in the next section, not all parts of the coastline are exposed to wave setup; indeed the location of the tide gauge within the sheltered deep water harbour at Apia is one such location. Without *a priori* information on wave setup, this supports the two-step approach adopted in this study of first evaluating the storm tide RIs (i.e., no waves) from the full population of stochastic

TCs, which are relevant for locations little affected by waves (e.g., the tide gauge), and second building in the additional contribution from the waves, through subsequent high-resolution modelling.

In summary, three different forcing regimes ("wind", "wave", and "all") are considered for each TC in the two 30-member ensembles, one corresponding to the 50-year RI storm tide levels and the other, to the 100-year RI levels, as defined by McInnes et al. [11]. Two different background sea levels are also considered: "baseline" and "SLR". This results in a total of 12 ensembles (each with 30 members), or 360 total simulations of the Apia model described in the next section (not including historical simulations used for model validation).

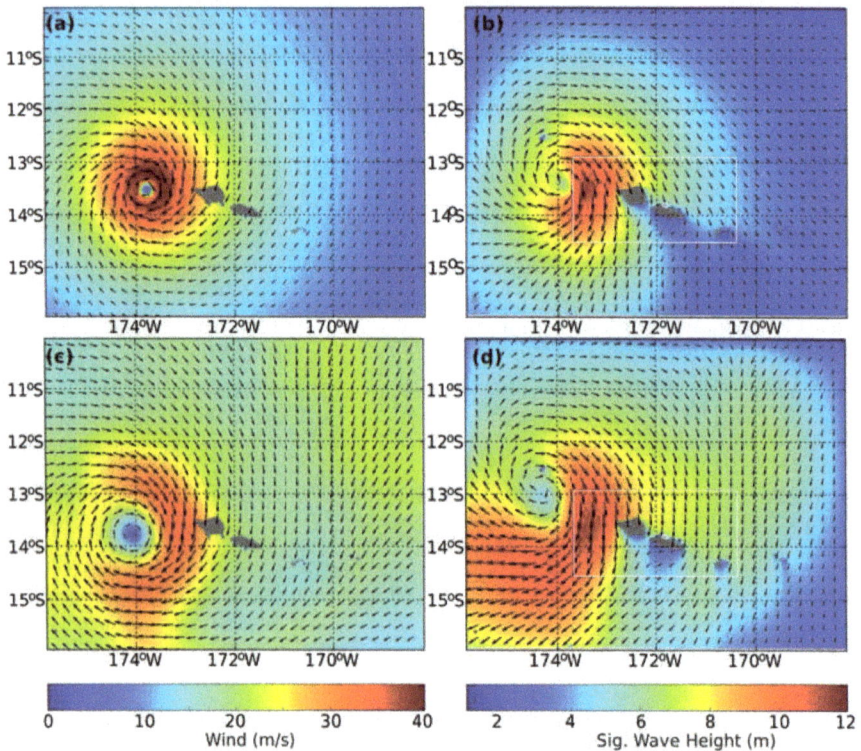

Figure 3. Example of Holland vortex winds (**a**) and simulated significant wave height and peak direction (**b**) during the historical TC Ofa (on 2 February 1990, 1200UTC). CFSR winds (**c**) and simulated waves (**d**) are included for comparison. The larger white box in (**b**) and (**d**) indicate the 1-km resolution nested SWAN model region, the smaller white box indicates the location and extents of the Apia Model.

2.3. Model Implementation

2.3.1. Archipelago Wave Model

In order to provide wave boundary conditions associated with the synthetic and the historic TCs, an archipelago-scale nested Simulating WAves Nearshore (SWAN) model (version 41.01) was implemented over the region. The SWAN model predicts the evolution in time and space of the wave action spectrum [23]; here, spectral resolution was as follows: 72 directional bins (5° resolution) and 24 frequency bins, logarithmically spaced between 0.042 and 0.411 Hz. A number of previous studies have shown SWAN's default bulk wind input formulation [24] may overestimate wind drag coefficients (C_d) in tropical cyclone wind conditions [25,26]; C_d is therefore capped at 2.5 m^2/s and the wind source term implementation is set to that reported in Janssen [27].

The SWAN model's outer boundaries were the same as the modelling boundaries described in [11], i.e., that of the TC wind fields (Figure 3a). Spatial resolution was set to 5 km, with a 1-km resolution nest centred on the islands of Upolu and Savai'i (Figure 3b). This model was validated by comparing wave fields generated by the SWAN model using both synthetic (Holland vortex) and Climate Forecast System Reanalyses (CFSR, [28]) winds with a multi-year wave hindcast for the region [15] for a number of historical TCs (e.g., Figures 3 and 4). Comparisons between CFSR and Holland vortex wave fields are covered briefly in the results section of this paper; however the validation is presented at some length in [12].

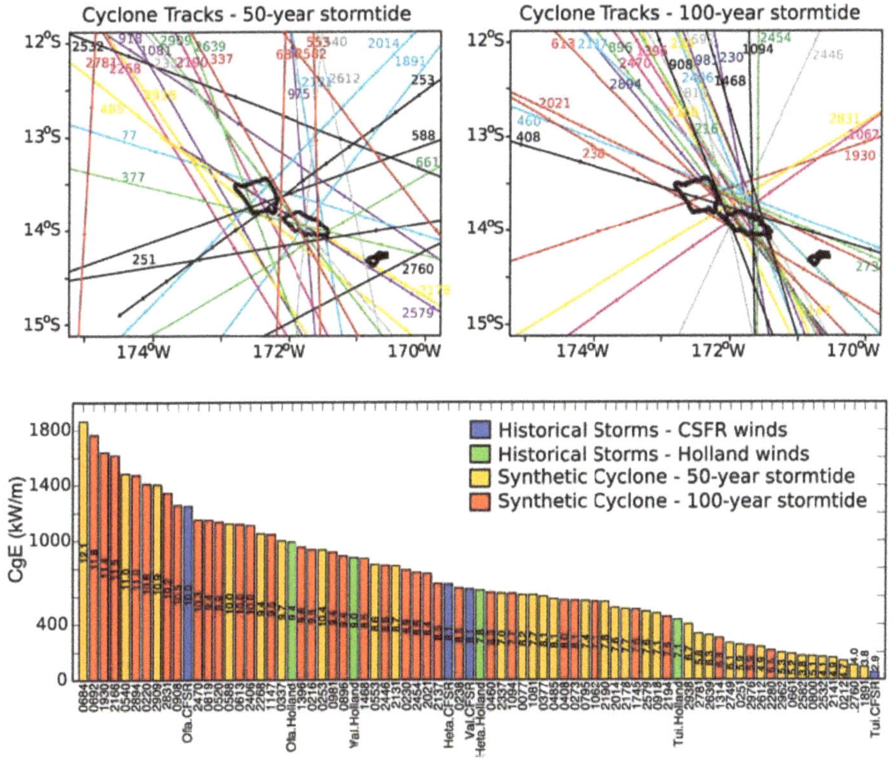

Figure 4. Upper panels: synthetic cyclone tracks used in this study: selected 50-year RI ensemble (**left**) and 100-year ensemble tracks (**right**). Lower panel: maximum wave energy flux (C_gE) calculated by the SWAN model at a point just north of the Apia Model's boundary for each of the events modeled in this study; labels on the bars indicate corresponding significant wave height.

2.3.2. Apia Model

For the simulation of extreme sea levels around Apia during TC conditions, the open-source version (6.01.15.5013) of the Delft3D flow-wave-coupled hydrodynamic modelling system [29] was used. This system has been widely applied to combined storm surge/wave setup processes, e.g., [30,31] and has been successfully applied to tropical fringing-reefs such as those found in Samoa, e.g., [32,33]. This system's "flow" module consists of a finite-difference solution to the Navier-Stokes equations for unsteady flow; the "wave" module is the previously described SWAN model. The two modules are iteratively coupled so total dissipation forces (radiation stresses) from the wave module are passed to the flow module to compute wave-induced residual flow and Stokes drift; the subsequent water levels and currents in the circulation module are passed back to the wave module to calculate an updated wave field. The Delft 3D model implementation used here (called the Apia model henceforth) was set up on a 2D curvilinear grid, which varied in spatial resolution from approximately 200 m near the northwest and southeast (lateral) boundaries, to approximately 10 m near the Mulinu'u Peninsula and Apia Harbor (Figure 1). This spatial resolution was made feasible by a high-resolution light detection and ranging (LiDAR) topography and bathymetry survey of the Apia area.

Astronomical tidal predictions, based on the Apia tide gauge, and background sea level were combined with linearly interpolated inverse barometer effect (IBE) from the archipelago storm tide model to produce a spatially- and temporally-varying offshore water level boundary; shore-normal lateral boundaries were calculated as water level gradients, i.e., Neumann boundary conditions [34]. Wave boundary conditions were supplied as directional spectra from the Archipelago wave model at points along the Apia model boundary; spectral resolution and source terms were set the same as for the archipelago SWAN model. Wave dissipation and subsequent wave setup over reefs have been shown to be sensitive to hydraulic bed roughness (f_w) and the wave breaking index (λ_b), defining the water depth at which waves will break, i.e., $H_b = h\lambda_b$ where H_b is breaking wave height, h is water depth. In keeping with a number of studies [8,33,35,36], we implement a spatially-varying f_w grid with roughness lengths equivalent to 0.15 m in reef areas and 0.02 in sandy areas [37], and set the λ_b to 1.0 based on mean fore-reef bottom slope according to the empirical relationship of Raubennheimer [38].

Water level observations at the Apia tide gauge and a nearby 11-month long mooring deployment were well correlated ($R \geq 0.98$) with those predicted by the Apia model; root mean square errors (RMSE) were less than 0.10 m at those locations. Reports of inundation extent and damage correspond well with that modelled, particularly for Cyclone Ofa, which experienced the worst coastal inundation of recent decades. Further details on validation, model parameters and other information are detailed in Hoeke et al. [12].

The three different forcing regimes were all simulated using the "flow" module with the same water level boundary forcing: in the "wind" regime the flow module was not coupled to the "wave" module. Thus, no wave radiation stresses were present; in the "wave" regime, the flow module was coupled to the wave module but assumed a surface wind stress of zero; in the "all" regime both wind and wave forcing was present.

3. Results

3.1. Archipelago Wave Model Results

Wave model simulations were performed for each member of the two synthetic TC ensembles (corresponding to 50 and 100-year storm tide RIs). The tracks of the cyclones (Figure 4: upper panels), indicated that those contributing to the 50-year storm tide levels were spread across the archipelago, while those producing the 100-year storm tide levels crossed in much closer proximity to Apia, consistent with these events leading to higher coastal sea levels through inverse barometer effect and wind setup (storm surge). An example of the modelled wind and wave field during a synthetic event is shown in Figure 3a,b. Maximum deepwater wave energy flux (C_gE, i.e., the power available to drive wave setup and other shallow water wave processes) for all ensemble members for a point on the Apia model's offshore boundary is shown in Figure 4 (lower panel). These maxima exhibit a broad range of values across the two ensemble members (60 storms in total); corresponding significant wave heights (H_s) range from 3.8 to 12.1 m. This upper range is within the empirical values found by Stephens and Ramsay [2] for Samoa: 10.6–13.7 m and 11.6–15.3 m for 50 and 100-year RI, respectively. The smaller values of many of the ensemble members found in this study is most likely due to the fact that Stephens and Ramsay [2] did not include any wave sheltering effects (e.g., of islands), whereas here they are dynamically included. The ranking of ensemble members by C_gE in Figure 4 (lower panel) indicates that while maximum wave heights generated by the 100-year ensemble are generally higher than those within the 50-year ensemble, this is not necessarily always the case; the maximum overall C_gE value in Figure 4 is actually from a member of the 50-year ensemble. This is a reflection of the fact that factors besides TC proximity to Apia and wind strength are important in determining maximum C_gE impingent on Apia, i.e., TC track orientation, radius of maximum winds and propagation speed. Furthermore this indicates that local maximum CgE is not necessarily well correlated with the local storm surge level (i.e., IBE and wind setup) simulated by McInnes et al. [11].

Historical TC waves simulated with either CFSR winds [28] or Holland vortex wind fields are also included in Figure 4 (lower panel) for comparison; these are scattered amongst the ensemble maxima, indicating no consistent bias in maximum C_gE between the synthetic TC approach and various historical approaches. In particular, the position of the TC Ofa's maximum C_gE in the middle (synthetic forcing) to upper end (CFSR forcing) of the rankings is broadly consistent with the TC's characteristics (radius of maximum winds and central pressure) relative to other historical TC information used to construct the synthetic cyclone population. However, as noted by a number of studies (e.g., [39,40]), CFSR wind fields may not be of sufficient spatial resolution to accurately simulate extreme waves near the TC eye wall; conversely synthetic vortex wind fields may underestimate far-field winds (and thus wave generation) by neglecting to take into account larger-scale meteorological conditions. An example of this is evident in Figure 3: an area of stronger winds and elevated H_s to the east of TC Ofa is evident in the CFSR simulation (Figure 3c,d) relative to the Holland vortex simulation (Figure 3a,b). Several researchers have attempted to ameliorate such issues by blending reanalysis winds with vortex winds (e.g., [40]); this is of course not an option for the synthetic cyclones used in this study, however. Sensitivity of extreme wave prediction to different input wind fields remains poorly understood; something exacerbated by an almost total lack of *in situ* TC wind and wave records in the insular Pacific [2,7]. Further consideration of this is beyond the scope of this study, however, and remains an area of future research. Comparisons of historic and synthetic cyclones and differences in wind forcing are discussed at greater length in [3,11,12,39].

3.2. Apia Model Results

Apia model simulations were typified by rapid wave dissipation through wave breaking and bed friction across the fore-reef slope of the fringing reefs, resulting in wave setup over the shoreward reef flats and rapid offshore return flow (on the order of 1 m/s) in reef passes and channels (Figure 5). This pattern is consistent with a number of studies at other locations, e.g., [13,33,41,42]. Figure 5 also shows the position of four locations ("offshore", "reef", "bay", and "tide") that are discussed in more detail below. It is noteworthy that the location of the tide gauge ("tide") in the harbour is relatively sheltered from both high waves and wave setup.

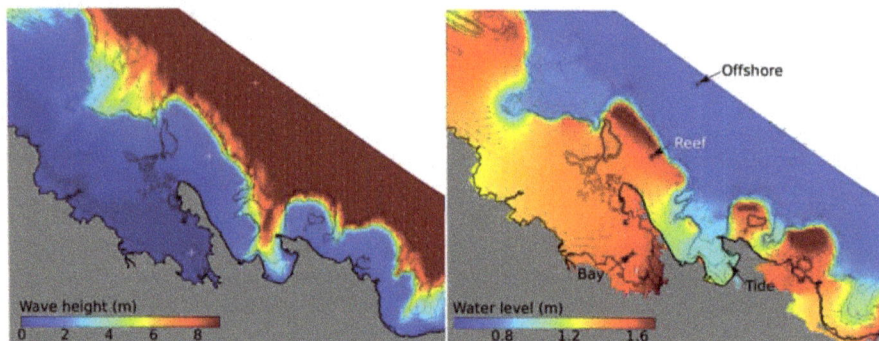

Figure 5. Apia Model output near the peak of local water levels during Cyclone Ofa. (**Left panel**): significant wave height and peak wave direction (**grey arrows**). (**Right panel**): water levels and depth-averaged current vectors (**grey arrows**). Locations discussed in the text are labelled in the right panel.

The spatial water level maxima of the Apia model simulations in the two TC ensembles exhibit a high degree of alongshore and cross-shore variability across all three forcing regimes ("wind", "wave", and "all"). These spatial patterns can be summarized by the median and 95th percentile values of the individual ensemble maxima; these are plotted in Figures 6 and 7 for the 100-year RI baseline and 100-year RI SLR scenario, respectively. (The spatial patterns of 50-year RI ensembles were similar albeit lower in height so are not plotted for brevity). Median values of the 100-year RI baseline ensemble range spatially between 0.9 and 1.9 m, except for wind forcing, which has a lower upper range, around 1.4 m (Figure 6a–c). The low end of the median values occur near the "tide" location in the harbour and are consistent with the 100-year RI for storm tide values (0.92 m) determined by McInnes et al. [11]. These results therefore support the idea that storm tide-only values (such as presented in McInnes et al. [11]) are relevant for the "tide" location, which is largely sheltered from wave-induced sea level contributions and local wind setup, but that large local excursions from this "reference" value occur due to bathymetric and shoreline variations in the local vicinity, which are accounted for in the high-resolution modelling. 95th percentile values from the baseline, 100-year RI ensemble vary between roughly 1.1 and 2.5 m for all three forcings (Figure 6d–f). While the ranges of ensemble values across the three forcings are similar, their spatial patterns are not. The wind forcing regime in particular shows that locally elevated water levels relative to the rest of the model grid are largely restricted to Vaiusu Bay (Figure 6a,d). Wave setup is significant throughout much of the grid in the wave forcing case, especially near the reef crest offshore from the Mulinu'u Peninsula, where values are 1 m or higher than elsewhere in the grid (Figure 6b,e); while lower than at the reef crest, wave forcing exhibits elevated water levels (relative to wind forcing) over much of the reef flat areas and Vaiusu Bay. Simulated "all" forcing regime water levels are qualitatively a combination the wind and wave forcing regime water levels (Figure 6c,f), though not a linear superposition of the two, particularly in areas close to shore. Reasons for this are discussed in the following sections. Patterns of the median and 95th percentile maximum water level values for the 100-year RI ensemble that includes the SLR scenario (Figure 7) are quite similar to the baseline, 100-year RI scenario, except being approximately 1 m higher. Here too, however, there are some significant departures from a simple linear increase between the two, particularly over the reef flat and Vaiusu Bay areas. This is also discussed in more detail in the following sections.

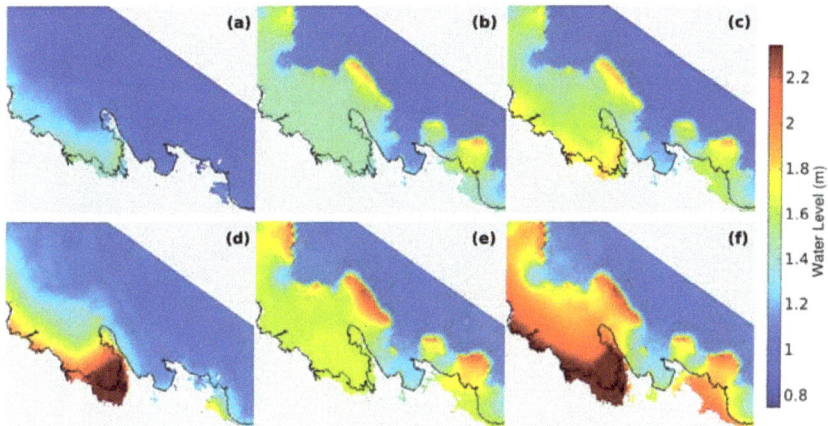

Figure 6. Water level maxima for the 100-year RI ensemble, "baseline" scenario. Median ensemble values for "wind" forcing (**a**); "wave" forcing (**b**); and "all forcing" (**c**). Subplots (**d–f**) are the same as (**a–c**) ensemble except are 95th percentile ensemble values.

It is apparent there is a large departure between the median and 95th percentile values within Vaiusu Bay for wind forcing (Figure 6a,d and Figure 7a,d) relative to other areas/other forcing. This is due in part to the high sensitivity of water levels within the bay to TC track: slight changes in maximum wind direction lead to large changes in local wind setup due to the bay's morphology. In fact, different areas within the grid show highly varying sensitivity to the different forcing regimes. This is summarized in Figure 8, which plots the range of water level maxima amongst ensemble members for each of all ensembles at the four locations ("offshore", "reef", "bay", and "tide") defined in Figure 5. The total range (variance) of maximum water levels among the ensemble members is relatively low, with only small differences between the forcing regimes not only at the offshore location (which would be expected), but also at the tide (gauge) location. At the reef location however, it is low only for the wind forcing regime; the wave and all forcing regimes exhibit more than three times the variance of the wind forcing, indicating the dominance of wave setup on extreme sea level processes at this location. At the bay location (within Vaiusu Bay), however, wind forcing ensembles exhibit a larger range of maximum water levels than the wave forcing ensembles at the "reef" location; closer inspection of these "wind" values, relative to the surrounding coast, reveals a strong relationship to the track of the TCs. Simulations that produced local maximum sea levels in the southern part of Vaiusu Bay tend to be those associated with north-to-south TC tracks and crossing in the immediate vicinity or to the west of Apia, maximising the northerly winds on the eastern flank of the cyclone (e.g., see 50-year RI tracks 684, 975, 2909, and 2190 and 100-year RI tracks 613, 692, 896, 908, 981, 1396 in Figure 4). In the simulations in which maximum setup occurred on the western side of Mulinu'u peninsula, the cyclones typically tracked from north to south on the eastern side of Apia (e.g., 50-year RI tracks 553 and 2131 and 100-year RI tracks 819, 981, and 2454) and produced strong westerly winds after the cyclone centre crossed Samoa. Water levels in Vaiusu Bay tended to be lower than surrounding coastal region (e.g., wind setdown occurred) for fast moving cyclones that followed a more east-west trajectory to the north of Samoa (e.g., 100-year RI tracks 460 and 1147) while cyclones whose track was at a greater distance from Samoa (e.g., 50-year RI track 1081) did not produce significant wind setup in Vaiusu Bay. Conversely, wave setup values across the Apia model domain did not exhibit nearly such local sensitivity to details of cyclone track; these values were more dependent on TC radius of maximum winds and central pressure.

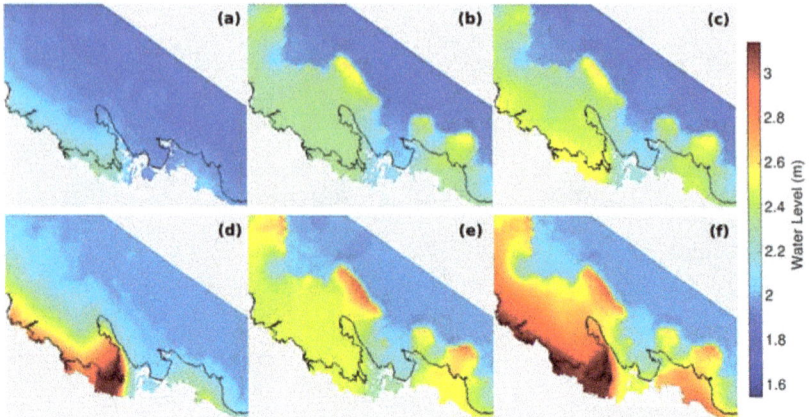

Figure 7. Water level maxima for the 100-year RI ensemble, "SLR" scenario. Median ensemble values for "wind" forcing (**a**); "wave" forcing (**b**); and "all forcing" (**c**); Subplots (**d**–**f**) are the same as (**a**–**c**) ensemble except are 95th percentile ensemble values.

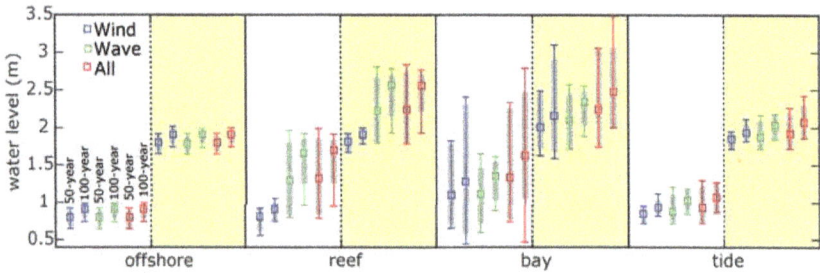

Figure 8. Distribution of maximum water levels amongst 50-year and 100-year RI ensemble members at "offshore", "reef", "bay", and "tide" locations (see Figure 5). Blue, green, and red indicate wind, wave, and all forcing, respectively (as indicated at upper left); median ensemble values are indicated with a square; 5th and 95th percentile values are indicated with grey bars, respectively, and ensemble minima and maxima by colored error bars. At each location indicated, the "baseline" scenario is on the left (white background) and the "SLR" scenario is on the right (yellow background).

4. Discussion

The results indicate that the amount of wind setup and wave setup varies locally by up to approximately a meter for both 50 and 100-year RI cyclones. Wave setup tends to be the dominant process, particularly at coastal locations adjacent to shore-parallel reef crests such as the Mulinu'u Peninsula. At such locations, the results presented here closely approximate a number of analytic/empirical solutions of wave setup which assume straight and parallel alongshore topography and wave conditions, e.g., [8,43,44]. In particular, water levels within several hundred meters shoreward of the reef crest fronting the Mulinu'u Peninsula (e.g., the "reef" location) simulated by the wave forcing ensemble are within approximately 10 centimetres of the solution described by Becker et al. [44], which is plotted for comparison with model output in Figure 9. Closer to shore, or at other locations without a neighbouring shore parallel reef crest of some length, these analytic methods greatly overestimate wave setup. This would be expected, since these solutions generally assume a complete momentum balance between wave radiation stress and pressure (water level) gradient, whereas the 2-dimensional numerical solution presented here allows for return flow dependent on morphological details and bed roughness, as would realistically be the case, particularly in the complex reef morphology typical of many tropical high islands.

While wave setup is the largest contributor to extreme water levels at most locations, that is not the case everywhere. This can be visualized by subtracting the wind forcing simulations from the wave forcing. Wave forcing is clearly dominant on the reef flats adjacent to exposed reef crests (negative values in Figure 10a), however, the high positive values in Figure 10a indicate the dominance of wind setup in Vaiusu Bay. Assuming that these processes are sufficiently locally independent so that separate (uncoupled) models for wind setup and wave setup can be combined (as may be desired) requires caution however. Adding the wind and wave forcing results leads to an overestimate of water levels relative to the "all" forcing in some areas, particularly near the shoreline (Figure 10b). This is primarily due to the increasing overall water levels (and thus depths) associated with the "all" forcing, which effectively decreases depth-integrated hydraulic roughness, allowing for greater volume transport of wind and wave setup offshore, resulting in decreased water levels relative to an independent combination of the two. This is also a reason that local increases in extreme sea levels are approximately 10%–30% less relative to background increases in sea level in the SLR *versus* the baseline scenario (Figure 11a), although decreased wave breaking (radiation stress) at the reef crest may also contribute. While this is potentially good news for coastal communities such as Apia, as it ameliorates the effect of extreme sea levels under conditions of SLR, it should be noted that this is accompanied by a relatively large increase in wave energy reaching the inner reef areas and the coastline as shown in Figure 11b; this is also visible in the higher SLR wave heights (compared to the baseline scenario) across the reef flat plotted in Figure 9.

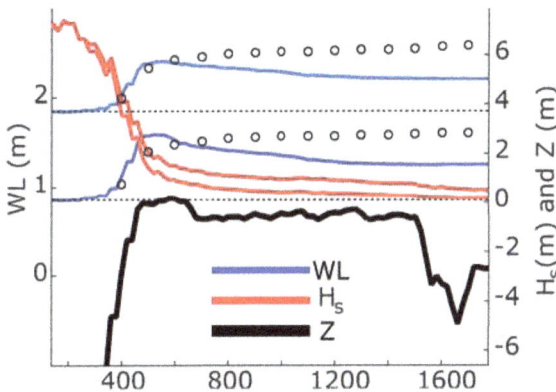

Figure 9. Ensemble (wave forcing, baseline, and SLR scenarios) significant wave heights (H_s, red) and water levels (WL, blue), along a transect through the "reef" location (shown in Figure 5); offshore is on the left side, landward direction towards the right; topography (Z) is indicated with a black line. The analytic estimate of wave setup described by Vetter et al. [8] and Becker et al. [44], based on H_s from the Apia model, is plotted with black circles, with the caveat that breaking parameter λ_b is set to 1.0 (consistent with the Apia model), rather than varying with water level.

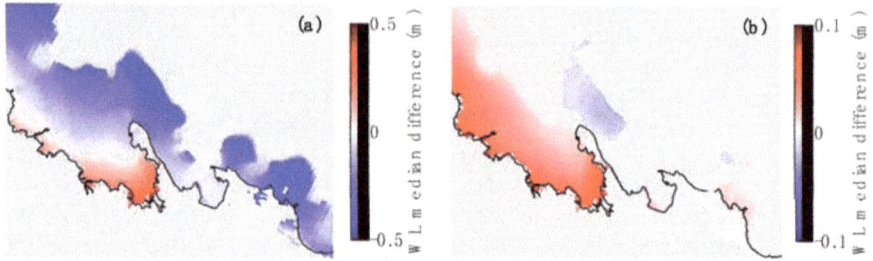

Figure 10. Differences in maximum water levels between "wind", "wave" and "all" forcing. (**a**): median difference between wind and wave maximas (wind minus wave); (**b**): median differences between the sum of wind and wave maximas (with a correction to prevent double counting tides and inverse barometer) and "all" forcing maximas.

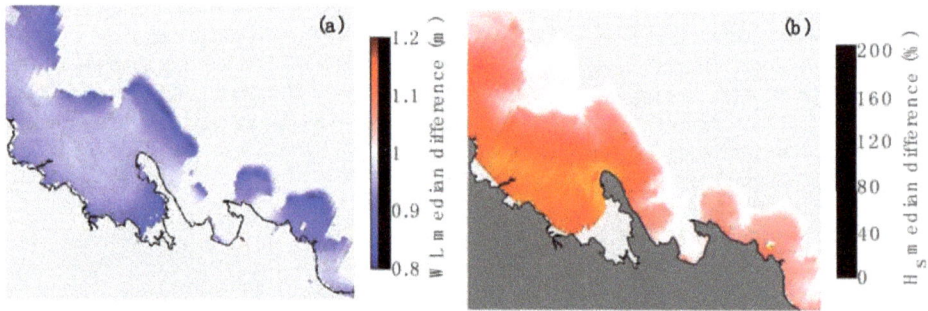

Figure 11. (**a**) Median differences between maximum water levels (WL) of the "SLR" and "baseline" ensemble members ("all" forcing); (**b**) median percent difference between maximum significant wave height (H_s) of the "SLR" and "baseline" ensemble members ("all" forcing). In (**b**) areas with median H_s less than 0.05 m or absolute differences of less than 5% are not plotted.

5. Conclusions

In this study, high-resolution numerical simulations of coupled storm surge, tides, and waves, forced by tropical cyclone (TC) ensembles drawn from a much larger stochastically generated population [11], were used to examine the overall local variance of extreme sea levels and the relative contribution of wind and wave setup. Local water level excursions were found to be very sensitive to local morphology and storm track, with median extreme sea levels (and total ranges) for each ensemble differing by a factor of approximately two or more within less than a kilometre. Wave setup dominates extreme sea level processes in many areas, meaning analyses that do not include wave processes may significantly underestimate extreme sea level likelihoods in such settings. Crucially, both wave and wind setup effects are minimal at some locations, such as near the Apia tide gauge. This is an illustration of how poorly the likelihood of coastal inundation predicated on tide gauge data may represent the coastal areas in the immediate vicinity, particularly in areas that lack continental shelves (e.g., oceanic islands such as those of Samoa), something noted by a number of previous studies [6,7,9].

To maintain tractable computational costs, a reduced member ensemble strategy was used in this study. Ensemble members were *a priori* based on cyclones that approximated the 50 and 100-year recurrence intervals (RI) storm tides at the Apia tide gauge, calculated from the archipelago-scale study of McInnes et al. [11], which did not consider the effects of wind-waves. While this approach is supported by the finding that both studies produced similar storm tide values near the tide gauge site (and other sites little effected by wave setup), the much larger overall spread of water level maxima among ensemble members at neighbouring reefs (such as fronting the Mulinu'u Peninsula,

Figure 1, Figure 8) found in this study indicates that local incident C_gE (and, thus, wave setup) is not necessarily well correlated with the local storm surge level (i.e., IBE + wind setup). This suggests that the same ensemble of TCs may not approximate the same local RI water levels at neighbouring locations. Therefore, future studies which utilize a pragmatic hybrid-downscaling approach similar to that used here would benefit from further investigation of the relationship between wind- and wave-induced sea level extremes, particularly when extrapolating to probabilistic statistics.

Comparisons of the high-resolution simulations show that 1 m of SLR relative to constant topography reduces the relative amount of combined wind and wave setup (on the order of 10%–20%) but leads to a large increase in nearshore wave energy (up to approximately 200%) in many areas. This is primarily due to decreased wave dissipation on outer reefs and is consistent with a number of other studies [32,44]. Seasonal and interannual sea-level variability, which is on the order of 30 cm at many Pacific islands, most likely similarly affects momentum balances in many locations. These local changes in momentum balances will result in changed patterns of inundation and sediment transport, particularly under progressively changing sea level. A problem with the assumption of static (constant) topography used in this and past studies is that it is probably unrealistic; however assuming alternate trajectories is far from straightforward as future changes in reef calcification rates remain unclear [44,45] and synergistic effects between water quality and sea level are likely, e.g., [46]. Changes in these processes may lead to large changes in reef morphology, complexity, and roughness, which would then also lead to large local changes in momentum balances. This is an area of future research important to understanding the future effectiveness of reefs as natural coastal defences in island nations.

Another important caveat is that the phase-averaged wave modelling in this and similar studies does not resolve individual waves or wave groups. As such it cannot resolve processes such as wave run-up, overtopping and infragravity frequency motions [47], which may be on the same order of magnitude as (sea-swell frequency) wind-waves near the shoreline in fringing reef systems [9,40,48,49]. Thus, transient extreme sea levels (e.g., time scales less than 30 min) and the extents of coastal inundation are likely to be significantly underestimated in some areas. Phase-resolved numerical simulation of these motions is yet again an order of magnitude more computationally expensive than the phase-averaged models used here. This highlights the importance of good techniques to minimize (reduce) the number of ensemble members for computationally efficient, yet statistically representative, dynamic downscaling simulations. Improving these statistical techniques, as well as increasing computational capacity and development of more efficient numerical approaches at scales relevant to coastal impacts, is essential to facilitate greater usage of stochastic prediction methods, such as those presented here.

Acknowledgments: Much of the work reported here was supported by CSIRO and the Pacific-Australia Climate Change Science Adaptation Planning program (PACCSAP, www.pacificclimatechangescience.org/), in particular the Samoa Parliament Complex Redevelopment Project. The authors would also like to thank Felix Lipkin and Frank Colberg who contributed to various aspects of this study and CSIRO's Advanced Scientific Computing group for their excellent support. This manuscript was improved by the comments of three anonymous reviewers.

Conflicts of Interest: The authors declare no conflict of interest.

References

1. Diamond, H.J.; Lorrey, A.M.; Knapp, K.R.; Levinson, D.H. Development of an enhanced tropical cyclone tracks database for the southwest Pacific from 1840 to 2010. *Int. J. Climatol.* **2011**, *32*, 2240–2250. [CrossRef]
2. Stephens, S.A.; Ramsay, D.L. Extreme cyclone wave climate in the Southwest Pacific Ocean: Influence of the El Niño Southern Oscillation and projected climate change. *Glob. Planet Chang.* **2014**, *123*, 13–26. [CrossRef]
3. McInnes, K.L.; Walsh, K.J.E.; Hoeke, R.K.; O'Grady, J.G.; Colberg, F.; Hubbert, G.D. Quantifying storm tide risk in Fiji due to climate variability and change. *Glob. Planet Chang.* **2014**, *116*, 115–129. [CrossRef]
4. Pugh, D.T. *Changing Sea Levels: Effects of Tides, Weather and Climate*; Cambridge University Press: Cambridge, UK, 2004.

5. Hubbert, G.D.; McInnes, K.L. A Storm Surge Inundation Model for Coastal Planning and Impact Studies. *J. Coast. Res.* **1999**, *15*, 168–185.

6. Kennedy, A.B.; Westerink, J.J.; Smith, J.M.; Hope, M.E.; Hartman, M.; Taflanidis, A.A.; Tanaka, S.; Westerink, H.; Fai, K.; Smith, T.; *et al.* Tropical cyclone inundation potential on the Hawaiian Islands of Oahu and Kauai. *Ocean Model.* **2012**, *52–53*, 54–68. [CrossRef]

7. Hoeke, R.K.; McInnes, K.L.; Kruger, J.C.; McNaught, R.J.; Hunter, J.R.; Smithers, S.G. Widespread inundation of Pacific islands triggered by distant-source wind-waves. *Glob. Planet Chang.* **2013**, *108*, 128–138. [CrossRef]

8. Vetter, O.; Becker, J.; Merrifield, M.; Pequignet, A.; Aucan, J.; Boc, S.; Pollock, C. Wave setup over a Pacific Island fringing reef. *J. Geophys. Res.* **2010**, *115*, 1–13. [CrossRef]

9. Merrifield, M.A.; Becker, J.M.; Ford, M.; Yao, Y. Observations and estimates of wave-driven water level extremes at the Marshall Islands. *Geophys. Res. Lett.* **2014**, *41*, 7245–7253. [CrossRef]

10. Walsh, K.J.E.; McInnes, K.L.; McBride, J.L. Climate change impacts on tropical cyclones and extreme sea levels in the South Pacific—A regional assessment. *Glob. Planet Chang.* **2012**, *80–81*, 149–164. [CrossRef]

11. McInnes, K.L.; Hoeke, R.K.; Walsh, K.J.E.; O'Grady, J.G.; Hubbert, G.D. Application of a Synthetic Cyclone Method for Assessment of Tropical Cyclone Storm Tides in Samoa. *Nat. Hazards* **2015**, *79*. [CrossRef]

12. Hoeke, R.; McInnes, K.; O'Grady, J.; Lipkin, F.; Colberg, F. *High Resolution Met-Ocean Modelling for Storm Surge Risk Analysis in Apia, Samoa*; CAWCR: Melbourne, Australia, 2014; Volume 71, pp. 1–80.

13. Mulligan, R.P.; Hay, A.E.; Bowen, A.J. Wave-driven circulation in a coastal bay during the landfall of a hurricane. *J. Geophys. Res.* **2008**, *113*. [CrossRef]

14. Australian Bureau of Meteorology and CSIRO. *Climate Variability, Extremes and Change in the Western Tropical Pacific: New Science and Updated Country Reports*; Australian Bureau of Meteorology and CSIRO: Melbourne, Australia, 2014.

15. Durrant, T.; Greenslade, D.; Hemer, M.; Trenham, C. *A Global Wave Hindcast focussed on the Central and South Pacific*; CAWCR: Melbourne, Australia, 2014.

16. Ready, S.; Woodcock, F. The South Pacific and southeast Indian Ocean tropical cyclone season 1989–1990. *Aust. Meteorol. Mag.* **1992**, *40*, 111–121.

17. Gill, J.P. The South Pacific and southeast Indian Ocean tropical cyclone season 1991–1992. *Aust. Meteorol. Mag.* **1994**, *43*, 181–192.

18. Solomon, S.M. *A Review of Coastal Processes and Analysis of Historical Coastal Change in the Vinicity of Apia, Western Samoa*; South Pacific Applied Geoscience Commission: Suva, Fiji, 1994; Volume 208, p. 62.

19. Bureau of Meteorology, Samoa. Available online: http://www.bom.gov.au/pacific/samoa (accessed on 18 September 2015).

20. Rearic, D.M. *Survey of Cyclone Ofa Damage to the Northern Coast of Upolu, Western Samoa*; South Pacific Applied Geoscience Commission: Suva, Fiji, 1990; Volume 104, p. 36.

21. Carter, R. *Design of the Seawall for Mulinu'u Point, Western Samoa*; South Pacific Applied Geoscience Commission: Suva, Fiji, 1987; Volume 78, p. 30.

22. Holland, G. A Revised Hurricane Pressure-Wind Model. *Mon. Weather Rev.* **2008**, *136*, 3432–3445. [CrossRef]

23. Booij, N.; Ris, R.C.; Holthuijsen, L.H. A third-generation wave model for coastal regions 1. Model description and validation. *J. Geophys. Res.* **1999**, *104*, 7649–7666. [CrossRef]

24. Wu, J. Wind-stress coefficients over sea surface from breeze to hurricane. *J. Geophys. Res.* **1982**, *87*, 9704. [CrossRef]

25. Huang, Y.; Weisberg, R.H.; Zheng, L.; Zijlema, M. Gulf of Mexico hurricane wave simulations using SWAN: Bulk formula-based drag coefficient sensitivity for Hurricane Ike. *J. Geophys. Res.* **2013**, *118*, 3916–3938. [CrossRef]

26. Zijlema, M.; van Vledder, G.P.; Holthuijsen, L.H. Bottom friction and wind drag for wave models. *Coast. Eng.* **2012**, *65*, 19–26. [CrossRef]

27. Janssen, P.A.E.M. Wave-Induced Stress and the Drag of Air Flow over Sea Waves. *J. Phys. Oceanogr.* **1989**, *19*, 745–754. [CrossRef]

28. Saha, S.; Moorthi, S.; Wu, X.; Wang, J.; Nadiga, S.; Tripp, P.; Behringer, D.; Hou, Y.-T.; Chuang, H.; Iredell, M.; *et al.* The NCEP Climate Forecast System Version 2. *J. Clim.* **2014**, *27*, 2185–2208. [CrossRef]

29. Lesser, G.R.; Roelvink, J.A.; van Kester, J.A.T.M.; Stelling, G.S. Development and validation of a three-dimensional morphological model. *Coast. Eng.* **2004**, *51*, 883–915. [CrossRef]

30. Mulligan, R.P.; Walsh, J.P.; Wadman, H.M. Storm surge and surface waves in a shallow lagoonal estuary during the crossing of a hurricane. *J. Waterw. Port. Coast. Ocean Eng.* **2015**, *141*, A5014001. [CrossRef]
31. Barnard, P.L.; van Ormondt, M.; Erikson, L.H.; Eshleman, J.; Hapke, C.; Ruggiero, P.; Adams, P.N.; Foxgrover, A.C. Development of the Coastal Storm Modeling System (CoSMoS) for predicting the impact of storms on high-energy, active-margin coasts. *Nat. Hazards* **2014**, *74*, 1095–1125. [CrossRef]
32. Taebi, S.; Pattiaratchi, C. Hydrodynamic response of a fringing coral reef to a rise in mean sea level. *Ocean Dyn.* **2014**, *64*, 975–987. [CrossRef]
33. Hoeke, R.K.; Storlazzi, C.D.; Ridd, P. V Drivers of circulation in a fringing coral reef embayment: A wave-flow coupled numerical modeling study of Hanalei Bay, Hawaii. *Cont. Shelf Res.* **2013**, *58*, 79–95. [CrossRef]
34. Roelvink, J.A.; Walstra, D.-J. Ro keeping it simple bu using complex models. *Adv. Hydro-Sci. Eng.* **2004**, *6*, 1–11.
35. Lowe, R.J.; Falter, J.L.; Bandet, M.D.; Pawlak, G.; Atkinson, M.J.; Monismith, S.G.; Koseff, J.R. Spectral wave dissipation over a barrier reef. *J. Geophys. Res.* **2005**, *110*, 1–16. [CrossRef]
36. Filipot, J.-F.; Cheung, K.F. Spectral wave modeling in fringing reef environments. *Coast. Eng. Proc.* **2012**, *67*, 67–79. [CrossRef]
37. Madsen, O.; Poon, Y.; Graber, H. Spectral Wave Attenuation by Bottom Friction: Theory. *Coast. Eng. Proc.* **1988**, *21*, 492–504.
38. Raubennheimer, B.; Guza, R.T.; Elgar, S. Field observations of wave-driven setdown and setup. *J. Geophys. Res.* **2001**, *106*, 4629–4638. [CrossRef]
39. Murakami, H. Tropical cyclones in reanalysis data sets. *Geophys. Res. Lett.* **2014**, *41*, 2133–2141. [CrossRef]
40. Li, N.; Roeber, V.; Yamazaki, Y.; Heitmann, T.W.; Bai, Y.; Cheung, K.F. Integration of coastal inundation modeling from storm tides to individual waves. *Ocean Model.* **2014**, *83*, 26–42. [CrossRef]
41. Hench, J. Episodic circulation and exchange in a wave-driven coral reef and lagoon system. *Limnol. Oceanogr.* **2008**, *53*, 2681–2694. [CrossRef]
42. Taebi, S.; Lowe, R.J.; Pattiaratchi, C.B.; Ivey, G.N.; Symonds, G.; Brinkman, R. Nearshore circulation in a tropical fringing reef system. *J. Geophys. Res.* **2011**, *116*, C02016. [CrossRef]
43. Gourlay, M.R.; Colleter, G. Wave-generated flow on coral reefs—An analysis for two-dimensional horizontal reef-tops with steep faces. *Coast. Eng.* **2005**, *52*, 353–387. [CrossRef]
44. Becker, J.; Merrifield, M.A.; Ford, M. Water level effects on breaking wave setup for Pacific Island fringing reefs. *J. Geophys. Res.* **2014**, *119*, 914–932. [CrossRef]
45. Andersson, A.J.; Gledhill, D. Ocean acidification and coral reefs: Effects on breakdown, dissolution, and net ecosystem calcification. *Ann. Rev. Mar. Sci.* **2013**, *5*, 321–348. [CrossRef] [PubMed]
46. Grady, A.E.; Moore, L.J.; Storlazzi, C.D.; Elias, E.; Reidenbach, M.A. The influence of sea level rise and changes in fringing reef morphology on gradients in alongshore sediment transport. *Geophys. Res. Lett.* **2013**, *40*, 3096–3101. [CrossRef]
47. Buckley, M.; Lowe, R.; Hansen, J. Evaluation of nearshore wave models in steep reef environments. *Ocean Dyn.* **2014**, *64*, 847–862. [CrossRef]
48. Pomeroy, A.; Lowe, R. The dynamics of infragravity wave transformation over a fringing reef. *J. Geophys. Res.* **2012**, *117*, C11022. [CrossRef]
49. Péquignet, A.-C.N.; Becker, J.M.; Merrifield, M.A. Energy transfer between wind waves and low-frequency oscillations on a fringing reef, Ipan, Guam. *J. Geophys. Res.* **2014**, *119*, 6709–6724. [CrossRef]

Journal of
*Marine Science
and Engineering*

MDPI

Article

A Flood Risk Assessment of the LaHave River Watershed, Canada Using GIS Techniques and an Unstructured Grid Combined River-Coastal Hydrodynamic Model

Kevin McGuigan, Tim Webster * and Kate Collins

Applied Geomatics Research Group, Nova Scotia Community College, Middleton, NS B0S 1M0, Canada; kevin.mcguigan@nscc.ca (K.M.G.); kate.collins@nscc.ca (K.C.)

* Author to whom correspondence should be addressed; timothy.webster@nscc.ca; Tel.: +1-902-825-2775; Fax: +1-902-825-5479.

Academic Editor: Rick Luettich
Received: 29 June 2015; Accepted: 15 September 2015; Published: 22 September 2015

Abstract: A flexible mesh hydrodynamic model was developed to simulate flooding of the LaHave River watershed in Nova Scotia, Canada, from the combined effects of fluvial discharge and ocean tide and surge conditions. The analysis incorporated high-resolution lidar elevation data, bathymetric river and coastal chart data, and river cross-section information. These data were merged to generate a seamless digital elevation model which was used, along with river discharge and tidal elevation data, to run a two-dimensional hydrodynamic model to produce flood risk predictions for the watershed. Fine resolution topography data were integrated seamlessly with coarse resolution bathymetry using a series of GIS tools. Model simulations were carried out using DHI Mike 21 Flexible Mesh under a variety of combinations of discharge events and storm surge levels. Discharge events were simulated for events that represent a typical annual maximum runoff and extreme events, while tide and storm surge events were simulated by using the predicted tidal time series and adding 2 and 3 m storm surge events to the ocean level seaward of the mouth of the river. Model output was examined and the maximum water level for the duration of each simulation was extracted and merged into one file that was used in a GIS to map the maximum flood extent and water depth. Upstream areas were most vulnerable to fluvial discharge events, the lower estuary was most vulnerable to the effect of storm surge and sea-level rise, and the Town of Bridgewater was influenced by the combined effects of discharge and storm surge. To facilitate the use of the results for planning officials, GIS flood risk layers were intersected with critical infrastructure, identifying the roads, buildings, and municipal sewage infrastructure at risk under each flood scenario. Roads were converted to points at 10 m spacing for inundated areas and appended with the flood depth calculated from the maximum water level subtracted from the lidar digital elevation model.

Keywords: storm surge; flood risk mapping; sea level rise; flexible mesh; GIS

1. Introduction

The LaHave River is an estuarine river in southern Nova Scotia that has a history of both fluvial and coastal flooding. Heavy rain and snowmelt can cause the river to overflow its banks, flooding the rural communities and roads within the watershed, while storm surge and sea-level rise can cause flooding near the coast; when a heavy rainfall event is combined with a storm surge, infrastructure within the watershed is especially at risk.

This project builds on a previous study which focused on the town of Bridgewater, Nova Scotia, which is located within the LaHave River watershed [1]. In this study we expand the domain to model the entire watershed, and present modifications to the modelling in order to best represent the large and complex system. Here, as in the Bridgewater study [1] we use a seamless digital elevation model (DEM) generated using high resolution lidar elevation data, bathymetric river data, and river cross-section information, along with river discharge and tidal elevation data to run a hydrodynamic model to produce flood risk predictions. The expanded study area necessitated augmenting our previous methodology of integrating the high resolution topography data with the various sources of bathymetric data. Like Merwade and Cook [2], we approached the generation of the continuous topographic-bathymetric surface using GIS techniques.

Instead of the nested-grid approaches used in the past, in this study we employ a two-dimensional flexible mesh model, a more sophisticated and more appropriate model for the complex LaHave watershed. We use the DHI Mike Flexible Mesh model which represents a two-dimensional model of the river channel and floodplain. Flexible meshes, also known as unstructured triangular grids, are composed of triangles that can be varied in size throughout the mesh allowing the user to refine the mesh density to be higher in more critical areas and lower in less sensitive regions. This has the advantage of having higher resolution results where required while reducing computational costs in areas where only a coarse resolution is required, such as in deeper bathymetries. Additionally, flexible meshes adapt easily to follow the often irregular features of coastal and riverine systems better than rectilinear grids.

Flexible mesh modeling is a widely used and desirable basis for coastal flood models [3–5] as well as river models [6–8]. Shubert et al. [3] use an unstructured mesh to model coastal flooding in Glasgow, Scotland using lidar terrain data; they note the benefits of the flexible mesh density in urban areas. Kliem et al. [4] simulated storm surge for the North Sea-Baltic Sea using the varying resolution irregular mesh 2D shallow water model MOG2D to overcome the limitations imposed by regular grids for flood modelling. Wang et al. [5] used the unstructured triangular finite volume model FVCOM to more accurately represent shorelines and bathymetry; the authors coupled the hydrodynamic model with a GIS to visualize and analyze the results and found that the simulations agreed well with observations. Hagen [6] and Gama et al. [7] used the Deltares-Flow-Flexible Mesh model to predict flood extent and both cited improvements in their ability to represent complex river geometries over structured grids.

Another improvement over Webster et al. [1] is the implementation of a spatially variable surface roughness map. Many have shown the importance of, and the differences between, including resistance parameters (e.g., Manning's n or M) through satellite imagery [8,9], landcover classification data [3], and lidar-based feature height data [10,11]. The collection of bathymetric data for rivers and shallow coastal areas is still a challenge, although new technology in the form of topo-bathymetric lidar provides the potential to acquire detail in shallow areas at a similar resolution and accuracy as traditional topographic lidar. Campana et al. [12] used bathymetric lidar to measure the topographic change of a river in the Italian Alps compared to manual river cross-section measurements. They conclude that bathymetric lidar offers great potential to monitor river evolution and to quantify morphological diversity. Kinzel et al. [13] compared the United States Geological Survey's hybrid topographic/bathymetric Experimental Advanced Airborne Research LiDAR (EAARL) sensor with detailed surveys of river channels collected using wading and sonar techniques for rivers in California and Colorado. They showed that the lidar elevation of the river bed matched the traditional survey data to on average within 30 cm. They note that the water clarity must be sufficiently clear and that the albedo river bottom must be sufficiently reflective to produce laser returns to the sensor.

1.1. Study Area

The LaHave River extends ~80 km from its headwaters at the southern base of South Mountain in the Annapolis Valley of Nova Scotia to its mouth on the Atlantic Ocean (Figure 1). The river passes through several rural communities, reaches the Town of Bridgewater approximately 20 km from the Atlantic Ocean, and passes through several more small communities before meeting the coast. The

LaHave River watershed drains many lakes and rivers, covers 1686 km^2, and encompasses sections of three counties and several municipalities, including the District of the Municipality of Lunenburg, for whom this study was completed. The watershed contains a mix of land use, including industrial and residential within the Town of Bridgewater, and shifting to mainly forested and agricultural in the majority of the watershed.

The LaHave watershed includes two Ecodistricts, as defined by the Nova Scotia Department of Natural Resources [14]. The LaHave Drumlins Ecodistrict is characterized by glacial till drumlins and coniferous forests, with soils that are mostly well-drained, except between the drumlins where soil is poorly drained. The LaHave River flows through the center of the Ecodistrict until it enters the South Shore Ecodistrict near the coast. The South Shore Ecodistrict is composed of a mixture of sandy beaches, lakes and streams, and coastal forests.

Tides are semi-diurnal in the Bridgewater area, with a tidal range of 2.5 m. The tidal influence extends ~20 km up the LaHave River, just to downtown Bridgewater. The river does not ice over during the winter, but ice does form upstream of the town and in the lakes and throughout the watershed.

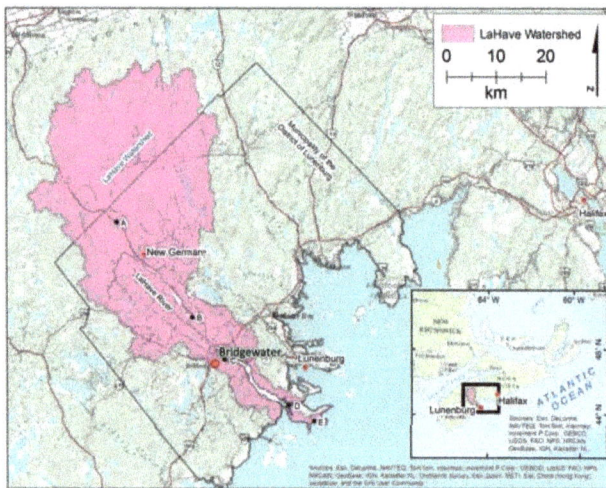

Figure 1. The LaHave River watershed extends from South Mountain, Nova Scotia, to the Atlantic Coast. (**A**) Applied Geomatics Research Group (AGRG) Cherryfield weather station; (**B**) Environment Canada water level gauge; (**C**) AGRG Marine Terminal tide gauge at Bridgewater; (**D**) AGRG Kraut Point tide gauge; (**E**) AGRG Hirtles Beach weather station.

1.2. Flood History

Storm surges can occur along the Atlantic coast when low pressure systems such as hurricanes, post-tropical storms, and Nor'easters cause high winds and heavy rain. Typically the coastal zone bears the brunt of these types of storms, but a 1 m storm surge can cause flooding and road closures 20 km inland as far as downtown Bridgewater, as occurred on 30 October No storm surge return period analysis has been completed for the LaHave Estuary directly because there is no tidal record for the area; instead, the Halifax tide gauge, located 60 km northeast along the Atlantic Coast is used for analysis (Figure 1).

Overland flooding of the LaHave River that results in property and infrastructure damage within the watershed occurs frequently and is caused by intense or prolonged rainfall, melting snow and ice, or a combination of these. Flow in the LaHave River follows a typical pattern, with maximum flow occurring in the spring and minimum flow occurring in the summer. An analysis of the 95-year LaHave River flow time series derived from the Environment Canada (EC) water level gauge in West Northfield (Figure 1) shows that almost 80% of the floods during that time period occurred during winter or spring, times when snow melt is likely to have contributed to the flood [15]. Maximum Instantaneous Peak Flow was highest on 10 January 1956 (1080 m^3/s) and second-highest on 31 March 2003 (663 m^3/s). Both of these floods were caused by heavy rain and melting snow and caused the highest water levels ever recorded (5.73 m and 5.17 m for 1956 and 2003, respectively), 1.5 to 2.0 m higher than any other flood event in the LaHave River watershed [15]. Two fatalities occurred upstream of Bridgewater in 2003 when a car was swept off the road and into the flooded river [16].

1.3. Climate Change

In the past ten years there have been 12 Storm Surge Warnings issued by EC for Lunenburg County [17]; of these, only two events (October 2011 and February 2013 [18]) resulted in a significant storm surge and flooding. Rainfall warnings are issued far more often: eight were issued in 2013, fourteen in 2014 and six within the first three months of 2015 [17]; several of these resulted in flooding with road closures, one of which was particularly dangerous [19]. Webster et al. [1] showed that combined storm surge and rainfall events can produce water levels higher than would be seen with a single event, and that risk to Bridgewater is increased as sea level rises.

Much attention has been devoted on a global scale to predicting sea-level rise [20–23]. The latest assessment of the Intergovernmental Panel on Climate Change (IPCC), AR5, projects a sea-level rise for 2046–2065 of 0.17 m to 0.38 m and for 2081–2100 of 0.26 m to 0.82 m, including the effects of melting ice sheets [22]. Rahmstorf et al. [24] have suggested a rise between 0.5 and 1.4 m from 1990 to In Maritime Canada many coastal areas have been deemed highly susceptible to sea-level rise [25]. In Nova Scotia, global sea-level rise is compounded by crustal subsistence [26] which contributes an estimated 0.16 m per century to relative sea-level rise [27]. A comprehensive, community-by-community report prepared by Richards and Daigle [28] provides estimated extreme total sea levels for Lunenburg, a neighboring community along the South Shore of Nova Scotia located northeast of the mouth of the LaHave estuary, using Rahmstorf et al. [24] as a basis for sea-level rise projections. Their report presents results for 10, 25, 50 and 100-year return periods, for years 2000, 2025, 2055, 2085 and 2100, and shows that a 10-year storm could result in a sea-level of 3.29 m by 2025, 0.86 m higher than HHWLT (Higher High Water Large Tide), and a 100-year storm in 2055 would increase water level to 3.80 m.

As is the case with temperature and sea-level, precipitation and river discharge patterns are changing with climate change. In Atlantic Canada there is evidence that heavy rainfall events are increasing in frequency [29,30] an observation that agrees with models and predictions [28,29,31,32]; the increase in rainfall is expected to occur in the winter and spring [28,31]. Increased and intensified rainfall is also observed and predicted for New England [33–35]. Studies of streamflow patterns during the last 50 years show that maritime rivers in the Atlantic provinces have been experiencing lower summer flows, but higher flows in early winter and spring [36,37]. Streamflow is expected to increase with temperature and precipitation in the Atlantic region [38], and spring flooding could become more common due to changes in late-winter early-spring precipitation patterns [39].

2. Experimental Section

This project shares data and some data analysis and modelling techniques with a previous study in the same area [1]. The cases of overlap in Section 2 are identified and discussed briefly and the reader is referred to Webster et al. [1], while new developments are discussed in detail. The former category includes hydrological data collection and model simulation design, while the latter category includes the GIS techniques employed to integrate topographic lidar data with coarser bathymetric data and the use of an image-derived surface roughness map. Additionally, we employ a different model, so Section 2.5 discusses the details of the flexible mesh generation, the model calibration, and boundary condition implementation.

2.1. Hydrology Data

The modelled factors driving the potential for flood risk in the LaHave Estuary include weather, river stage and tides; these data are required to run the hydrodynamic model and predict flood risk and were obtained and used in the previous Bridgewater study [1]. Discharge data for the river runoff model are from the Environment Canada flow sensor [15] (Figure 1B), additional water level data were observed near Bridgewater (Figure 1C) and near the mouth of the river (Figure 1D) using AGRG pressure sensors. Air pressure data used to compensate the water level sensors came from AGRG weather stations in the center of the watershed (Figure 1A) and the Atlantic coast (Figure 1E).

2.2. DEM Development

An accurate representation of bathymetry and topography is essential for successful hydrodynamic flood risk modelling. In this study bathymetry data from a suite of data sources were combined with topographic lidar data to generate a continuous DEM that was used in the model.

2.2.1. Bathymetric Survey

A bathymetric grid was compiled from a variety of sources to accurately represent seabed, lakebed and river channel bathymetry as well as their geometries (Figure 2). Lake bathymetry was obtained from the Nova Scotia Department of Fisheries and Aquaculture [40] (Figure 2a). Upstream of Bridgewater a combination of depths were measured using a depth sounder mounted on a canoe (Figure 2b) and for extremely shallow locations RTK GPS and depth measurements were obtained manually by walking across the river (Figure 2c), and from Bridgewater to Upper LaHave the depth sounder was mounted on a 15 foot aluminum boat. The Canadian Hydrographic Survey (CHS) nautical chart Chart 4381 at a scale of 1:38,900 was digitized to obtain soundings between Upper LaHave and Riverport (Figure 2d) and the digital chart covered the area from Riverport to the coast, including the mouth of the LaHave River. Areas where no bathymetric information existed were modelled to artificially form a channel using a method discussed in Section 2.2.3 (Figure 2e–g).

2.2.2. Lidar Survey

Lidar data for this project were obtained for the coastal region in 2009 [41] and for the remainder of the watershed in 2012 [1] at ~1 m spacing. Two surface models were constructed from these data; a Digital Surface Model (DSM) which incorporates all the points and a bare-earth DEM which incorporates only the classified ground points.

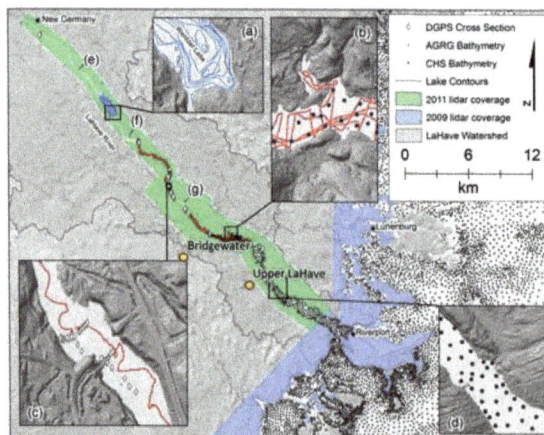

Figure 2. Topo-bathymetric data sets amalgamated in this study include (**a**) lake bathymetry contours; (**b**) AGRG bathymetry from canoe survey; (**c**) AGRG RTK GPS cross sections; (**d**) CHS bathymetry from paper chart; (**e–g**) areas of no bathymetric information where the channel was modified. Blue area represents the lidar coverage from 2009, green area represents the lidar coverage from 2012.

2.2.3. Topographic-Bathymetric Data Integration

The lidar survey provides sufficient detail to model the floodplain but it does not penetrate the water surface and must be combined with the bathymetry data to generate a seamless topo-bathymetric DEM that represents the topography above and below the water line. The water surface extent was delineated from the combined 2011 and 2009 lidar datasets on a 2 m spatial resolution using a custom built ArcGIS tool that is based on user input cross-sections, as described in Webster et al. [1]. The water surface was refined with additional cross-sections and manual digitization where necessary.

The process of integrating the bathymetry data with the lidar data differs from and improves upon the previous study [1]. Here, all bathymetric data from New Germany to Riverport (Figure 2) were averaged by depth into a 12 m grid; the digital CHS chart covering the coastal area was dealt with separately and is discussed later. The river bank was identified using the water surface extent and a 4 m landward buffer. Bathymetry points that were within a specified distance to the river bank were removed to reduce severe oscillations in subsequent bank-bathymetry interpolations. The river bank then contained only lidar points at 2 m resolution, and these were assigned a value of 0 m depth. The bank points were then integrated with the now 12 m gridded bathymetry points at 16 m resolution using a spline interpolation method, and resampled to a 4 m resolution water depth grid ("observed spline"). A buffer extending from the river bank points to 10 m landward of the lidar bank points was used in the spline to eliminate interpolation calculations occurring between river segments.

Where bathymetry data were nonexistent (Figure 2e–g) an experimentally determined depth model was developed based on Euclidian distance from the banks in areas of known bathymetry, such that:

$$h = \frac{\left(x^{0.4} + 0.1x\right)}{3.5} \tag{1}$$

where h is water depth relative to the bottom and x is the Euclidian distance. Modelled depths were sampled to 12 m spacing and points within a 4 m distance of the river bank were removed. A spline was generated at a 4 m resolution using modelled and observed bathymetry points, lidar bank points, and a 10 m landward bank-buffered spline barrier ("modelled spline"). The resulting grid indicated that modelled depth values tended to over-estimate channel depth in some areas, specifically near the town of Bridgewater, producing an artificial scouring effect (Figure 3a). This is to be expected as the depth model was designed to predict bathymetry further upstream where channel banks are steeper and the river is shallower. To combine the modelled and observed bathymetry more smoothly, a 1 bit raster mask was used to remove modelled bathymetry points in areas where the "modelled spline" depth points were more than 0.50 m deeper than the "observed spline" depth points (Figure 3b). Furthermore, a manual estimation of water depth was included as points where some narrow sections of the LaHave water surface were unaccounted for by this technique (Figure 3c) and the spline interpolation was recomputed to generate a final river bathymetry grid (Figure 3d).

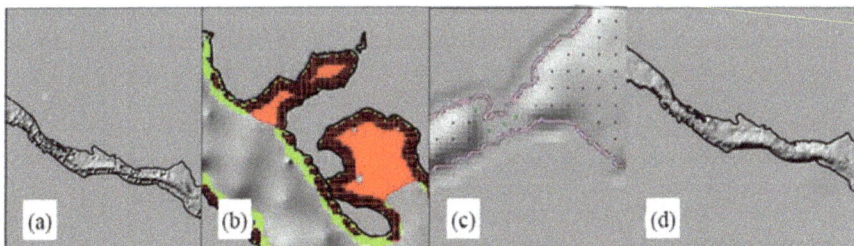

Figure 3. (**a**) The initial modelled-observed bathymetry interpolation; (**b**) the mask used to remove poorly modelled bathymetric points: red points were kept and green points discarded; (**c**) the resultant spline with no artificial scouring; (**d**) additional bathymetry was added manually as required (green points).

The river bathymetry grid was bi-linearly resampled to a 2 m spatial resolution and subtracted from the 2 m resolution lidar DEM with the water surface included (Figure 4); this generated a DEM of lidar topography and river bathymetry for the lidar extent. Sections where noise, offsets, or artifacts existed in the water surface of the lidar were selected and smoothed using a low-pass filter, or removed. The tidal portion of the study area extended ~15 km from Riverport into the Atlantic Ocean (Figure 2). This area was outside of the lidar extent and so was modelled using 20 m Nova Scotia Topographic Database (NSTDB) elevation data up to 10 m relative to the Canadian Geodetic Vertical Datum of 1928 (CGVD28) merged with the CHS digitized chart bathymetry; the resulting coastal DEM was merged with the river DEM to generate a seamless topographic-bathymetric model of the LaHave River estuary relative to CGVD28.

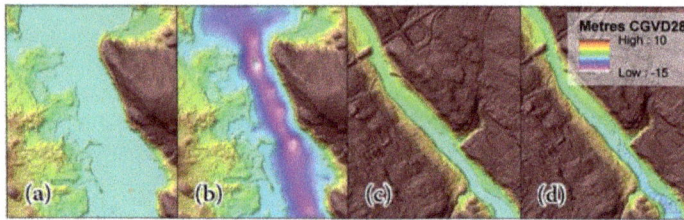

Figure 4. Lidar data before (**a,c**) and after (**b,d**) bathymetric integration; An area upstream of Bridgewater is shown in (**a,b**); downtown Bridgewater is shown in (**c,d**).

2.3. Unstructured Mesh Generation

An unstructured grid, or flexible mesh, was generated from the seamless 2 m topographic-bathymetric DEM using the Mike DHI mesh generator toolset (Figure 5). Before computation of the flexible mesh, the modelled domain was limited to areas where flooding was possible by calculating a maximum flood extent. To accomplish this, the slope of the LaHave River level was determined through the use of a flow accumulation calculation, and an approximate maximum flood level (z_{max} relative to CGVD28) along the river channel was calculated such that:

$$z_{max} = 0.89z + 10 \tag{2}$$

where z refers to vertical elevation relative to CGVDThis function was developed such that the maximum flood level would be limited to 10 m above areas with a river level of 0 m CGVD28 (which can accommodate a storm surge of approximately 8.5 m if necessary) and flood levels upstream at the head of the river (62 m CGVD28) would have a maximum flood level of ~65 m CGVD28, or a maximum flood depth of ~3 m. The maximum flood extent was generated by buffering points along the river by a factor based on the flow accumulation, calculating the maximum flood level using Equation (2), and intersecting the resultant surface with the seamless DEM. The maximum flood extent was used to generate the outer boundary of the flexible triangular mesh grid (Figure 5). The boundary was smoothed using the Polynomial Approximation with Exponential Kernel (PAEK) algorithm with a smoothing tolerance of 20 m to eliminate sharp angles of the boundary originating from the vectorization of the maximum water level raster.

The boundary was also used to clip the lidar elevation grid to reduce the input of elevation data into the Mike DHI mesh building tool. Topographic data density was reduced further using a key-point analysis to eliminate redundant elevations while preserving geometry to within 0.10 m; this technique reduces the number of points in a mesh but still provides an accurate representation of the topography. The mesh density was further customized to increase model stability and reduce simulation time by separating it into sections of varying mesh density. Mesh density near the river banks was defined to increase linearly with slope and elevation, and the remainder of the mesh was designed following the advancing-front method [42] so that it varied in density from 50,000 m^2 maximum area per triangle element in the coastal region (Figure 5c); to 500 m^2 in the densest regions upstream (Figure 5a,b). All triangles were built with a minimum allowable angle of 26 degrees.

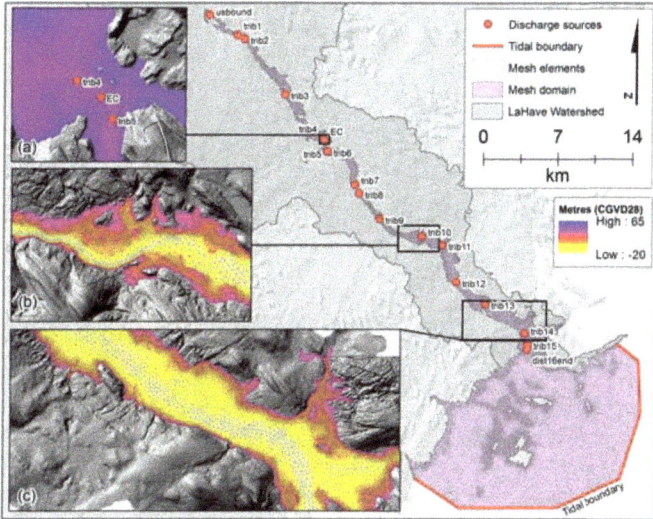

Figure 5. Flexible mesh modelling domain with tidal and discharge (tributary) boundary locations. (a) Inset of the mesh at the EC gauge; (b) variable mesh density and elevation midway along the estuary; (c) variable mesh density and elevation midway near the mouth of the estuary.

2.4. Surface Roughness Map

The stability and calibration of coupled coastal-fluvial hydrodynamic flood models depends in large part on the accurate representation of dynamic flow characteristics over varying surfaces such as cobbled streams, vegetation and concrete. As such, bed roughness indices were employed to model the observed relationship between depth-averaged flow velocity and total water depth across varying bed types. To represent this relationship spatially the model included a variable Manning's M index grid, for which small numbers represent the roughest surfaces and most resistance to flow, and the largest numbers represent the smoothest surfaces and least resistance to flow; Manning's M is the reciprocal of the commonly used Manning's n. Here, we used a single Band 5 (near infrared) 30 m resolution Landsat 5 image (taken 27 July 2008) to develop a 10 m resolution surface roughness grid based on vegetation density and land use (Table 1).

Table 1. Derivation of Manning's M spatial roughness grid.

Feature				Manning's M
	Heavy		16–54	10
Vegetation	Medium	Digital Number	55–69	15
	Light–none		>70	20
Initial value for river channel and coast				25
Roads				90

2.5. Hydrodynamic Modelling

There are many approaches to the hydrodynamic modelling of estuarine systems, all with various levels of detail and requirements. This model was designed using the Mike DHI software wherein a 2-d depth averaged incompressible Reynolds averaged Navier-Stokes shallow-water equation was solved over the constructed flexible mesh domain with a variable time step scheme. Operationally the simulations ran at an average time step of approximately 0.4 s. The shallow water equations used allowed for variation in temperature, salinity, and density, but all were assumed constant in this study to decrease processing time. Ideally, discharge data would be available for each tributary entering the LaHave River to incorporate river discharge from the whole watershed into the downstream tidal model (Figure 5). In the absence of these data, our approach was to include discharge from all major tributaries as discharge point sources, each scaled relative to a single long term observed discharge record which exists in the mid-section of the model domain. Catchments for the 17 tributaries were defined using a flow accumulation calculation for the entire LaHave watershed executed on a 20 m spatial resolution DEM using available NSTDB data. Each source point was assigned a scale factor based on the area of its catchment relative to the drainage area of the LaHave River EC gauge (Table 2). Additional discharge sources intermediate to defined tributaries were included in tributary point sources downstream. The scale factors were applied to the EC daily discharge record [15] resulting in a discharge time series for each tributary. This approach allowed the model to be linked to a large amount of discharge data for statistical purposes while not being required to analyze rainfall rates or the complexities of snow melt, both of which are inherent to the discharge record; additionally, we were able to incorporate contributions from all catchments without deploying sensors at each tributary location. This method is limited by the assumption that all catchments in the watershed have the same hydrological properties. The scaled discharge time series were used as boundary conditions that were varied for each model simulation and area discussed in Section 2.5.2.

Table 2. Major LaHave River tributaries, catchment contributions, and scale factor applied to the EC gauge discharge.

Name	Catchment Contribution (Number of 20 × 20 m Pixels)	EC Gauge Ratio
EC Gauge	3,113,163	1.00000
usbound	1,781,664	0.57230
trib3	1,148,538	0.38428
trib6	459,567	0.14796
dist16end	14,972	0.00481

2.5.1. Model Calibration

The model was calibrated by comparing observed water levels at the EC gauge, near the town of Bridgewater, and at the mouth of the river (locations shown on Figure 1) to modelled water levels extracted from the model at those locations. All water level recording equipment were referenced vertically to the datum of the model (CGVD28) using survey grade RTK GPS (with a precision of 2.5 cm or better). Tidal calibration was done for 8–15 August 2012 which was selected as a period where the predicted and gauged tides were most congruent and did not exhibit any high frequency variations observed during the winter (Figure 6a,b). At Kraut Point (Marine Terminal) the mean difference between modelled and observed water level was −1.5 cm (−5.2 cm), standard deviation 24 cm (15 cm). This very low mean difference between the modelled and observed water levels in the southern portion of the model domain (Kraut Point, Figure 1) indicates that the tidally dominated portion of the flow model operates very accurately, specifically during non-storm events. A discharge event on 1 November 2012 was used for the fluvial model calibration as the river water level conditions before the event were stable and the curve of the event was smooth and thus near the signature desired to be used for the various flood simulations (Figure 6c).

In the calibration phase, all hydrodynamic parameters were determined and remained consistent for all subsequent flood scenario simulations (Table 3). Principal parameters which contributed to the fluvial calibration include the relative timing of discharge sources upstream of the EC gauge as well as the bed roughness of the river. It was experimentally determined that each of the discharge sources upstream of the EC gauge (usbound, trib1, trib2, trib3) should be shifted by several hours (7, 5, 5, 2, respectively). A Manning M value of 33 was determined for the river channel and the coastal domain of the model using a trial and error process of model and observation time series comparison. A small section on the edge of the tidal boundary was assigned a very rough Manning M value of 5 to eliminate oscillations which existed along the boundary and thus improved model stability.

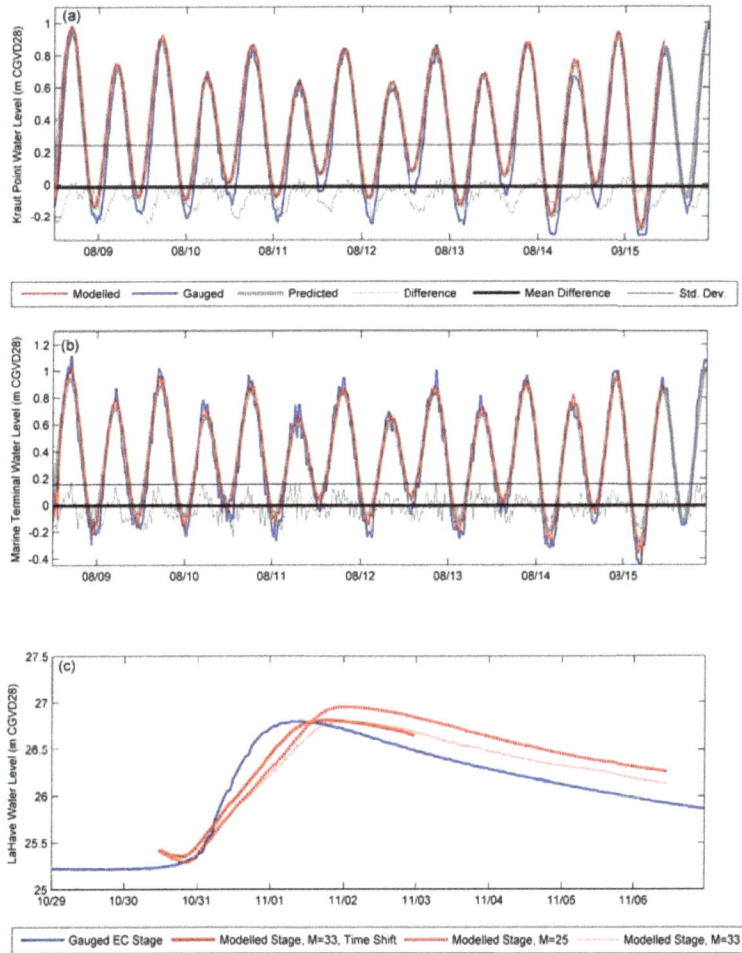

Figure 6. (a) Comparison of tide gauge observations with model results at Kraut Point (mean difference −1.5 cm, standard deviation 24 cm); (b) comparison of tide gauge observations with model results at the Marine Terminal (mean difference −5.2 cm, standard deviation 15 cm). Predicted tide used for the model boundary is plotted on both (a,b); (c) shows gauged and modelled stage of the LaHave River at the Environment Canada Gauge location. Three different model results are shown with variable Manning's M; the final model version (solid red line) uses M = 33 and a time shift. Note that (a,b) share the same legend; (c) references the legend at the bottom of the figure.

Table 3. Final model parameter values.

Parameter	Value
Time step	Variable, 0.01–30 s
Critical CFL number	0.8
Drying Depth	0 m
Flooding Depth	0.05 m
Wetting Depth	0.1 m
Density Type	Barotropic
Smagorinsky eddy viscosity coefficient	0.28
Final Manning M of river channel and most of coast	33
Final Manning M at tidal boundary	5

2.5.2. Model Simulations

2.5.2.1. Extreme Value Analysis

In this study we are examining the risk of flooding from two possible sources which can interact to compound the problem: river runoff and storm surges or long-term sea-level rise. The model scenarios follow Webster et al. [1], who use extreme value models (EVMs) to determine the return periods of extreme events using the discharge and sea-level time series to examine how often such events have occurred in the past. The time series of measured discharge of the LaHave River (1915–2012) was used to determine the annual probability of extreme events and also the return period of extreme high flow events. Table 4a summarizes the return period in years and the associated discharge for a 65% probability of occurrence, which equals at least one occurrence. Re-examining the historical high discharge events of January 1956 and March 2003, we see that the 1956 event has a recurrence interval greater than 100 years (1080 m³/s), while the 2003 event (663 m³/s) is approximately equivalent to a 50 year return period event with a 65% probability of occurrence.

Table 4. Model simulations. Simulation naming scheme format: first 3 digits indicate a discharge return period of 0, 50 or 100 years, last digit indicates level of storm surge (0 for maximum high tide conditions, 2 for a 2.0 m surge on top of the predicted tide or 3 for a 3.0 m surge on top of the predicted tide).

(a) LaHave River Scenario			(b) Sea Level Scenario		
Probability	Return Period	Discharge (m³/s)	Max. High + 0.0 m	Max. High + 2.0 m	Max. High + 3.0 m
	0 year	210	Sim000_0	Sim000_2	Sim000_3
65%	50 year	652	Sim050_0	Sim050_2	Sim050_3
	100 year	741	Sim100_0	Sim100_2	Sim100_3

The Halifax tide gauge record (1920–2010) was used to estimate the risk or probability of flooding related to high water events along the coast [1,41]. The 100-year return period flood level under current RSL conditions is 2.2 m [1], which is 10 cm below previous high water level maximum observed during Hurricane Juan in Halifax in 2003 which had an associated storm surge of 1.63 m [27]. If RSL increases to a rate of 0.73 m/century [21], the 100 year water level increases to 2.5 m CGVD28 which further inundates areas. If RSL increases to a rate of 1.46 m/century [24], the 100 year water level increases to 3.1 m CGVDIn this study, we use these estimates as guidelines, and use storm surge levels of 0.0 m, 2.0 m and 3.0 m (Table 4b) added to the maximum predicted high tide for These values can be thought of as extreme storm surges today or moderate surges in the future considering RSL.

Three sets of model simulations were executed for each different return period discharge: one set based on variable discharge under maximum high tide conditions in 2012, and two other sets based on different storm surge or long-term sea-level rise conditions (Table 4).

2.5.2.2. Boundary Conditions

The discharge and storm surge scenarios were implemented using boundary conditions (Figure 7). The river discharge boundary condition was scaled proportionally to the drainage area for the LaHave River as described above and varied according to the 0, 50 and 100 year return period discharge values (Figure 7a,b). We obtained predicted water level time series from the DHI Global Tidal Model, which includes the 10 major tidal constituents [43]. The storm surge scenarios (0.0 m, 2.0 m, and 3.0 m water levels) were added to the predicted maximum 2012 high tide boundary (Figure 7c) at the mouth of the LaHave River (Figure 5). The timing of the peak river discharge relative to the storm surge peak was determined experimentally, after calibrating for surface roughness, to ensure the worst case flooding condition.

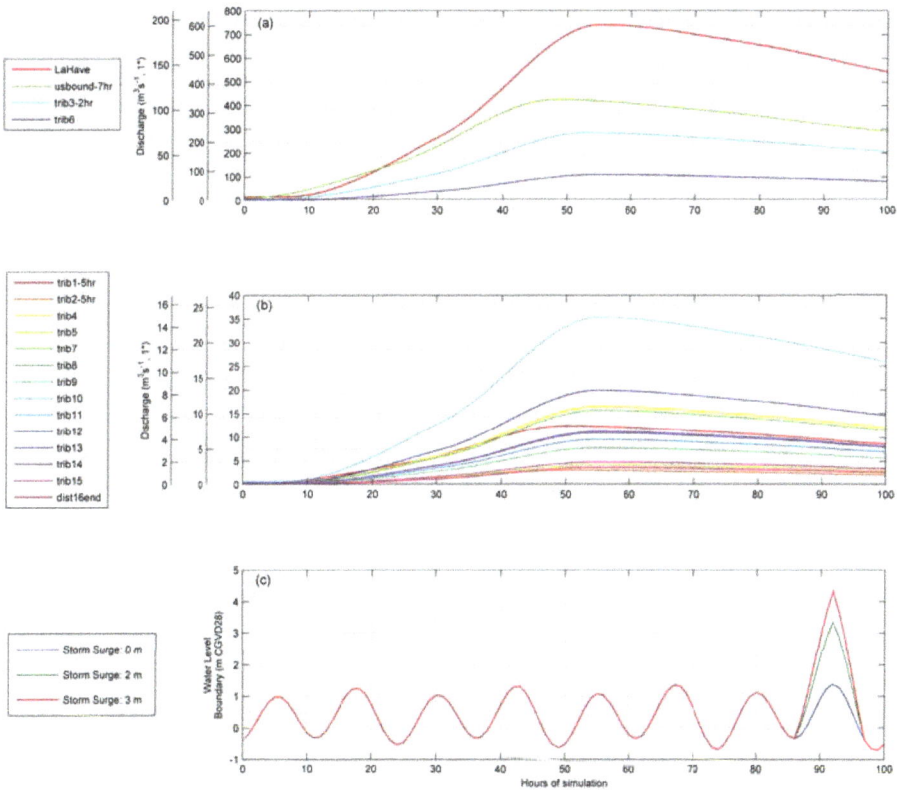

Figure 7. Model Boundary Conditions for (**a**) discharge at major tributaries for 0, 50, and 100 year return level flows; (**b**) discharge at minor tributaries for 0, 50, and 100 year return level flows; (**c**) tidal boundaries for storm surge levels 0 m, 2 m, and 3 m.

3. Results and Discussion

3.1. Simulation Results

Maximum water level from each of the nine simulations were output from Mike as flexible mesh elements and interpolated to a surface which was intersected with the DEM. The simulation naming scheme format is as follows: the first 3 digits indicate a discharge return period of 0, 50 or 100 years, the last digit indicates level of storm surge (0 for maximum high tide conditions, 2 for a 2.0 m surge on top of the predicted high tide or 3 for a 3.0 m surge on top of the predicted high tide). Analysis of the results of all nine simulations revealed that the area upstream of Bridgewater is dominated by discharge events with limited flooding from elevated sea-level (Figure 8), the area around Bridgewater is influenced by the combined effects of discharge and sea-level (Figure 9), and downstream of Bridgewater is dominated by elevated sea-levels and less by major discharge events (Figure 10). We present results of three simulations, Sim000_0, Sim050_2, and Sim100_3, which represent the range of conditions from typical annual events to rare extreme events.

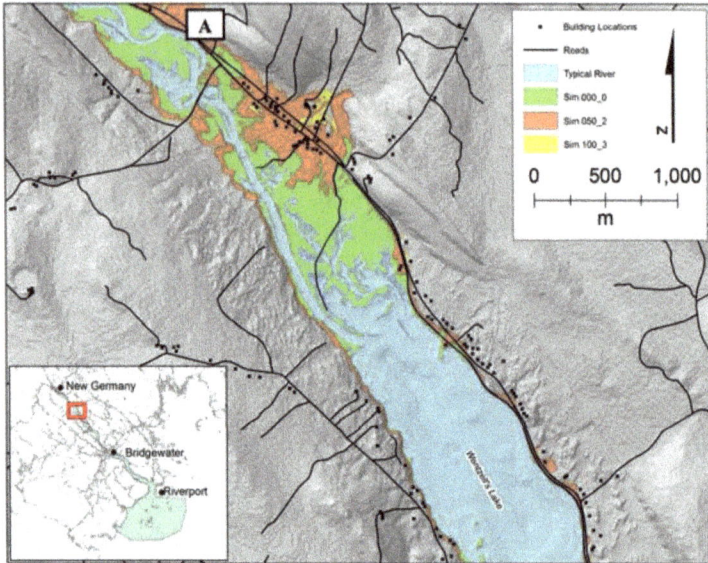

Figure 8. Maximum flood extents from Sim000_0 (green), Sim050_2 (orange) and 100_3 (yellow) south of New Germany. The normal river extent is shown in blue and the road (black lines) and buildings (black squares) effected. The background image is a combination of the lidar and 20 m DEM. "A" marks the location of two deaths in 2003 during a March flood event.

Although Sim050_2 shows flooding around New Germany (Figure 8) from the 1 in 50 year return period discharge with a 2 m storm surge on high tide in orange, a similar flood extent is observed for this area when we plot Sim050_0 which represents the 1 in 50 year return period discharge with high tide indicating there is virtually no influence of the sea-level at the mouth of the LaHave estuary for this area. The 100 year discharge event only increases the maximum flood extent slightly (yellow Figure 8). There are significant impacts from flooding for the 50 year discharge event including overtopped road and abandoned rail lines and flooded buildings on the east side of the river. The discharge values for Sim 050_2 and the March 2003 flood event are similar, and the simulation shows flood extents that agree with observations of road overtoppings in this area, including the area where two people died because they could not see the inundated road and drove into the ditch and drowned (Figure 8A).

The influence of the 1 in 50 year discharge scenario dominates the flooding resulting from Sim 050_2 until immediately upstream of Bridgewater. Downstream, the effects of storm surge dominate the flooding from Sim 050_2; Figure 9 shows the Bridgewater Mall parking lot flooded due to the 2 m storm surge. Farther downstream of Bridgewater the 2 and 3 m storm surge water levels flood sections of the highway, parks, and buildings. In October 2011 a 2 m storm surge was observed to flood the Bridgewater waterfront, especially at Shipyards Landing Park (Figure 9A).

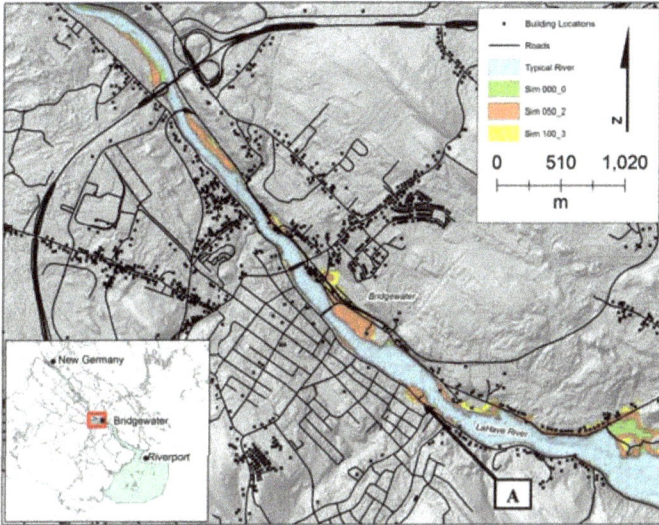

Figure 9. Maximum flood extents from Sim000_0 (green), Sim050_2 (orange) and 100_3 (yellow) at Bridgewater. The normal river extent is shown in blue and the road (black lines) and buildings (black squares) effected. The background image is a combination of the lidar and 20 m DEM. "A" shows the location of flooding during the 2 m storm surge in October 2011.

The area from Upper LaHave downstream to the mouth of the river at Riverport is vulnerable to elevated sea-level scenarios where the 2 and 3 m storm surge levels inundate and overtop the roads on both sides of the river (Figure 10), matching photographic evidence of flooding in these locations during the October 2011 storm surge event.

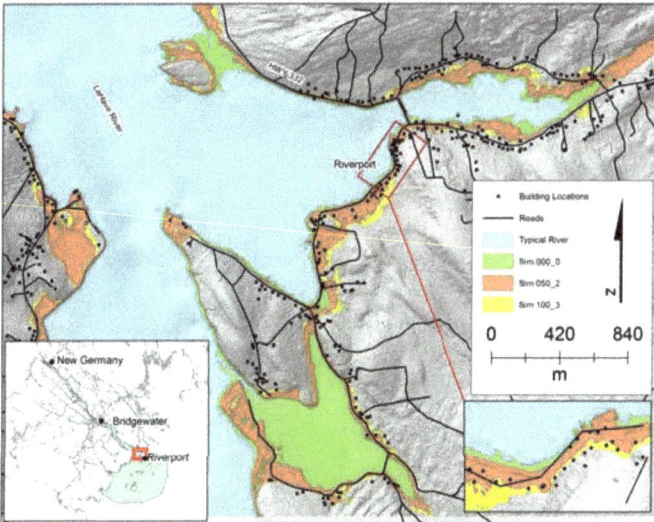

Figure 10. Maximum flood extents from Sim000_0 (green), Sim050_2 (orange) and 100_3 (yellow) at Riverport. The normal river extent is shown in blue and the road (black lines) and buildings (black squares) effected. The background image is a combination of the lidar and 20 m DEM.

3.2. Infrastructure Risk Assessment

An infrastructure risk assessment was conducted based on the nine simulations. Road vectors were segmented at 10 m intervals into points and overlaid on the flood depth maps from the model simulations along with points representing buildings, manholes and pump stations. Maximum water depths were tabulated for each flood scenario per point where flooding had occurred. The infrastructure exhibited a range of impact levels varying by location along the river and by the various flood scenarios (Figure 11). This point analysis approach enables a quick assessment of the at risk infrastructure along the river. Infrastructure in the northern portion of the model domain, which flood as the result of fluvial runoff, exhibit only minor flooding in annual maximum discharge conditions (Figure 11a,c). Roads, though sometimes bounded by flood waters on either side, are typically not overtopped in this case. Flooding becomes much more severe and infrastructure is widely affected and even overtopped upstream during a 50 year discharge event and more severely so during a 100 event. Roadways which may have been inundated yet traversable in the 50 year event can become impassable due to an increase in water depth during a 100 year discharge event. A similar relationship can be observed in the depth over roads in the 2 m and 3 m storm surge simulations in the tidal flooding dominated southern portion of the study area (Figure 11b,d).

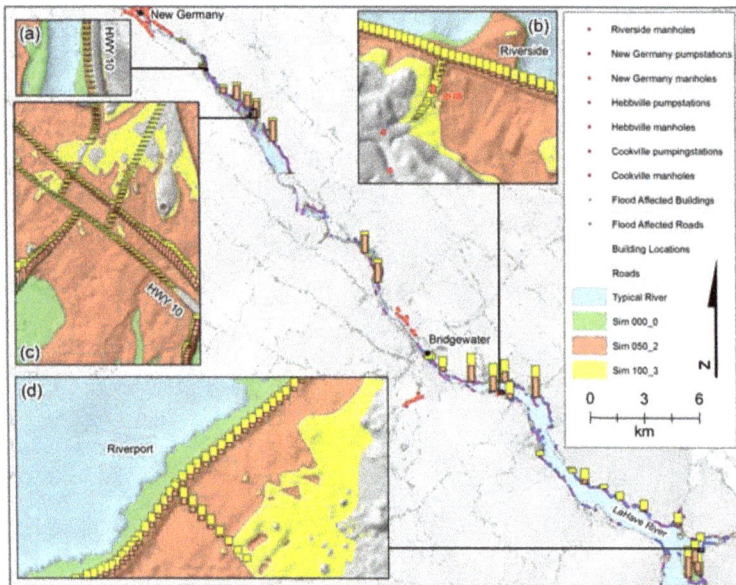

Figure 11. The infrastructure assessment of flood risk along the LaHave river study area indicates a wide range of risk from both discharge (north of Bridgewater, (**a,c**)) and storm surge (south of Bridgewater, (**b,d**)). At-risk roads (purple), buildings (blue), and sewers (red) are indicated. Relative water depths over infrastructure are indicated as bar graphs and color coded by flood simulation (Sim 000_0, Sim 050_2, and Sim 100_3).

3.3. Discussion

3.3.1. Interpretation and Implications

The results of this study agree with and improve upon the findings of Webster et al. [1] who studied a limited portion of the LaHave watershed to estimate flood risk only within the town of Bridgewater. Here, we do a far more in-depth examination of flood risk in the entire watershed. This required a more complex integration of high resolution lidar topography data with several additional sources of different resolution bathymetry; but most importantly, expansion of the model domain allowed us to indicate which areas were most at risk to different discharge or sea-level flood scenarios. The precise division of areas at risk to each type of flooding provided invaluable information to municipal planners and emergency response officials as it allows them to customize their Emergency Response Operations and develop their urban planning strategies with a more directed spatial approach. Once an integrated fluvial-coastal hydrodynamic analysis is completed, at-risk infrastructure mapping can be relatively trivial and is an effective tool for the municipality to prioritize allocation of infrastructure funding.

Perhaps the most original contribution of this study is the specific approach developed to smoothly amalgamate coarse bathymetry points with the much finer resolution topographic lidar data, most notably the use of a combination of measured and predicted water depths to seamlessly fill large gaps in measured bathymetry and populate the flexible mesh.

The approach taken in this study to derive discharge contributions from along the river (tributaries) simply by way of cumulative catchment areas scaled using a single gauged location was an efficient and effective method to incorporate total watershed flow; however, it presumes that all catchment runoff characteristics throughout the watershed are identical, or perform similarly to the gauged catchment. The assumption is valid because the catchments share similar geographic, geologic, and landuse characteristics. Analysis of an additional water level gauge downstream supported this presumption. Additionally, specific critical infrastructure such as bridge decks and inflow culvert dimensions were not surveyed or incorporated into the hydrodynamic modelling given the large scale of the study area. Such infrastructure are typically "weakest links" for fluvial discharge as they can restrict or alter flow and contribute to flooding. This omission reduces the accuracy of our model but considering the extreme flood events being modelled and the large geographic area of the watershed these limitations are acceptable. Finally, this study does not consider the potential for increased flooding from wave run-up as the result of wind, nor does it include additional atmospheric forcings such as the effect of ice build-up in the channel and snow melt in isolation of rainfall.

3.3.2. Future Work

The methods developed here to integrate many varied sources of bathymetry data with topographic lidar data were innovative and successful. However, data collection and analysis time would be reduced, and model results improved, if an area were surveyed with topographic-bathymetric lidar, which would inherently provide a high resolution seamless land-sea DEM. We anticipate conducting future flood risk studies that consider the combined effects of fluvial and coastal flooding at NSCC with our recently acquired topographic-bathymetric lidar system, and we expect to build even further on the modelling methodologies developed here using lidar-derived seamless bathymetry-land DEMs.

4. Conclusions

This study modelled combined river discharge and ocean tide-surge hydrodynamics to map flood risk using a combination of innovative GIS tools and methodologies and a sophisticated hydrodynamic model that was established for the study area using a flexible mesh representation of the hydrography and topography.

Discharge was simulated for events that represent a typical annual maximum runoff and extreme events with at least one occurrence within a 50 and 100 year return period based on the EC gauge time-series data measured upstream of Bridgewater. Tide and storm surge events or the equivalent of long term sea-level rise were simulated by using the predicted maximum high tide during 2012 and adding a 2 and 3 m storm surge to the ocean level seaward of the mouth of the river. In total nine simulations were carried out using a combination of three discharge levels coupled with three tidal-surge levels. For extreme events that are dynamic in nature, the maximum water level can occur at different places and at different times during the simulation. The maximum flood extent for each simulation was examined and the maximum water level for the duration of the simulation was extracted and merged into one file and used in the GIS to map the maximum flood extent and water depth.

Areas upstream of Bridgewater appear most vulnerable to fluvial discharge events and maps demonstrate the areas that are most vulnerable. Areas downstream of Bridgewater appear to be most vulnerable to storm surge and sea-level rise. Infrastructure including roads, buildings and the municipal wastewater (e.g., lift stations) were intersected with the flood layers to map areas at greatest risk. The road information was converted to points every 10 m and analyzed along with the building points and wastewater infrastructure where the flood depth information for each for each of the nine simulations was appended to their respective attribute tables. There are areas where the combined effects of large discharge with elevated sea-levels produce higher water levels than any single event on its own. No adaptation measures have been considered as that was beyond the scope of this current project. However, the information provided from this study will allow the municipality and provincial departments such as the Nova Scotia Department of Transportation and Infrastructure Renewal the ability to evaluate the risk of flooding from large river discharge events and storm surge or longer term sea-level rise and their combined effects and begin considering adaptation measures to mitigate flooding.

Acknowledgments: We thank the Municipality of the District of Lunenburg for funding this study, and the Town of Bridgewater for contributions and for funding the previous study [1].

Author Contributions: T.W. and K.M. conceived and designed the experiments; K.M. performed the experiment; T.W., K.M. and K.C analyzed the data; K.C. and T.W. contributed analysis; K.M., K.C and T.W. wrote the paper.

Conflicts of Interest: The authors declare no conflict of interest.

References

1. Webster, T.; McGuigan, K.; Collins, K.; MacDonald, C. Integrated River and Coastal Hydrodynamic Flood Risk Mapping of the LaHave River Estuary and Town of Bridgewater, Nova Scotia, Canada. *Water* **2014**, *6*, 517–546. [CrossRef]
2. Merwade, V.; Cook, A.; Coonrod, J. GIS techniques for creating river terrain models for hydrodynamic modeling and flood inundation mapping. *Environ. Model. Softw.* **2008**, *23*, 1300–1311. [CrossRef]
3. Schubert, J.E.; Sanders, B.F.; Smith, M.J.; Wright, N.G. Unstructured mesh generation and landcover-based resistance for hydrodynamic modeling of urban flooding. *Adv. Water Resour.* **2008**, *31*, 1603–1621. [CrossRef]
4. Kliem, N.; Nielsen, J.W.; Huess, V. Evaluation of a shallow water unstructured mesh model for the North Sea—Baltic Sea. *Ocean Model.* **2006**, *15*, 124–136. [CrossRef]
5. Wang, L.; Zhao, X.; Shen, Y. Coupling hydrodynamic models with GIS for storm surge simulation: Application to the Yangtze Estuary and the Hangzhou Bay, China. *Front. Earth Sci.* **2012**, *6*, 261–275. [CrossRef]
6. Ten Hagen, E. Hydrodynamic River Modelling with D-Flow Flexible Mesh: Case Study of the Side Channel at Afferden and Deest. Master's Thesis, University of Twente, Enschede, The Netherlands, 18 September 2014.
7. Gama, M.C.; Popescu, I.; Mynett, A.; Shenyang, L.; van Dam, A. Modelling extreme flood hazard events on the middle Yellow River using DFLOW-flexible mesh approach. *Nat. Hazards Earth Syst. Sci. Discuss.* **2013**, *1*, 6061–6092. [CrossRef]
8. Gallegos, H.A.; Schubert, J.E.; Sanders, B.F. Two-dimensional, high-resolution modeling of urban dam-break flooding: A case study of Baldwin Hills, California. *Adv. Water Resour.* **2009**, *32*, 1323–1335. [CrossRef]
9. Mtamba, J.; van der Velde, R.; Ndomba, P.; Zoltán, V.; Mtalo, F. Use of Radarsat-2 and Landsat TM Images for Spatial Parameterization of Manning's Roughness Coefficient in Hydraulic Modeling. *Remote Sens.* **2015**, *7*, 836–864. [CrossRef]
10. Mason, D.C.; Horritt, M.S.; Bates, P.D.; Hunter, N.M. Improving models of river flood inundation using remote sensing. In *New Developments and Challenges in Remote Sensing*; IOS Press: Amsterdam, The Netherlands, 2007.

11. Cobby, D.M.; Mason, D.C.; Davenport, I.J. Image processing of airborne scanning laser altimetry data for improved river flood modelling. *ISPRS J. Photogramm. Remote Sens.* **2001**, *56*, 121–138. [CrossRef]

12. Campana, D.; Marchese, E.; Theule, J.I.; Comiti, F. Channel degradation and restoration of an Alpine river related morphological changes. *Geomorphology* **2014**, 230–241. [CrossRef]

13. Kinzel, P.; Legleiter, C.; Nelson, J. Mapping River Bathymetry with a Small Footprint Green LiDAR: Applications and Challenges. *JAWRA J. Am. Water Resour. Assoc.* **2013**, *49*, 183–204. [CrossRef]

14. Neily, P.D.; Quigley, E.; Benjamin, L.; Stewart, B.; Duke, T. *Ecological Land Classification for Nova Scotia Volume 1—Mapping Nova Scotia's Terrestrial Ecosystems*; Nova Scotia Department of Natural Resources Renewable Resources Branch: Halifax, Canada, 2003; p. 83.

15. Government of Canada, E.C. Real-Time Hydrometric Data—Environment Canada. Available online: http://www.wateroffice.ec.gc.ca/index_e.html (accessed on 8 November 2013).

16. Brown, L. Flooded LaHave Claims Two Lives. South Shore Now, 2003.

17. Talbot, T. *Environment Canada Storm Surge and Heavy Rainfall Warnings*; Meteorological Service of Canada, Environment Canada: Gatineau, Canada, 2015.

18. CBC News Storm Surge Floods Coastal Nova Scotia. Available online: http://www.cbc.ca/news/canada/nova-scotia/storm-surge-floods-coastal-nova-scotia-1.1300403 (accessed on 7 April 2015).

19. CBC News Nova Scotia Flooded Roads Slowly Reopening. Available online: http://www.cbc.ca/news/canada/nova-scotia/nova-scotia-flooded-roads-slowly-reopening-1.2870854 (accessed on 7 April 2015).

20. Church, J.A.; Gregory, J.M.; Huybrechts, P.; Kuhn, M.; Lambeck, K.; Nhuan, D.; Qin, D.; Woodworth, P.L. Changes in Sea Level. In *Climate Change 2001: The Scientific Basis: Contribution of Working Group I to the Third Assessment Report of the Intergovernmental Panel*; Douglas, B.C., Ramirez, A., Eds.; Cambridge University Press: Cambridge, UK, 2001.

21. Meehl, G.A.; Stocker, T.F.; Collins, W.D.; Friedlingstein, P.; Gaye, A.T.; Gregory, J.M.; Kitoh, A.; Knutti, R.; Murphy, J.M.; Noda, A.; *et al.* Global Climate Projections. In *Climate Change 2007: The Physical Science Basis. Contribution of Working Group I to the Fourth Assessment Report of the Intergovernmental Panel on Climate Change*; Solomon, S., Qin, D., Manning, M., Chen, Z., Marquis, M., Averyt, K.B., Tignor, M., Miller, H.L., Eds.; Cambridge University Press: Cambridge, UK; New York, NY, USA, 2007.

22. IPCC. Summary for Policymakers. In *Climate Change 2013: The Physical Science Basis. Contribution of Working Group I to the Fifth Assessment Report of the Intergovernmental Panel on Climate Change*; Stocker, T.F., Qin, D., Plattner, G.K., Tignor, M., Allen, S.K., Boschung, J., Nauels, A., Xia, Y., Bex, V., Midgley, P.M., Eds.; Cambridge University Press: Cambridge, UK, 2013.

23. Rahmstorf, S. A Semi-Empirical Approach to Projecting Future Sea-Level Rise. *Science* **2007**, *315*, 368–370. [CrossRef] [PubMed]

24. Rahmstorf, S.; Cazenave, A.; Church, J.A.; Hansen, J.E.; Keeling, R.F.; Parker, D.E.; Somerville, R.C.J. Recent Climate Observations Compared to Projections. *Science* **2007**, *316*, 709–709. [CrossRef] [PubMed]

25. Shaw, J.; Taylor, R.B.; Forbes, D.L.; Ruz, M.H.; Solomon, S. *Sensitivity of the Coasts of Canada to Sea-level Rise*; Natural Resources Canada: Calgary, AL, Canada, 1998; pp. 1–79.

26. Peltier, W.R. Global Glacial Isotasy and the Surface of the Ice-Age Earth: The ICE-5G (VM2) Model and GRACE. *Annu. Rev. Earth Planet. Sci.* **2004**, *32*, 111–149. [CrossRef]

27. Forbes, D.L.; Manson, G.K.; Charles, J.; Thompson, K.R.; Taylor, R.B. *Halifax Harbour Extreme Water Levels in the Context of Climate Change: Scenarios for a 100-Year Planning Horizon*; Natural Resources Canada: Calgary, AL, Canada, 2009; p. 22.

28. Richards, W.; Daigle, R. *Scenarios and Guidance for Adaptation to Climate Change and Sea Level Rise—NS and PEI Municipalities*; Atlantic Climate Adaptation Solutions Association: Halifax, Canada, 2011; p. 87.

29. Bruce, J.; Burton, I.; Martin, H.; Mills, B.; Mortsch, L. *Water Sector: Vulnerability and Adaptation to Climate Change*; Final Report; The Government of Canada Climate Change Action Fund: Ottawa, Canada, 2000.

30. Mekis, E.; Hogg, W.D. Rehabilitation and analysis of Canadian daily precipitation time series. *Atmos. Ocean* **1999**, *37*, 53–85. [CrossRef]

31. Government of Canada, Natural Resources Canada. Canada in a Changing Climate: Atlantic Canada. Available online: http://www.nrcan.gc.ca/earth-sciences/climate-change/community-adaptation/830 (accessed on 6 November 2013).

32. *Toward a Greener Future: Nova Scotia's Climate Change Action Plan*; Nova Scotia Department of Environment: Halifax, Canada, 2009.

33. Madsen, T.; Willcox, N. *When It Rains, It Pours Global Warming and the Increase in Extreme Precipitation from 1948 to 2011*; Environment America Research & Policy Center: Washington, DC, USA, 2012.
34. Singh, D.; Tsiang, M.; Rajaratnam, B.; Diffenbaugh, N.S. Precipitation extremes over the continental United States in a transient, high-resolution, ensemble climate model experiment. *J. Geophys. Res. Atmos.* **2013**, *118*, 7063–7086. [CrossRef]
35. Toreti, A.; Naveau, P.; Zampieri, M.; Schindler, A.; Scoccimarro, E.; Xoplaki, E.; Dijkstra, H.A.; Gualdi, S.; Luterbacher, J. Projections of global changes in precipitation extremes from Coupled Model Intercomparison Project Phase 5 models. *Geophys. Res. Lett.* **2013**, *40*, 4887–4892. [CrossRef]
36. Whitfield, P.H.; Cannon, A.J. Recent Variations in Climate and Hydrology in Canada. *Can. Water Resour. J.* **2000**, *25*, 19–65. [CrossRef]
37. Zhang, X.; Harvey, K.D.; Hogg, W.D.; Yuzyk, T.R. Trends in Canadian streamflow. *Water Resour. Res.* **2001**, *37*, 987–998. [CrossRef]
38. Najjar, R.G.; Walker, H.A.; Anderson, P.J.; Barron, E.J.; Bord, R.J.; Gibson, J.R.; Kennedy, V.S.; Knight, C.G.; Megonigal, J.P.; O'Connor, R.E.; *et al.* The potential impacts of climate change on the mid-Atlantic coastal region. *Clim. Res.* **2000**, *14*, 219–233. [CrossRef]
39. Intergovernmental Panel on Climate Change Working Group II. *The Regional Impacts of Climate Change: An Assessment of Vulnerability*; Cambridge University Press: Cambridge, UK, 1998.
40. Nova Scotia Department of Fisheries and Aquaculture Lake Inventory Maps. Available online: http://novascotia.ca/fish/programs-and-services/industry-support-services/inland-fisheries/lake-inventory-maps/ (accessed on 16 April 2015).
41. Webster, T.; McGuigan, K.; MacDonald, C. *Lidar Processing and Flood Risk Mapping for Coastal Areas in the District of Lunenburg, Town and District of Yarmouth, Amherst, County Cumberland, Wolfville and Windsor*; Atlantic Climate Adaptation Solutions Association: Halifax, Canada, 2011; p. 130.
42. Pirzadeh, S. Structured background grids for generation of unstructured grids by advancing-front method. *Am. Inst. Aeronaut. Astronaut.* **1993**, *31*, 257–265. [CrossRef]
43. DHI MIKE 21 Toolbox. *Global Tide Model—Tidal Prediction*; Aerospace Research Central: Reston, VA, USA, 2014.

Journal of
Marine Science and Engineering

MDPI

Article

Application of a Coupled Vegetation Competition and Groundwater Simulation Model to Study Effects of Sea Level Rise and Storm Surges on Coastal Vegetation

Su Yean Teh [1,†], Michael Turtora [2,†], Donald L. DeAngelis [3,*], Jiang Jiang [4], Leonard Pearlstine [5], Thomas J. Smith III [6] and Hock Lye Koh [7]

[1] School of Mathematical Sciences, Universiti Sains Malaysia, Penang 11800, Malaysia; syteh@usm.my
[2] U.S. Geological Survey, Caribbean-Florida Water Science Center, 4446 Pet Lane, Suite #108, Lutz, FL 33559-630, USA; mturtora@usgs.gov
[3] U.S. Geological Survey, Southeast Ecological Science Center, Coral Gables, FL 33124, USA
[4] Jiang Jiang, Forestry College of Nanjing Forestry University, Key Laboratory of soil and water conservation and Ecological Restoration, Nanjing Forestry University, Nanjing 210037, China; ecologyjiang@gmail.com
[5] Leonard Pearlstine, Everglades National Park, South Florida Natural Resources Center, 950 N Krome Ave, Homestead, FL 33030, USA; Leonard_Pearlstine@nps.gov
[6] U.S. Geological Survey, 600 Fourth Street South, St. Petersburg, FL 33701, USA; tom_j_smith@usgs.gov
[7] Hock Lye Koh, Sunway University Business School, Jalan Universiti, Bandar Sunway, Selangor 47500, Malaysia; hocklyek@sunway.edu.my
*   Author to whom correspondence should be addressed; don_deangelis@usgs.gov; Tel.: +1-305-284-1690.
†   These authors contributed equally to this work.

Academic Editor: Rick Luettich
Received: 31 July 2015; Accepted: 21 September 2015; Published: 25 September 2015

Abstract: Global climate change poses challenges to areas such as low-lying coastal zones, where sea level rise (SLR) and storm-surge overwash events can have long-term effects on vegetation and on soil and groundwater salinities, posing risks of habitat loss critical to native species. An early warning system is urgently needed to predict and prepare for the consequences of these climate-related impacts on both the short-term dynamics of salinity in the soil and groundwater and the long-term effects on vegetation. For this purpose, the U.S. Geological Survey's spatially explicit model of vegetation community dynamics along coastal salinity gradients (MANHAM) is integrated into the USGS groundwater model (SUTRA) to create a coupled hydrology–salinity–vegetation model, MANTRA. In MANTRA, the uptake of water by plants is modeled as a fluid mass sink term. Groundwater salinity, water saturation and vegetation biomass determine the water available for plant transpiration. Formulations and assumptions used in the coupled model are presented. MANTRA is calibrated with salinity data and vegetation pattern for a coastal area of Florida Everglades vulnerable to storm surges. A possible regime shift at that site is investigated by simulating the vegetation responses to climate variability and disturbances, including SLR and storm surges based on empirical information.

Keywords: coupled hydrology–vegetation model; salinity; coastal Everglades; hardwood hammock; mangroves; vadose zone; groundwater

1. Introduction

Sea Level Rise (SLR) is one of the most significant predicted consequences of global climate change and has the potential for severe effects on the vegetation of low-lying coastal areas and islands [1]. Rising sea level will also mean higher storm surges [2], even if the frequencies do not change. Mean SLR will have a gradual effect on shoreline retreat and subsequent loss of ecosystem area in these locations. However, large-scale marine water intrusion through storm surges may affect large areas on a short time scale, including the inundation of whole low-lying islands. The immediate effect will be on the freshwater lenses that sit on top of saline groundwater in these areas. Such effects on available fresh water may have negative consequences for the ecological and human populations of coastal areas and, particularly, islands, as they depend critically on fresh water stored in the lenses [3–5]. Longer-term consequences may involve large-scale vegetation regime changes, which can pose risks to conservation and restoration efforts in coastal national parks and preserves. For example, in southern Florida, USA, the beneficial effects of increased freshwater flow resulting from the Comprehensive Everglades Restoration Plan (CERP) [6] may be compromised in some places by increased saltwater intrusion and salinity overwash events.

In tropical and subtropical coastal areas, increase in salinity of the vadose zone (unsaturated soil zone) induced by storm surges might reduce or eradicate the salinity-intolerant (glycophytic) species and promote rapid landward migration of salinity-tolerant (halophytic) species such as mangroves. Inland expansion of mangroves at the expense of glycophytic vegetation has been noted in coastal ecosystems, e.g., see [7,8]. The effect of a disturbance may cause a rapid shift in the transition zone between vegetation types, also called the ecotone [9]. While many ecotones between floristic types are broad and diffuse, some are remarkably narrow. An example of the sharpening of ecotones in coastal areas involves halophytic vegetation (mangrove vegetation in tropical and sub-tropical regions) and glycophytic vegetation (including tropical hardwood trees forming "hammocks", and freshwater marsh) in southern Florida coastal areas [10]. Typically, these are not interspersed. Stability of the ecotone is promoted by self-reinforcing positive feedback as follows [11]. During the dry season, plant transpiration can lead to infiltration by highly brackish underlying ground water into the vadose zone. Hardwood hammock trees reduce transpiration when the salinity in the vadose zone increases. This limits the salinization of the vadose zone. Meanwhile, the transpiration of mangroves can continue even at relatively high salinities, sustaining salt-water infiltration. Thus, through self-reinforcing positive feedback, each type of vegetation has a tendency to promote the salinity condition that is favorable to itself in competition.

However, both SLR over decades and the acute effects of large storm surges may upset that stability. If a large enough pulse of salinity from overwash remains in the soil for a long period, it may overwhelm the feedback maintaining the favorable conditions for glycophytic vegetation, such that the ecotone can no longer be maintained; that is, a regime shift could occur. Large areas of glycophytic vegetation could be replaced by halophytic vegetation, which may lead to permanent salinization of the vadose zone. For example, Baldwin and Mendelssohn [12] studied the effects of salinity and inundation coupled with clipping of aboveground vegetation on two adjoining plant communities, *Spartina patens* and *Sagittaria lancifolia*. The study reported that that the levels of flooding and salinity at the time of disturbance determined the potential shift of vegetation to a salt tolerant species. Large storm surges created by Hurricanes Katrina and Rita (2005) affected the coastal areas of Louisiana, USA. Subsequently, both freshwater and brackish communities exhibited changes in vegetation [13]. It was noted that in the central region of their study area, marsh composition changed to a more saline classification, and high mean salinities exceeded mean pore-water salinity levels tolerated by the previous dominant species.

Hindcasting the effects of previous storm surges and on coastal vegetation and forecasting the future effects of both storm surges and SLR requires modeling. In particular, forecasting these effects on the halophyte-glycophyte ecotone requires that the competition of these two vegetation types be modeled along with hydrology and salinity dynamics. To investigate the dynamics of the ecotone between halophytic and glycophytic vegetation, including the possibility of a regime shift, a spatially explicit computer simulation model, MANHAM (MANgrove HAMmock model), has been developed [14].

MANHAM simulates the competition of hardwood hammock trees and mangrove trees on a grid of spatial cells; each a few square meters in area. Vegetation of each type may be present in a given cell, and growth and competition are modeled on this local scale. Dispersal of propagules of each species is also modeled. In cells where vadose zone salinity is low, hammock trees grow faster and outcompete mangroves, but in higher salinity cells, hammock tree growth is slowed and mangroves can outcompete the hammock trees.

MANHAM also models water flow and salinity in the vadose zone, which depends on precipitation, tides, evaporation, plant transpiration, and groundwater infiltration. The dynamics of hydrology and salinity are modeled on a time resolution of less than a day, whereas the dynamics of vegetation is modeled on monthly time steps. The key mechanism in the model is the self-reinforcing positive feedback relationship between each vegetation type and vadose zone salinity described above. In each spatial cell these feedbacks help maintain dominance of the current vegetation type in the cell, hammock or mangrove, against invasion by the other type. However, a large external impact on the vadose zone salinity of a cell dominated by hammock trees, such as from storm surge overwash, could lead to decline in hardwood growth and favor growth of mangrove propagules, or seedlings, into trees. Through positive feedback between the growing mangrove vegetation and the vadose zone salinity, the cell could eventually shift to mangrove domination. Shifts from mangrove to hardwood hammock are also possible if there is a sufficient external forcing of freshwater. Mathematical details of MANHAM are described in Appendix 5.

MANHAM has examined the impact of SLR on southern Florida coastal forests [15], showing that buttonwood forest (*Conocarpus erectus*), could be squeezed out by red mangrove (*Rhizophora mangle*). Simulations have also indicated that a significant one-day storm surge event could feasibly initiate a vegetation shift from hardwood trees to mangroves in areas initially dominated by the former. Mangroves in the model were able to take over large areas when storm surges saturated the vadose zone with over 0.015 kg/kg salinity, if the salinity was not quickly washed out of the vadose zone by precipitation, and if a sufficient density of mangrove propagules (seedlings, by which mangroves are spread) were present. It was observed that such a shift might not be conspicuous at first, but would be inevitable once a tipping point had been passed. These findings are relevant to many coastal areas, including the coastal Everglades. They have motivated us to apply the modeling to forecasting future changes in coastal vegetation and developing plans to meet the challenges of these changes.

MANHAM simulates the vadose zone as a uniform compartment and does not model underlying groundwater dynamics [14] and assumes that ground water is a constant boundary condition. These assumptions are certainly violated in real systems. For example, in low-lying coastal areas and small atoll islands, tides cause diurnal fluctuations in groundwater, and storm surges can cause major changes in groundwater that can last for years. A deficiency caused by these assumptions in MANHAM is that it does not consider the freshwater lens, which in coastal areas and atolls typically overlies deeper water salinity levels of the neighboring seawater. This freshwater lens is an important constituent of the water balance for the overlying vegetation through transpiration and plays a key role on the salinity balance as well. Because of these deficiencies we developed MANTRA (MANhamsuTRA), which builds in more detailed hydrological and salinity dynamics. Below we describe the USGS's Saturated–Unsaturated TRAnsport (SUTRA) groundwater model and its coupling with MANHAM to form MANTRA. We apply MANTRA to a coastal area of Everglades National Park (ENP), southern Florida, and demonstrate how it can be used to project the effects of both storm surges and SLR on coastal vegetation.

2. Methods

2.1. MANTRA Model

To overcome the limitation due to MANHAM's lack of a freshwater lens, MANHAM has been integrated with an established groundwater hydrology and salinity model, the United States Geological Survey (USGS)'s Saturated–Unsaturated TRAnsport (SUTRA) groundwater model [16,17]. The fluid pressure and salinity gradients in the transition zone between glycophytic and halophytic vegetation associated with the freshwater lens are quantified by the variable density flow simulated by SUTRA. SUTRA also simulates the unsaturated zone of the soil, and so can substitute for the hydrodynamics and salinity dynamics of the soil and groundwater. However, SUTRA does not include vegetation competition dynamics. By combining MANHAM with SUTRA, forming MANTRA (MANhamsuTRA), we provide an integrated model that simulates the possible effects of gradual SLR, as well as both short- and long-term effects of a single or a sequence of overwash events on a coastal area or small island, containing zones of glycophytic and halophytic vegetation.

MANTRA input data are (1) vegetation type and (2) groundwater conditions (fluid pressure and salinity) and it simulates the changes in vegetation type biomass over time subject to the groundwater conditions. Because MANTRA is an extension of SUTRA, it also delivers the output of SUTRA, including fluid pressure and solute concentration. The primary variable upon which the groundwater model SUTRA is based is fluid pressure, which varies spatially and temporally. Variations in fluid density and fluid pressure differences drive flow of groundwater, which is a fundamental mechanism upon which the solute transport model is based. MANTRA employs spatial discretization called mesh by quadrilateral finite elements. In a cross sectional form, the elements are organized in rows and columns with each element having four nodal points. Nodal points (or nodes) are shared by the elements adjoining the node. A cell is centered on a node, not an element. Cell boundaries are half way between opposite sides of an element as shown in Figure 1. Further details of MANTRA are described in Appendix 5.

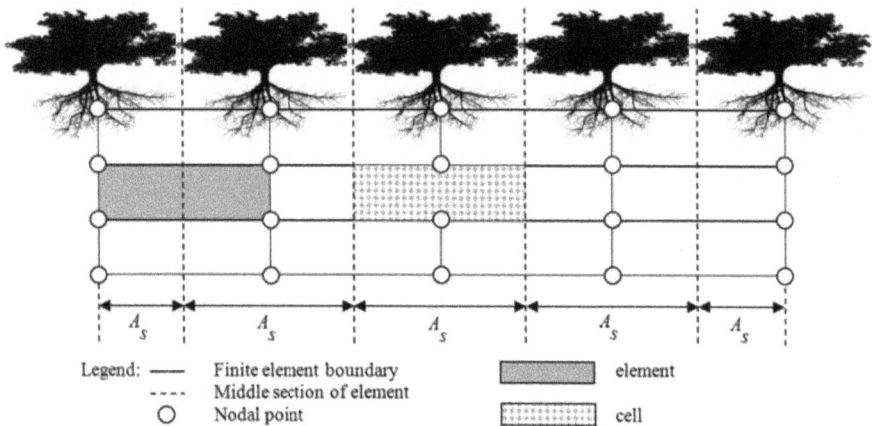

Figure 1. Schematic sketch of a hypothetical simulation case for illustration purposes.

Here we describe an important aspect of how the hydrology of SUTRA and the vegetation dynamics of MANHAM interact, which is through water uptake by plants. The fluid mass source/sink term Q [M/s] (where M = mass and s = second) in SUTRA, which accounts for external addition/subtraction of fluid including pure water mass plus the mass of any solute dissolved in the source fluid, can be used to characterize the uptake of water by plants (Q_p) [18–20]. This term in a mass balance equation is used to represent the addition (source) or extraction (sink) from the mass balance system. As a function of salinity, the total water uptake $R = f$ (C) [L/s] (where L here is the dimension of vertical distance or depth) by plants is determined by the salinity concentration C [M_s/M_f] (where M_s = mass of solute and M_f = mass of fluid) calculated by SUTRA. The salinity C is derived from the solute mass balance equation that includes processes such as fluid flow and diffusion. Then, assuming a closed canopy so that transpiration can be assumed constant for the vegetation types and evaporation ignored, the fluid mass per unit time (Q_p) required by the plants in a certain horizontal cell for transpiration can be estimated by:

$$Q_p = R \cdot A_s \cdot \rho \ [\text{M/s}] \tag{1}$$

Here, A_s represents the surface along the depth dimension [L^2] and ρ is the fluid density of fresh water [M/L^3]. For cross-sectional model, the width of each cell is assumed to be 1.0 m. Thus, A_s depends on the length or horizontal grid size (Figure 1).

It may happen that the fluid mass required by the plant for transpiration is more than what is available (unsaturated flow). Hence, there should be a relation between the fluid mass required by transpiration and the fluid mass available. It is assumed that actual fluid mass being subtracted from a cell due to transpiration depends on the saturation S_w and porosity ε in the cell, leading to the following relation:

$$Q_{IN} = -Q_p \cdot \varepsilon \cdot S_w \ [\text{M/s}] \tag{2}$$

where, Q_{IN} = total mass sink (due to plant transpiration) [M/s]; ε = porosity [V_v/V]; S_w = water saturation [V_w/V_v] with V = total volume, V_v = volume of voids, V_w = volume of water. Suppose porosity ε is kept constant at 1.0. When the void space is completely filled with fluid and is said to be saturated, that is $S_w = 1$, the actual water uptake by plants, Q_{IN}, will be equivalent to the amount of water required by the plants, Q_p. When the void space is only partly water filled and is referred to as being unsaturated, that is $S_w < 1$, the actual water uptake by plant will decrease as a factor of the saturation, S_w.

To simulate fluid inflow to the soil pores due to seawater inundation during a storm surge, Equation (3) below [21] was used to specify the pressure at surface nodes inundated by seawater.

$$p = \rho_{sea} \cdot g \cdot h \tag{3}$$

Here, p = pressure at top layer nodes [$M/(Ls^2)$], ρ_{sea} = fluid density of seawater [M/L^3], g = gravity [L/s^2], h = inundation depth [L].

2.2. Study Transect

To evaluate MANTRA's effectiveness in a coastal Everglades setting, we chose a site and a set of scenarios related to past and possible future events. MANTRA was used on a coastal site that has been exposed to storm surges. The main objective of MANTRA is to project possible future changes in hardwood hammocks in southern Florida under conditions of gradually rising sea level and/or major storm surges. As an example, we apply MANTRA to a specific hardwood hammock along the southwestern coast of ENP, bordering Florida Bay (25°12′24.13″ N, 80°55′47.48″ W). This hammock has been described by Saha et al. [22], where it is referred to as the Coot Bay Hammock (see aerial view in Figure 2). The hammock is in the middle of a low ridge with north-south orientation. Along a 370 m transect from west to east across the hammock (see Figures 2 and 3), vegetation changes from black mangroves (*Avicennia germinans*) at the low elevation (0.4–0.5 m above mean sea level) western end to mixed halophyte coastal prairie (*Batis/Salicornia*) (0.5–0.7 m) with individual buttonwoods (*Conocarpus erectus*) (0.7–0.8 m), to mixed-hammock, and hardwood hammock communities (0.9–1.5 m) at the peak in elevation in the middle of the ridge. As one continues farther along the transect, the associations occur in reverse order, except for a change from white to black mangroves. The transitions are relatively sharp. The geology is marl mixed with peat to a depth of 2–3 m over karst bedrock. Table 1 shows assumed ranges of conductivities. The site has a tropical climate with an average of 1570 mm annual precipitation, nearly 60% of which is from June through September. Mean January and July temperatures are, respectively, 22 °C and 30 °C. Hourly salinity data of water at 0.5 m below land surface were available at two locations (blue dots near transect on Figure 2) and monthly point measurements of salinity and water depth at five locations (other blue dots on Figure 2) from in

February 2011, through July 2012. Coot Bay Hammock is an ideal study transect because hardwood hammocks occur at the highest, and mangroves and coastal prairies occur at the lowest end of the gradient in elevation.

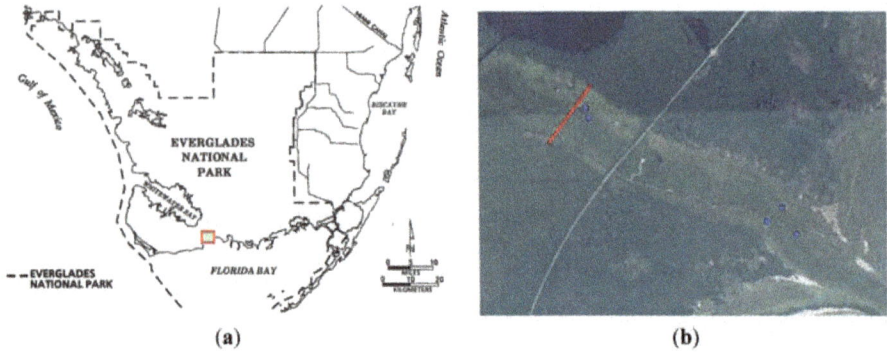

(a) (b)

Figure 2. (a, **Left**) Southern Florida with site shown in red box (from Light and Dineen 1994 [23]). (b, **Right**) Aerial view of Coot Bay Hammock, with 370 m transect, shown in red. Blue dots indicate well sites. l. (Figure 2a adapted from reference [23] with copyright permission).

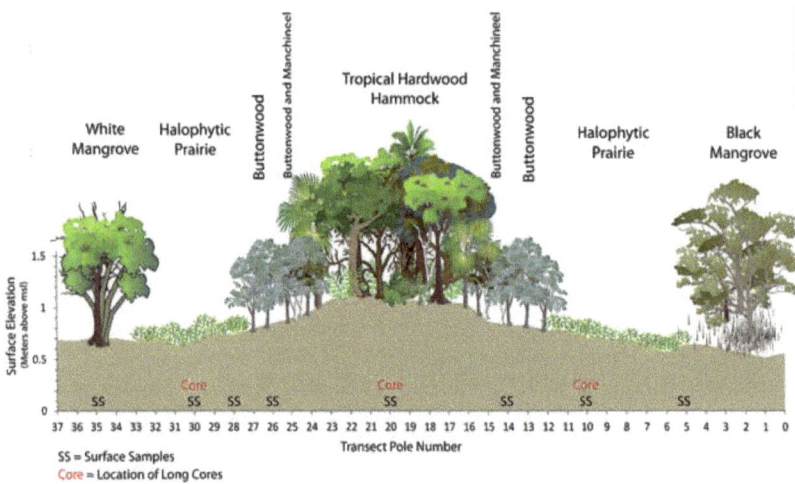

Figure 3. West-to-east transect of about 370 m across the Coot Bay Hammock showing the sharp gradations between vegetation types. Transect poles mark 10 m distances, and locations where surface samples and core samples were taken. Figure courtesy of Brandon Gamble, National Park Service. Graphic symbols courtesy of the Integration and Application Network, University of Maryland Center for Environmental Science.

Saha et al. [22] noted an increase in salinity over the last decade (data from hydrological station maintained by ENP). Saha et al. [22] suggest that SLR can induce a rising water table, which will cause a shrinking of the vadose zone and an increase in salinity in the bottom portion of the freshwater lens, subsequently increasing brackishness of plant-available water. For these reasons, the Coot Bay Hammock was selected as a first site to test MANTRA. The 370 m transect was modeled as two-dimensional, with the horizontal dimension along the transect, and the vertical axis for depth.

Table 1. Stratigraphy and hydraulic conductivities used for the Coot Bay Hammock. Parameter values are assumed from other sources.

Layer Name	Lower Boundary (m)	Hydraulic Conductivity (Qualitative)	(m/Day)
Marl plus peat	~2–3	low	<0.3
karst	~6	high	>1000
sand	~9	medium	10 to 100
noflow	~9		

Storm surge/SLR would most likely occur from the lower right of Figure 2, right panel. This happened in November 2005 due to Hurricane Wilma [24]. A storm surge could also enter the area through a canal system. The park road through ENP to the town of Flamingo on Florida Bay (visible in Figure 2) cuts across the southern tip of the tropical hardwood forest that extends to the NW. The ridge extends further SE with vegetation that was identified in 1981 as "collapsed hammock" [22] and is now referred to as a transitional buttonwood hardwood hammock.

The Coot Bay Hammock area is vulnerable to both wind damage and storm surges from hurricanes. Hurricanes have been important in shaping the vegetation of the region. The "Labor Day" hurricane of 1935 killed many buttonwoods, while Hurricane Donna (1960) produced a storm surge of 4 m in this area, causing 90% mortality of trees on lower ground, and 25%–50% mortality buttonwoods and hardwood hammock trees on coastal hammocks such as the Coot Bay Hammock. These hurricanes were factors that "produced new vegetation mosaics of white, black and red mangroves and buttonwoods" [25]. Buttonwoods have not recovered in some areas they once dominated. While fires have been suggested as a possible factor in their loss, Olmstead and Loope [25] discount this, which leaves open the possibility that a storm surge induced regime shift was the cause. Two hurricanes that more recently have affected the Coot Bay Hammock area were Hurricane Andrew (1992) and Hurricane Wilma (2005). The storm surge from Andrew measured 1.2–1.5 m at Flamingo, near Coot Bay Hammock, while that from Wilma measured 0.7 m at Black Forest, also near Coot Bay Hammock, see [26]. Hurricane Andrew caused very little structural damage to trees that far south, while Wilma caused moderate though not severe damage.

2.3. Model Simulations

A 2-D model of the transect across the Coot Bay Hammock was developed using MANTRA. For simplicity, buttonwoods were aggregated with the hardwood hammock, and hardwood hammock and mangroves were modeled as competing vegetation types. Coastal hammocks, tree islands, and buttonwood forests of Florida Bay experience tidal amplitude of only ~15 cm [27], so tides were ignored in the model. Model scenarios are described below.

2.3.1. Scenario 1: Existing Conditions

The model was first applied to the existing conditions of the Coot Bay Hammock. The aim was to calibrate the model to produce results that are consistent with the observed data; that is, with the observed sharp boundary and with virtually no mangroves in areas dominated by hardwood hammocks and *vice versa*.

2.3.2. Scenarios 2 and 3: Storm Surges

Preliminary MANTRA simulations (not shown here) showed that neither of the storm surges plus the associated light damage to the hammock from Hurricanes Andrew and Wilma would be sufficient to cause a regime shift from hardwood hammock to mangroves. However, sites such as Coot Bay Hammock have been struck by larger hurricane disturbances in the more distant past, and will be in the future. Therefore, to project the effects of greater disturbances, two storm surge scenarios were applied that used inundation depths consistent with those caused by Hurricane Andrew, i.e., about

1 m, but which had some additional factors that would amplify the effects of the surge. Scenario 2 assumes that the storm inflicted heavy damage to the hardwood hammock trees, reducing their living biomass to below the level of mangrove seedlings. Scenario 3 consisted of a storm surge in which only moderate damage was done to the hardwood hammock vegetation, reducing the initial vegetation by one-half. However, the storm surge was followed immediately by a severe four-year drought in which mean precipitation was reduced by half. Although a four-year drought starting immediately following a hurricane would be unusual, periodic droughts are part of the climate of southern Florida, and a drought of this magnitude is plausible.

2.3.3. Scenario 4: Gradual SLR

A final scenario (Scenario 4) of gradual SLR was applied over a period of 150 years without storm surge events.

3. Results

3.1. Scenario 1: Existing Conditions

The results for the distribution of mangroves and hardwood hammock trees for the calibrated MANTRA are shown in Figure 4. Consistent with the vegetation distribution observed at the Coot Bay Hammock, our results show the hardwood hammocks occupy the slightly elevated ridge, where a freshwater lens of about 0.5 m is maintained by precipitation and feedback from hardwood hammocks, whereas the mangroves occupy the lower elevated areas on either side of the hardwood hammock area.

3.2. Scenarios 2 and 3: Storm Surges

Next, storm surge Scenario 2 was applied using MANTRA. The results of the simulation are shown in Figure 5. The process of the positive feedback involving increased mangrove invasion causing increased soil salinity, allowed the mangroves to take over in about 16 years. Scenario 3 consisted of a storm surge in which only moderate damage was done to the hardwood hammock vegetation, but was followed by a severe four-year drought. The results are shown in Figure 6, where again the mangroves took over the entire site within about 16 years. Again, the positive feedback loop of mangrove invasion and increasing salinity drove the transition.

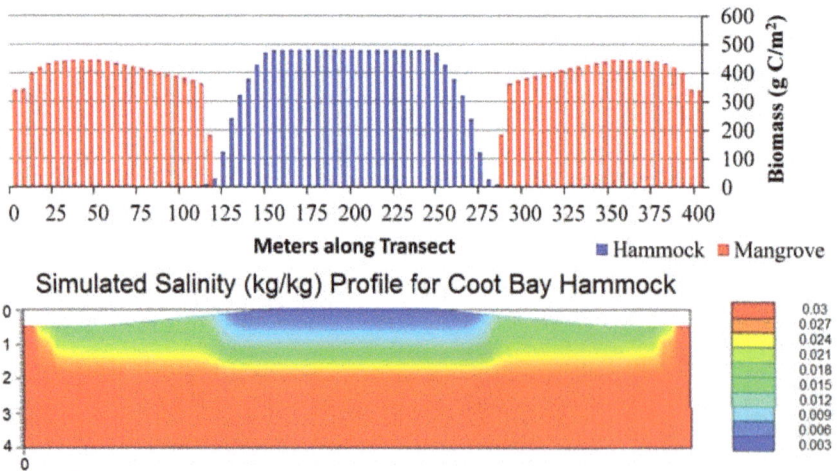

Figure 4. (**Top**) Simulated distribution of mangroves (red) and hardwood hammock (blue) trees along a 370-meter transect (rounded to 400 m in the model) across the Coot Bay Hammock. (**Bottom**) Simulated salinity profile with ground depth in meters and salinity in kg/kg.

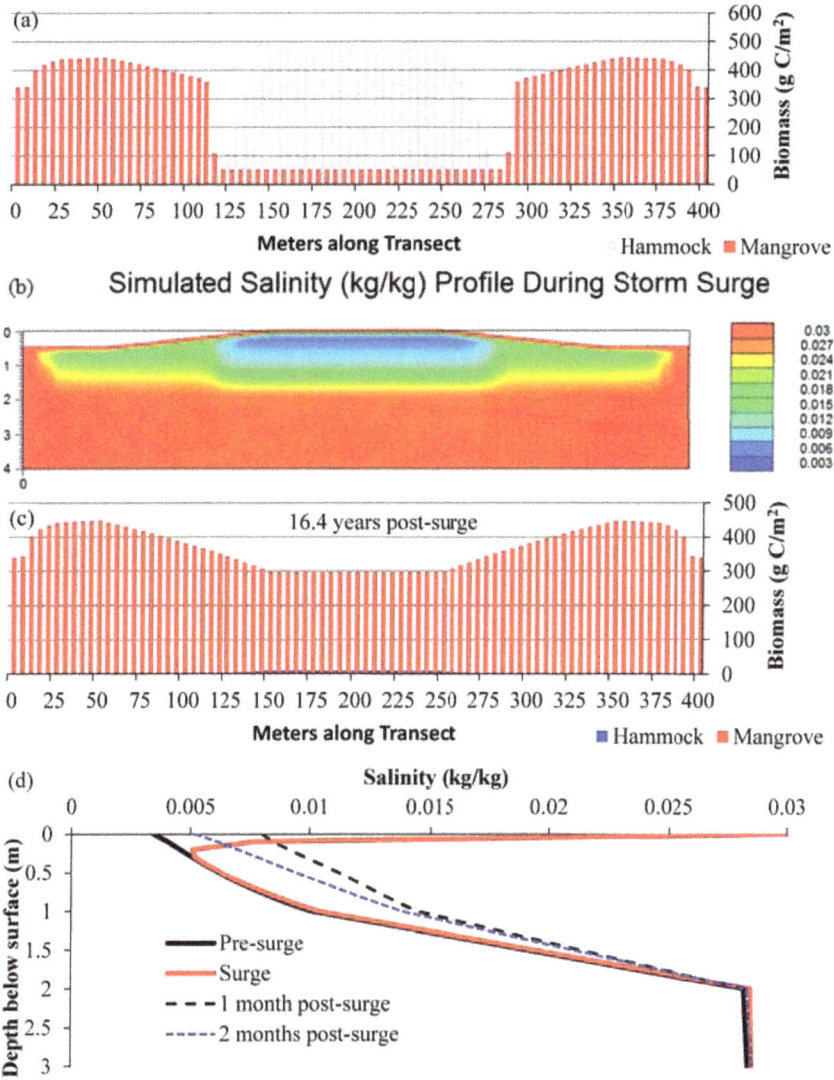

Figure 5. (**a, Top**) Depiction of initial conditions following storm surge, with almost complete elimination of hardwood hammock living biomass through knockdown of trees. Faded blue colors indicate destruction of initial trees. (**b, Second from top**) Simulated salinity profile during the storm surge. (**c, Third from top**) Simulated distribution of mangrove (red) vegetation 16.4 years (6000 days) after the storm surge. (**d, Bottom**) Sequence of salinity profiles starting before the surge until two months after, as predicted by MANTRA, with the particular precipitation pattern.

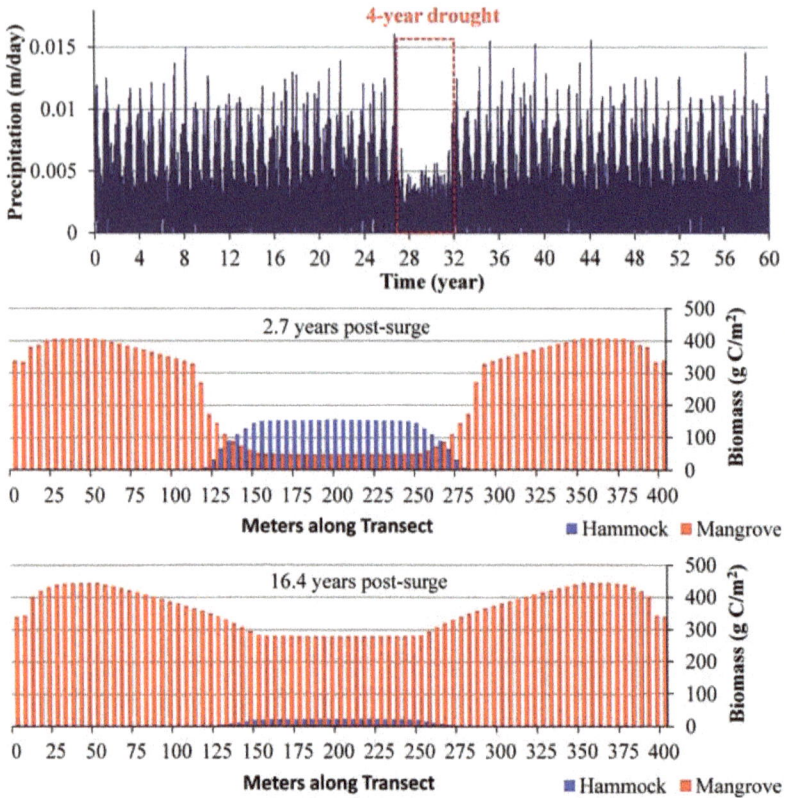

Figure 6. (**Top**) Depiction of precipitation over time used in the scenario, with four-year drought. (**Middle**) Simulated distribution of mangrove (red) and hardwood hammock (blue) vegetation 2.7 years (1000 days) after the storm surge, which occurs at 27.3 years into the simulation. (**Bottom**) Simulated distribution of mangrove (red) vegetation 16.4 years (6000 days) after the storm surge.

3.3. Scenario 4: Gradual SLR

Figure 7 illustrates Scenario 4, of SLR effect on the vegetation distribution at the Coot Bay Hammock transect. We consider here only the effects of sea level rise and ignore any possible effects of major storm surges. These simulation results indicate the mangroves will encroach into the areas of hardwood hammock, confining the freshwater vegetation to a smaller area. Hardwood hammock would persist on the elevated ridge. Further simulations (results not shown) indicate that the hardwood hammocks appear to persist at the elevated ridge unless the sea level rises to a level where the ridge is frequently inundated; e.g., every 20 years or so, with seawater.

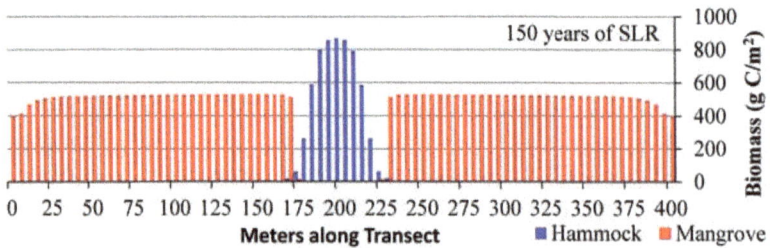

Figure 7. *Cont.*

Simulated Salinity (kg/kg) Profile - 150 years of SLR

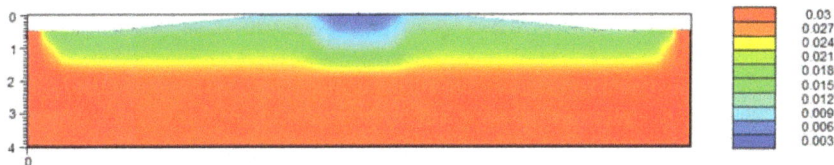

Figure 7. (Top) Simulated distribution of mangroves (red) and hardwood hammock (blue) trees along a 400-meter transect across the Coot Bay Hammock after 150 years subject to plausible SLR scenario (3 mm/year). **(Bottom)** Simulated salinity profile.

4. Discussion

Application of MANTRA to the Coot Bay Hammock transect provides insights on the potential vulnerability of the vegetation to storm surges and SLR. Simulations (not shown) that assumed storm surges of about 1 m, but little damage to trees and no prolonged drought did not produce regime shifts of the Coot Bay Hammock to halophytic vegetation. This is consistent with observations in the field showing no recent signs of the start of a shift. The scenarios that we presented that included damage or prolonged drought indicated that a shift might occur under those circumstances. Simulation results for Scenario 2 indicate that mangroves might be able to take over the slightly elevated ridge previously dominated by hardwood hammocks after a storm surge if the surge inflicts heavy damage to the hardwood hammock trees. It is possible that the effects of earlier hurricanes, Hurricane Donna (1960) in particular, may have led to permanent changes over parts of the Coot Bay hammock area. That hurricane produced a 4 m storm surge in the area and destroyed many buttonwood trees, which have not returned. This could be explained by an event like Scenario 2. Precipitation is the source for groundwater lens recharge. Scenario 3, a moderate storm surge followed by a prolonged period of drought, could also cause the shift from hardwood hammocks to mangroves, even though the hardwood hammocks were not badly damaged by the surge.

MANTRA improves greatly over MANHAM on the resolution with which hydrology and salinity are simulated, particularly along the vertical axis. The previous results of MANHAM suggested the possibility of a regime shift from a storm surge [14], but it was cautioned that that it was only a hypothetical result based on simple assumptions of water budget for hydrological dynamics. SUTRA is a highly detailed hydrologic model that has been tested in many contexts and can reliably predict hydrology and solute dynamics if provided good parameter values. Therefore, in MANTRA, the vadose zone is now connected seamlessly with the ground water, rather than the latter being treated as a boundary condition. A freshwater lens emerges naturally above the groundwater in simulations. Importantly, this more realistic treatment of hydrology and salinity dynamics does not change the emergence of sharp boundaries between glycophytic and halophytic vegetation, which was observed in MANHAM [11,14]. This gives us confidence that the self-reinforcing positive feedbacks hypothesized to be acting between each vegetation type and its local soil environment are a reasonable explanation for the sharp ecotone observed, and that these feedbacks may provide resilience to storm surge disturbances that are not too strong. MANTRA shows, like MANHAM, that a storm surge can cause a regime shift, but it is more conservative, as it shows salinity washing out faster unless there is a drought. So the results of MANTRA show that wind damage to the freshwater vegetation must be severe enough that the mangrove seedlings washed in have a high chance of not being outcompeted by the remaining freshwater vegetation.

Our simulations underscore that three conditions are necessary for a hardwood hammock to undergo a regime shift leading to a mangrove community; sufficiently severe damage to the existing hammock to open a gap to allow growth of invading seedlings, a large input of salinity persisting for a long enough period of time to favor growth of mangrove seedlings in competition remaining freshwater vegetation, and an input of enough mangrove seedlings to allow mangroves to be present in sufficient number to influence the future soil salinity. Surveys of hurricane damage to Everglades hardwood hammocks from Hurricane Andrew (1992) provide an estimate of what a major hurricane can inflict. Studies show heavy damage to 85% of all stems >2 cm diameter and loss of almost all leaves [28,29]. As pointed out in another study [30], the damage provided opportunities for invasive seedlings, which negatively influenced regrowth of the native vegetation. If such damage were inflicted on a coastal hammock, the damage would likely be accompanied by storm surge overwash. Hydrodynamic simulations have been performed of the effects of hurricanes on southern Florida [31]. In particular, those authors performed a hindcast of the "Great Miami Hurricane" of 18 September 1926, including the subsequent meteorological conditions. Simulations of surface water and groundwater salinity in an on shore area showed that salinity levels above 5 g/kg could remain in the soil for close to three years (until July 1929 in their Figure 10), and longer in the upper layer of groundwater. Of course, persistence of soil salinity conditions will vary from location to location with geology and freshwater influxes. However, even short-term exposure to salinity would kill much salinity-intolerant vegetation [32], and the persistence of the high salinity levels for two or three years shown in the modeling of [31] is more than sufficient in MANTRA to favor mangrove seedlings over freshwater vegetation regrowth. Input of mangrove seedlings by storm surges has rarely been studied, but on the basis of one study [33], Jiang et al. [34] estimated that up to 2000 propagules per ha could be input from nearby mangrove forest. It has also been suggested that a strategy of mangroves is to constantly produce a large number of seedlings [35] that can be spread by high tides, wind, or animals to provide a "sit-and-wait" seedling bank. Mangrove seedlings and small plants are commonly observed in nearby freshwater areas (*personal observation*).

MANTRA was developed as a tool to study the potential impact of SLR and storm surges on competing halophytic and glycophytic vegetation and, in particular, to investigate the hypothesis that a large input of salinity to a community such as hardwood hammock could result in a regime shift to a halophytic community, such as mangroves. The scenario simulations of MANTRA indicate the feasibility of such shifts, but it can be asked what the evidence is for occurrences in the past. Solid evidence for past regime shifts in southern Florida may be lacking, but that may reflect that up until now there have been few studies focused on vegetation changes following storm surges. A regime shift of vegetation would also take at least a decade to be noticed, and so might appear to be ordinary gradual change rather than an irreversible transition.

Nonetheless, there are some additional examples where regime shifts may be inferred to have occurred in southern Florida. A report on Cape Sable at the southwestern tip of Florida documented the shift in much of this region from freshwater to marine marsh [36]. This shift started to occur suddenly in the 1930s and shows no signs of returning to its original state, so it appears to be an irreversible change, perhaps a regime shift. It is possible that the Labor Day Hurricane of 1935, which produced a nearly four-meter storm surge over Cape Sable, was at least a partial cause of this shift. As another case, Ross et al. (2009) [37], studying vegetation changes in the Florida Keys, noted that "Once sea level reaches a critical level, the transition from a landscape characterized by mesophytic upland forests and freshwater wetlands to one dominated by mangroves can occur suddenly, following a single storm-surge event. We document such a trajectory, unfolding today in the Florida Keys. With sea level projected to rise substantially during the next century, ex-situ actions may be needed to conserve individual species of special concern".

In the Introduction, we noted general evidence from outside of southern Florida that salinity input to a glycophytic community could lead to apparent long-term vegetation shifts [12,13]. In addition, the role of overwash salinity pulses in causing long-term effects on vegetation has been noted in islands of the southern Pacific, where widespread sea flooding by storm surges around the coastlines of South Pacific atolls is a serious hazard during tropical cyclones. For example, in 2005, tropical cyclone Percy inundated the three atolls of Tokelau. The high surge allowed waves to sweep across the low-lying atoll islands. It also inundated the Pukapuka Atoll in the Northern Cook Islands. The immediate effect was on the freshwater lenses that sit on top of saline ground water in these areas [38,39]. Both [38] and [40] stressed that recovery from an overwash event may be prolonged, depending on the amount of seawater that accumulates in the central depression of the atolls. The reason that we extended our original MANHAM model to be combined with SUTRA in MANTRA was to be able to simulate the changes in the freshwater lens and groundwater from storm surges, which could have a long-lasting effect on vegetation.

4.1. Relevance of MANTRA for Management

A goal of the CERP for restoration of the Everglades is to bring additional fresh water south into ENP to restore freshwater habitats. It has been challenging, however, to deliver historic quantities of fresh water sufficient to improve conditions all the way to the coast and Florida Bay, and climate change and SLR will complicate this further [41]. Coastal hardwood communities provide unique habitat for a high diversity of species from plants to mammals [42,43] and protect fresh marsh communities behind them from storm surges. Mangrove migration inland along the west coast, at the expense of hardwood communities and freshwater marsh, has often kept pace with current rates of SLR, although the coastal forests are further stressed by the historic and current reductions of fresh water flow to the Everglades. Along the southern coast, including the study transect of this paper, there is an elevation dip inland of the mangrove/buttonwood/hardwood zone that isolates these hardwood forest communities from simple migration to higher elevations inland, e.g. see [15]. Projections of increasing rates of SLR heighten concerns for maintaining a fresh water hydrologic head that slows salt-water intrusion and allows coastal hardwood communities to have critical time to adapt to changing conditions. MANTRA provides information on how SLR and storm surges may affect vulnerable hardwood hammocks.

4.2. Future Plans

MANTRA development and application is documented in this paper to provide a start in developing a robust model for projecting the effects of overwash and climate change events on groundwater salinity as well as potential changes in vegetation composition. In this paper, the halophytic plant species at Coot Bay Hammock are generally grouped together as mangroves. However, in fact, there were other halophytic plants that have different dynamics than mangroves along the Coot Bay Hammock transect. Therefore, a better representation of the plant community at Coot Bay can be obtained by including more vegetation types in the model. MANTRA will be improved by revising the plant root network horizontally and vertically, which allows water uptake farther from the main stem and deeper into the ground. Future uses of MANTRA will be extended to a three-dimensional environment.

Precipitation is a major source of groundwater lens recharge. Hence, changes in precipitation pattern will affect groundwater lens recharge and vegetation distribution. Precipitation interception by plant foliage is not modeled explicitly in this model but it should be noted that rainfall interception by plant foliage is an important component in hydrological studies. Zinke [44] reported that interception loss is commonly 10% to 20% in hardwoods. A simulation study assuming this common interception loss will not change the conclusion of this paper.

Based upon the findings of a recent study [45], MANTRA may be revised by using SUTRA-MS to simulate oxygen isotope transport in addition to salt transport, because ^{18}O may be an early indicator of salinity stress on trees. MANTRA shall be applied to study sites along the Waccamaw River, South Carolina, USA. Potential applications to study sites in Malaysia and Mekong River, Vietnam, where floodwaters brought devastating damage to crops like paddy, are also planned. Transitioning to the three-dimensional version of SUTRA will be necessary for some of these projects.

5. Conclusions

The object of this paper was to describe a new model, MANTRA, that combines hydrologic and salinity dynamics based on USGS's SUTRA model, with MANHAM, which models two vegetation types, glycophytic and halophytic, having different transpiration properties with respect to soil salinity. The purpose of the model is to accurately describe effects of SLR and storm surges on the ecotone between these types, given that positive feedbacks between vegetation and soil salinity are important components of this system. The application to Coot Bay Hammock shows consistency with historical data showing that the last major hurricanes, Andrew and Wilma, did not cause a major change (regime shift) of the ecotone, but indicates that larger disturbances, which cause substantial damage to existing vegetation, might have such an effect.

Acknowledgments: S.Y.T. was supported in part by the USGS's Across Trophic Level System Simulation program. Financial support by Grants 305/PMATHS/613418 and 203/PMATHS/6730101 to S.Y.T. and H.L.K. is gratefully acknowledged. M.T. and D.L.D. were supported in part by the USGS's Natural Resources Preservation Project. J.J. was supported in part by National Basic Research Program of China (No. 31200534). Use of trade or product names does not imply endorsement by the U.S. Government. We greatly appreciate the many suggestions and edits of two reviewers for the journal and a USGS reviewer.

Appendix

Appendix 1. Description of MANHAM Model

MANHAM is spatially explicit model of two competing vegetation types, mangroves and hardwood hammocks (though it is adaptable to other competing vegetation types). The vegetation types compete for light and have different tolerances of salinity. The basic assumptions are similar to those in Sternberg et al. [11], in which the formation of a sharp boundary between the vegetation types was modeled. Both vegetation types use water from the vadose zone, which overlies a saturated zone of brackish groundwater. Hammock species are assumed to be better competitors in low salinity areas, but cannot grow well under high salinity, where they are out-competed by mangroves. If enough water is withdrawn from the vadose layer by plant water uptake or evaporation, groundwater will infiltrate by capillary action into the vadose layer and increase its salinity. On the other hand, if precipitation exceeds evaporation plus the transpiration of water, then salinity in the vadose layer is percolated towards the underlying ocean water layer and salinity decreases [46]. High vadose zone salinity that develops during Florida's dry season is considered to be the major determinant of vegetation distribution in Florida in this model.

The main mechanism in the model is based on the feedback relationship between the two vegetation types and vadose zone salinity mentioned above. For example, consider a microsite and assume the vadose zone has a particular average salinity during the dry season, which is not sufficiently high to decrease the complete domination of hammock species in an area. During the dry season, as freshwater hammock species continue to transpire water from the vadose layer, ocean water tends to infiltrate and to increase the salinity of the groundwater of the microsite. Because freshwater plants are sensitive to salinity [47], they decrease their transpiration rates, reducing further infiltration of ocean water. In this way the salinity of the vadose layer may be stabilized at low concentrations that are not lethal to freshwater plants. Conversely, consider the alternate equilibrium state where mangroves dominate an area. As mangroves transpire the water in the vadose layer, underlying groundwater with ocean water salinity infiltrates upwards into the vadose zone, but unlike the hammock species, mangroves will continue to transpire and continue to increase the salinity of the vadose layer to levels which would not be tolerated by freshwater hammock species. Thus there will be a tendency for one or the other vegetation type to stabilize itself in a given area, by reinforcing salinity conditions favorable to itself.

This mechanism may explain observations at the landscape level. Our model conceptualizes the landscape as a grid of microsites, or spatial cells, and assumes each grid cell is occupied by a closed canopy of a small number of plants, which can include both mangrove and hammock individuals. Each cell, whether currently dominated by mangrove or hammock species, always contains at least some small fraction of the other type, which can act as 'seeds' for growth under more favorable conditions. Each cell is exposed to precipitation, soil evaporation, tidal deposition of saline water (depending on the cell's elevation in the landscape) and transpiration, which produce vertical fluxes of water in a cell and either increase or decrease the salinity of the vadose zone of that cell. (Evaporation of intercepted water is not considered in this version of the mode). The evapotranspiration depends on the fractions of each of the vegetation types in the cell. The vadose layer of the cell is also assumed coupled to neighboring cells through lateral movement of salinity. The strongest mechanism for this transport may be water uptake by the roots of plants in the neighboring cells, which redistributes water and salinity between cells. Thus there is some tendency for adjacent cells to approach over time the same vadose zone salinity, allowing the possibility for each vegetation type to spread horizontally from one cell to dominate adjacent cells.

We hypothesize that a model landscape of mangrove and hardwood hammock trees, initially randomly mixed and then subjected to these abiotic factors, will self-organize into a pattern similar to those observed in nature, having strong aggregation into areas of either solid hammock or mangrove vegetation (vegetation clumping), such that there can be rapid changes between the vegetation types along gradual clines in microtopography. However, we also hypothesize that a large enough disturbance can change this pattern. For example, a storm surge that deposits a large amount of saline water across the landscape may cause hammock trees to slow their growth sufficiently to be outcompeted by mangroves over a sufficiently long time period to allow mangroves take over. Thus a large area may "switch" vegetation type quickly.

This model was implemented quantitatively as a two-dimensional grid of square spatial cells, where the sides of each cell were assumed to be in the range of a few to several meters.

Hydrology and salinity: The salinity in a given spatial cell is determined first of all by the difference between the precipitation, P, which brings in fresh water to the top of the vadose zone, and the evaporation, E, and plant uptake of water, R. This difference is called the infiltration rate, I_{NF};

$$I_{NF} = E + R - P \text{ (mm day}^{-1}) \tag{A1.1}$$

and the dynamics of salinity in the vadose zone are given by the equations

$$nz\frac{dS_V}{dt} = I_{NF}S_{wt} \text{ for } I_{NF} > 0 \tag{A1.2}$$

$$\varepsilon z\frac{dS_V}{dt} = I_{NF}S_V \text{ for } I_{NF} < 0 \tag{A1.3}$$

where z (mm) is the depth of the vadose zone of a given cell, ε is the porosity, and S_V and S_{wt} are the salinities of the pore water in the vadose zone and of the underlying saline groundwater, respectively. Positive values of infiltration (A1.2) occur when precipitation is less than the water demanded by evaporation and transpiration; then water from the underlying saline groundwater infiltrates upward into the vadose zone. Note that when $I_{NF} > 0$, salt is deposited in the vadose zone by evapotranspiring water, so concentrations can build up to high levels. Negative values occur when precipitation exceeds evaporation and transpiration demands; then water percolates downward into the underlying groundwater table.

The assumption of a groundwater table with fixed salinity is a useful first approximation, but it is also possible that the salinity dynamics of at least the surface layer of groundwater is more complex. The effect of an upper layer of groundwater that is affected both by precipitation and flow of groundwater from higher elevations is examined in an appendix (see on-line Appendix 5), while here we restrict ourselves to examining the model with the simpler assumption.

Evaporation was assumed to be small compared with transpiration and was neglected, since we are assuming a dense canopy in each cell, which inhibits evaporation. Moreira, et al. [48] and Harwood, et al. [49] both observed that in forests transpiration dominates as the vapor generator compared to evaporation. R_{TOTAL} depends on the transpiration and gross productivity of each vegetation type in the spatial cell (see A1.13 below). The maximum possible water uptake rate by freshwater hammocks is assumed to be 2.6 mm/d. This value is based on previous studies indicating that transpiration in tropical forests lies within this range [38]. Uptake of water as a function of salinity by the hardwood hammock $R_1(S_v)$ and mangrove species $R_2(S_v)$ is given by the respective empirical relations (Figure 2):

$$R_1(S_v) = 2.6 \left(1 - \frac{S_v}{3.14 + S_v} \right) \text{ mm day}^{-1} \tag{A1.4a}$$

$$R_2(S_v) = 4.4 \left(\frac{100 - S_v}{15 + 100 - S_v} \right) \text{ mm day}^{-1} \tag{A1.4b}$$

in which hammocks reduce their transpiration by $\frac{1}{2}$ when the salinity of the pore water is 3.14 ppt, while mangrove transpiration is not reduced by $\frac{1}{2}$ until the salinity of the pore water is 85 ppt.

In addition to the above hydrologic processes, tidal effects were imposed on all spatial cells at elevations low enough to be affected. The effect of tides on the salinity of spatial cells was calculated as follows. On each day a single high tide was assumed. The height of the tide above the surface of each spatial cell was generated as a function of the mean and a randomly generated variation within the observed limits of tidal flux of the empirical data, so that the number of spatial cells covered by the tide on a given day varied in number. The amount of salt contained in the volume of water above the cell, assumed to have a salinity of 30 ppt, was allowed to mix homogeneously with the vadose zone below. In all model simulations precipitation and effects of tides were prescribed on a daily basis. Means and standard deviations of daily precipitation (NOAA, National Weather Services Forecast Office, Florida) and daily tidal height (NOAA, Tide & Current Historic data base, Key West Station) for each month were derived from 162 and 5 years of empirical data respectively. Daily values were determined using a normal random number generator, with values truncated at zero.

We assume there is also horizontal diffusion of salinity between cells. We used a diffusion constant of $D = 0.0005$, which is about seven times the theoretical value used by Passioura, et al. [50] and ten times the laboratory values of [51]. However, we assume that classical diffusion is not the only process causing mixing of solute between cells. The extension of roots across cell boundaries can contribute to the mixing among cells, and we believe our value is reasonable.

Vegetation dynamics: A given cell could be occupied by the two types of vegetation simultaneously, and, in fact, even in cells dominated by one type, small amounts of the other type tended to persist. The biomasses of each species in a given spatial cell were explicitly modeled, as well as the mechanism of competitive dominance of the hammock vegetation over mangrove vegetation under very low soil salinity conditions. We used an approach similar to that of Herbert, et al. [52] for competition between species of different functional types, with a slight difference. This is explained in detail in Teh et al. [14] and Herbert, et al. [52]. Herbert et al. [52] assumed that the plant types differed in their abilities to compete for light and nutrients. Here we assumed that the plants differed only in their ability to compete for light, with the additional assumption that the hardwood species were superior in low salinity. We assumed that in any particular microsite the equations for the different vegetation types (hardwood hammock and mangrove in this case) were

$$\frac{dB_{Ci}}{dt} = U_{Cvi} - M_{Cvi} - L_{Cvi} \ (i = 1, 2) \tag{A1.5}$$

where B_{ci} is carbon in plant biomass (gCm^{-2}), U_{Cvi} is gross productivity (gCm^{-2}day^{-1}),

$$U_{Cvi} = \frac{Q(S_v) g_{Ci} w_{Ci} I (1 - e^{-k_l S_{CT}})}{w_{C1} + w_{C2}} \ (i = 1, 2) \tag{A1.6}$$

where w_{C1} and w_{C2} incorporate competition for light, depending on how much of the canopy of a spatial cell each occupies (see Herbert et al. [52] for details);

$$w_{Ci} = \frac{2(1 - e^{-k_l S_{Ci}})}{(1 + e^{-k_l S_{Ci}})} \prod_{j=1}^{2} \left(\frac{f_{Cj} e^{-k_l S_{Ci}} + f_{Ci}}{f_{Ci} + f_{Cj}}\right) \quad (i = 1,2) \tag{A1.7}$$

where the $f_{Ci} = 0.58/c_{ii}$ parameters are measures of canopy dominance (e.g., the relative degree to which trees of one species shade another due to height differentials). S_{Ci} (m^2 m^{-2}) is the leaf area index of species i in a spatial cell,

$$S_{Ci} = b_{Ci} B_{Ai} \quad (i = 1,2) \tag{A1.8}$$

B_{Ai} (gCm^{-2}) is active tissue carbon of each plant species,

$$B_{Ai} = \frac{c_{ii} B_{Amax,i} B_{Ci}}{B_{Amax,i} + c_{1i} B_{C1} + c_{2i} B_{C2}} \tag{A1.9}$$

where the parameters c_{ij} are allometric parameters governing the amount of energy allocated to active tissue (leaves). Note that the biomass of species j can affect the allocation of biomass of species i. We assumed the effect of salinity on productivity of species i, $Q_i(S_v)$, occurs through its effect on the water uptake rate, normalized by the maximum possible rate;

$$Q_i(S_v) = \frac{R_i(S_v)}{R_i(0)} \quad (i = 1,2) \tag{A1.10}$$

M_{Cvi} (g C m^{-2} day^{-1}) is the respiration of each plant;

$$M_{Cvi} = m_{Ai} B_{Ai} + m_{wi}(B_{Ci} - B_{Ai}) \quad (i = 1,2) \tag{A1.11}$$

where the rates differ between dead and living matter. L_{Cvi} (g C m^{-2} day^{-1}) is litterfall of each plant,

$$L_{Cvi} = l_{Ai} B_{Ai} + l_{wi}(B_{Ci} - B_{Ai}) \quad (i = 1,2) \tag{A1.12}$$

Total evapotranspiration from a cell is linearly related to the evapotranspiration of each species, multiplied by its fraction of the primary production in that spatial cell;

$$R_{TOTAL} = \frac{U_{Cv1}}{U_{Cv1} + U_{Cv2}} R_1 + \frac{U_{Cv2}}{U_{Cv1} + U_{Cv2}} R_2 \tag{A1.13}$$

We do not have parameter values related to light competition for these vegetation types, but made estimates that allowed hardwood hammock vegetation to dominate for salinities below 7 ppt.

To improve the fit of the model to the boundary between hardwood hammocks and mangroves of the Coot Bay Hammock, some moderate modifications were made to the original MANHAM formulation and the parameter value. Notably, we varied the light extinction coefficients, k_i, for the two species, as well as the parameters and equation for leaf area index of each species i in a spatial cell. We gave the hardwood hammock an advantage in light use shading out of the mangroves, which would allow hardwood hammock to outcompete the mangroves at lower salinity areas. The equation for leaf area index, S_{Ci}, of species i in a spatial cell is revised from

$$S_{Ci} = b_{Ci} B_{Ai} \tag{A1.14a}$$

(where b_{Ci} is the leaf area index per unit active tissue and B_{Ai} is the active tissue carbon, see Equation (8) in Teh et al. [14]) to Equation (A1.14b) below to include the plant water uptake effort $Q_i(C)$, following Herbert et al. [52].

$$S_{Ci} = b_{Ci} B_{Ai} \cdot Q_i(C) \tag{1}$$

where C = solute concentration $[M_s/M_f]$. The changes to (A1.14a) would cause the hardwood hammocks to diminish in biomass quickly at higher salinities, as the water uptake effort $Q_i(S_v)$ is less than halved when the salinity is higher than 3.14 ppt. The revised formulation (A1.14b) alone results in a situation where the mangroves outcompete the hardwood hammocks at the elevated ridge, which also is not realistic. However, the increase of light extinction factor k_i for hardwood hammocks ensures that hardwood hammocks outcompete the mangroves at lower salinities. These revisions allow the mangrove and hardwood hammocks to evolve, in a more realistic manner, into the distribution observed at Rowdy Bend.

Appendix 2. MANTRA Version 1 Manual: August 1, 2014

This report briefly describes the coupled hydrology-salinity-vegetation model, MANTRA, for analysis of the feedback effect of competing glycophytes (hardwood hammocks) and halophytes (mangroves) on the groundwater flow and salinity regime. MANTRA can be further revised to include more competing plants [15] by incorporating the dynamics of these plants in the MANHAM module. MANTRA was developed by integrating the USGS spatially explicit models of vegetation community dynamics along coastal salinity gradients (MANHAM) into the USGS groundwater models (SUTRA). Table A1 lists the files needed to run MANTRA. The MANHAM [11,14] module is incorporated into SUTRA [16] in sutra_2_2.f by accounting for the exchange of fluid and solute mass between the models. A storm surge event is incorporated into usubs_2_2.f as time-dependent specified pressure.

In MANTRA, the total amount of water required for plant transpiration is subtracted from the SUTRA cells covered by plants. The fluid mass source/sink term Q_{IN}, originally available in SUTRA, is used to characterize the uptake of water by plant [18–20]. This source/sink term accounts for external addition/subtraction of fluid including pure water mass plus the mass of any solute dissolved in the source fluid. This fluid uptake by plant reduces the fluid mass in the cells, which in turn increases the solute (salt) concentration. Plant growth in MANHAM depends on the actual fluid available in the cells for transpiration and the solute concentration. Units are presented in square brackets [ ] with M being the unit of mass, L the unit of length and s the unit of time in seconds. The relations are formulated for simulations of a 2D cross-sectional fishnet domain with saturated-unsaturated, variable-density (pressure) with single species (solute) transport. It is assumed at present that the plants will only withdraw water from the uppermost layer of cells/nodes and that the plant roots cover the entire surface area of the cells (Figure A1). Currently, the vegetation and groundwater modules operate at the same time step. This is expected to slow down the computation, particularly when there are large numbers of element. As the hydrological processes and vegetation dynamics operate at different time scale, a method similar to those employed in the SEHM model by Jiang, et al. [53] can be employed to optimize the computation time.

Table A1. List of files needed to run MANTRA.

Filename	Remarks
SUTRA.FIL SUTRA.inp SUTRA.ics	These are the files that are also needed to run the original version of SUTRA. These files contain the input parameters for the groundwater flow and solute transport simulation. These files can be created by using ArgusONE.
MANTRA.exe	This is the executable file of MANTRA. This executable file is built from fmods_2_2.f, sutra_2_2.f, ssubs_2_2.f, usubs_2_2.f. Modifications have been made to incorporate the MANHAM module into sutra_2_2.f. A storm surge event is incorporated into usub_2_2.f as time-dependent specified pressure.
MANHAM.DAT	This is the input file for the MANHAM module in MANTRA. This file contains the input parameters related to the vegetation.

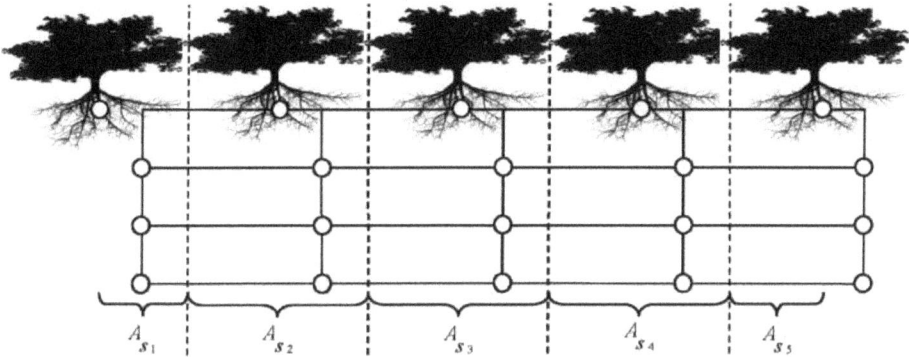

Figure A1. Schematic sketch of a simulation case. A_{si} = Surface area for node i at uppermost layer of cells.

The uptake of water as a function of salinity for hardwood hammock (R_1) and mangrove (R_2) are estimated by the empirical relations Equations (A2.1) and (A2.2).

$$R_1(C) = R_{max,1}\left(1 - \frac{C}{S_{half,1} + C}\right) \text{ [L/s]} \tag{A2.1}$$

$$R_2(C) = R_{max,2}\left(\frac{0.1 - C}{S_{half,2} + 0.1 - C}\right) \text{ [L/s]} \tag{A2.2}$$

with C = solute concentration $[M_s/M_f]$ calculated by SUTRA. Here, Fluid mass per unit time (Q_p) required by the plants in a certain cell is then estimated by (A2.3).

$$Q_p = (R_1 + R_2) \cdot A_s \cdot \rho \text{ [M/s]} \tag{A2.3}$$

Here, Q_p = fluid mass per unit time extracted in a certain cell for plant transpiration [M/s]; A_s = cell surface area $[L^2]$; ρ = fluid density $[M/L^3]$. For cross-sectional model, the width of each cell is assumed to be 1.0. Thus, the cell surface area depends mainly on the length or horizontal grid size (Figure A1). The actual fluid mass being subtracted from a cell due to evapotranspiration depends on the saturation S_w and porosity ε in the cell, leading to Equation (A2.4).

$$Q_{IN} = -Q_p \cdot \varepsilon \cdot S_w \text{[M/s]} \tag{A2.4}$$

Here, Q_{IN} = total mass sink (due to plant transpiration) [M/s]; ε = porosity $[V_v/V]$; S_w = water saturation $[V_w/V_v]$ with V = total volume, V_v = volume of voids, V_w = volume of water. Suppose porosity ε is kept constant at 1.0. When the void space is completely filled with fluid and is said to be saturated, that is $S_w = 1$, the actual water uptake by plant will be equivalent to the amount of water required by the plants Q_p. When the void space is only partly water filled and is referred to as being unsaturated, that is $S_w < 1$, the actual water uptake by plant will decrease as a factor of the saturation S_w. Similar to transpiration, precipitation is implemented as a source term in MANTRA. The precipitation rate is varied stochastically on daily basis. Daily values are determined using a normal random number generator, with values truncated at zero.

Since the storm surge event is incorporated into usubs_2_2.f as time-dependent specified pressure, time-dependent specified pressure for the surface nodes should be indicated in the SUTRA input file (sutra.inp) so that the usubs_2_2.f routine will be called for implementation. The storm surge event is simulated by allowing the surface nodes to be inundated with seawater of certain depth and salinity. This form of inundation will change the hydrostatic pressure at the surface nodes. Equations (A2.5) and (A2.6) [21] are respectively used to specify the pressure and concentration at surface nodes inundated by seawater during a storm surge event.

$$p = \rho_{sea} \cdot g \cdot (h_{surge} - y) \left[M/(L \cdot s^2) \right] \tag{A2.5}$$

$$C = S_{sea} \; [M_s/M_f] \tag{A2.6}$$

Here, p = pressure at top layer nodes [$M/(L \cdot s^2)$], ρ_{sea} = fluid density of seawater [M/L^3], h_{surge} = inundation depth [L], y = node height [L], and S_{sea} = seawater salinity [M_s/M_f]. Figure A2 shows an example of input file (MANHAM.DAT) for MANHAM module in MANTRA for the example case of Rowdy Bend. Table A2 summarizes the list of parameters in MANHAM.DAT with their description, type, value and unit.

```
INPUT FILE FOR MANHAM

---+----+----+----+----+----+----+----+----+----+----+----+----+
  SUTRA Node Control
---+----+----+----+----+----+----+----+----+----+----+----+----+
  NSNODE              1
  NFIRST              1
  NLAST            1601
  NDIFF              20

---+----+----+----+----+----+----+----+----+----+----+----+----+
  Storm Surge Control
---+----+----+----+----+----+----+----+----+----+----+----+----+
  IT_Surge        60000
  SDepth           10.5
  SSalinity       0.030

---+----+----+----+----+----+----+----+----+----+----+----+----+
  Plant IC Control
---+----+----+----+----+----+----+----+----+----+----+----+----+
  NSPEC               2
  NRAND            1234
  BEGINHAM         0.50
  BCO          20000.00
  PERSPEC          0.50

---+----+----+----+----+----+----+----+----+----+----+----+----+
  Light Parameters
---+----+----+----+----+----+----+----+----+----+----+----+----+
  SI              0.010
  EKI             0.600       0.400
  GC(NSPEC)       520.0       380.0
  BAMAX(NSPEC)    350.0       350.0

---+----+----+----+----+----+----+----+----+----+----+----+----+
  Plant Parameters
---+----+----+----+----+----+----+----+----+----+----+----+----+
  bc(NSPEC)       0.0355      0.0170
  RA(NSPEC)       4.0000      4.0000
  RW(NSPEC)       0.0296      0.0296
  RMA(NSPEC)      1.7000      1.7000
  RMW(NSPEC)      0.0148      0.0148
  RFWMAX(NSPEC)   0.0026      0.0088
  SATK(NSPEC)     3.1400     15.0000
  C(NSPEC,NSPEC)  0.1000      0.1000
                  0.5000      0.5000  ( C11,C21,C12,C22)

---+----+----+----+----+----+----+----+----+----+----+----+----+
  Environmental Parameters
---+----+----+----+----+----+----+----+----+----+----+----+----
+----+----+----+----+----+----+----+----+----+----+----+----+
  VPRE(12)        1.590       1.360    1.320    1.500    2.730    3.820
                  3.090       4.050    5.480    4.640    2.020    1.570
  VPRESD(12)      1.870       1.200    1.290    1.800    2.220    2.710
                  1.970       2.130    2.710    3.470    2.840    1.520
```

Figure A2. Example input file (MANHAM.DAT) for MANHAM module in MANTRA for the example case of Rowdy Bend.

Table A2. List of parameters in MANHAM.DAT with their description, type, value and unit.

Variable Name (Input File)	Eqn	Type	Description	Value	Unit
SUTRA Node Control					
NSNODE	—	Integer, I10	Number of sets of surface nodes	1	—
NFIRST	—	Integer, I10	First surface node number in SUTRA domain	1	—
NLAST	—	Integer, I10	Last surface node number in SUTRA domain	1601	—
NDIFF	—	Integer, I10	Interval between surface node numbers	20	—
Storm Surge Control					
IT_Surge	—	Integer, I10	Iteration time of a storm surge event	60,000	—
SDepth	h_{surge}	Real, F10.2	Storm surge inundation depth in relation to the height of SUTRA computational domain	10.5	m
SSalinity	S_{sea}	Real, F10.3	Storm surge water salinity	0.030	kg/kg
Plant Initial Condition (IC) Control					
NSPEC	—	Integer, I10	Number of plant species	2	—
NRAND	—	Integer, I10	Seed for random number generator	1234	—
BEGINHAM	—	Real, F10.2	Ratio of cells dominated by hardwood hammock	0.50	—
BC0	—	Real, F10.2	Total initial biomass in a cell	20,000.00	$g\,C\,m^{-2}$
PERSPEC	—	Real, F10.2	Ratio of plant i in a cell	0.50	—
Light Parameters					
SI	I	Real, F10.3	Solar irradiance	0.010	$GJ\,m^{-2}day^{-1}$
EKI	k_i	Real, F10.4	Light extinction factor	0.600 / 0.400	—
GC(NSPEC)	g_{Ci}	Real, F10.1	Light-use efficiency	520.0 / 380.0	$g\,C\,GJ^{-1}$
BAMAX(NSPEC)	$B_{Amax,i}$	Real, F10.1	Maximum value attainable by B_{Ai}	350.0 / 350.0	$g\,C\,m^{-2}$
Plant Parameters					
bc(NSPEC)	b_{ci}	Real, F10.4	Leaf area per unit carbon	0.0355 / 0.0170	$m^2/g\,C$
RA(NSPEC)	r_{Ai}	Real, F10.4	Active tissue respiration rate	4.0000 / 4.0000	$year^{-1}$
RW(NSPEC)	r_{Wi}	Real, F10.4	Woody tissue respiration rate	0.0296 / 0.0296	$year^{-1}$
RMA(NSPEC)	m_{Ai}	Real, F10.4	Active tissue litter loss rate	1.7000 / 1.7000	$year^{-1}$
RMW(NSPEC)	m_{Wi}	Real, F10.4	Woody tissue litter loss rate	0.0148 / 0.0148	$year^{-1}$
RFWMAX(NSPEC)	$R_{max,i}$	Real, F10.4	Maximum water uptake for plant NSPEC	0.0026 / 0.0088	$mm\,day^{-1}$
SATK(NSPEC)	$S_{half,i}$	Real, F10.4	Half saturation constant for maximum water uptake for plant NSPEC	3.1400 / 15.0000	ppt
C(NSPEC, NSPEC)	c_{ii}	Real, F10.4	Parameters for plant allometry	0.1000 / 0.5000	—
Environmental Parameters					
VPRE(12)	M_{pre}	Real, F10.4	Means precipitation rate for twelve months	1.590 / 1.360 ⋮	$mm\,day^{-1}$
VPRESD(12)	SD_{pre}	Real, F10.4	Standard deviations of precipitation rate	1.870 / 1.200 ⋮	$mm\,day^{-1}$

References

1. Nicholls, R.J.; Cazenave, A. Sea-level rise and its impact on coastal zones. *Science* **2010**, *328*, 1517–1520. [CrossRef] [PubMed]
2. Gornitz, V. *Rising Seas: Past, Present, Future*; Columbia University Press: New York, NY, USA, 2013; p. 344.
3. Anderson, W.P., Jr. Aquifer salinization from storm overwash. *J. Coast. Res.* **2002**, *18*, 413–420.
4. Anderson, W.P., Jr.; Lauer, R.M. The role of overwash in the evolution of mixing zone morphology within barrier islands. *Hydrogeol. J.* **2008**, *16*, 1483–1495. [CrossRef]
5. Terry, J.P.; Falkland, A.C. Responses of atoll freshwater lenses to storm-surge overwash in the Northern Cook Islands. *Hydrogeol. J.* **2010**, *18*, 749–759. [CrossRef]
6. Sklar, F.H.; Chimney, M.J.; Newman, S.; McCormick, P.; Gawlik, D.; Miao, S.; McVoy, C.; Said, W.; Newman, J.; Coronado, C. The ecological-societal underpinnings of Everglades restoration. *Front. Ecol. Environ.* **2005**, *3*, 161–169.
7. Ross, M.S.; O'Brien, J.J.; Flynn, L.J. Ecological site classification of Florida Keys terrestrial habitats. *Biotropica* **1992**, *24*, 488–502. [CrossRef]
8. Smith, T.J.I.; Foster, A.M.; Tiling-Range, G.; Jones, J.W. Dynamics of mangrove–marsh ecotones in subtropical coastal wetlands: fire, sea-level rise, and water levels. *Fire Ecol.* **2013**, *9*, 66–77. [CrossRef]
9. Gosz, J.R. Ecotone hierarchies. *Ecol. Appl.* **1993**, *3*, 369–376. [CrossRef]
10. Snyder, J.R.; Herndon, A.; Robertson, W.B.J. South Florida rockland. In *Ecosystems of Florida*; Myers, R.L., Ewel, J.J., Eds.; The University of Central Florida Press: Orlando, FL, USA, 1990; pp. 230–279.
11. Sternberg, L.D.L.; Teh, S.Y.; Ewe, S.M.L.; Miralles-Wilhelm, F.; DeAngelis, D.L. Competition between hardwood hammocks and mangroves. *Ecosystems* **2007**, *10*, 648–660. [CrossRef]
12. Baldwin, A.H.; Mendelssohn, I.A. Effects of salinity and water level on coastal marshes: An experimental test of disturbance as a catalyst for vegetation change. *Aquatic Botany* **1998**, *61*, 255–268. [CrossRef]
13. Steyer, G.D.; Cretini, K.F.; Piazza, S.; Sharp, L.A.; Snedden, G.A.; Sapkota, S. *Hurricane Influences on Vegetation Community Change in Coastal Louisiana*; U.S. Geological Survey: Reston, VA, USA, 2010.
14. Teh, S.Y.; DeAngelis, D.L.; Sternberg, L.D.L.; Miralles-Wilhelm, F.R.; Smith, T.J.I.; Koh, H.L. A simulation model for projecting changes in salinity concentrations and species dominance in the coastal margin habitats of the Everglades. *Ecol. Model.* **2008**, *213*, 245–256. [CrossRef]
15. Saha, A.; Saha, S.; Sadle, J.; Jiang, J.; Ross, M.; Price, R.; Sternberg, L.; Wendelberger, K. Sea level rise and South Florida coastal forests. *Clim. Chang.* **2011**, *107*, 81–108. [CrossRef]
16. Voss, C.I.; Provost, A.M. *SUTRA, A Model for Saturated-Unsaturated Variable-Density Ground-Water Flow with Solute or Energy Transport*; U.S. Geological Survey Water-Resources Investigations Report 02–4231; U.S. Geological Survey: Reston, VA, USA, 2010; p. 291.
17. Voss, C. USGS SUTRA code—History, practical use, and application in Hawaii. In *Seawater Intrusion in Coastal Aquifers—Concepts, Methods and Practices*; Practices, J., Bear, A., Cheng, H.D., Sorek, S., Ouazar, D., Herrera, I., Eds.; Kluwer Acdemic Publishers: Dordrecht, Netherlands, 1999; pp. 249–313.
18. Vrugt, J.; Wijk, M.V.; Hopmans, J.W.; Šimunek, J. One-, two-, and three-dimensional root water uptake functions for transient modeling. *Water Resour. Res.* **2001**, *37*, 2457–2470. [CrossRef]
19. Zhu, Y.; Ren, L.; Skaggs, T.H.; Lü, H.; Yu, Z.; Wu, Y.; Fang, X. Simulation of Populus euphratica root uptake of groundwater in an arid woodland of the Ejina Basin, China. *Hydrol. Process.* **2009**, *23*, 2460–2469. [CrossRef]
20. Tian, W.; Li, X.; Wang, X.-S.; Hu, B. Coupling a groundwater model with a land surface model to improve water and energy cycle simulation. *Hydrol. Earth Syst. Sci. Discuss.* **2012**, *9*, 1163–1205. [CrossRef]
21. Kooi, H.; Groen, J.; Leijnse, A. Modes of seawater intrusion during transgressions. *Water Resour. Res.* **2000**, *36*, 3581–3589. [CrossRef]
22. Saha, A.; Moses, C.; Price, R.; Engel, V.; Smith, T., III; Anderson, G. A hydrological budget (2002–2008) for a large subtropical wetland ecosystem indicates marine groundwater discharge accompanies diminished freshwater flow. *Estuaries Coasts* **2012**, *35*, 459–474. [CrossRef]
23. Light, S.S.; Dineen, J.W. Water control in the Everglades: A historical perspective. In *Everglades: The Ecosystem and its Restoration*; Davis, S.M., Ogden, J.C., Eds.; St. Lucie Press: Florida, FL, USA, 1994; pp. 47–84.
24. Sadle, J.; Everglades National Park. Homestead, Florida, FL, USA. Personal communication, 2014.
25. Olmstead, I.C.; Loope, L.L. *Vegetation along a Microtopographic Gradient in the Estuarine Region of Everglades National Park, Florida, USA*; South Florida Research Center: Florida, FL, USA, 1981; p. 41.

26. Smith, T.; Anderson, G.H.; Tiling, G. *Science and the Storms: The USGS Response to the Hurricanes of 2005*; Farris, G.S., Smith, G.J., Crane, M.P., Demas, C.R., Robbins, L.L., Lavoie, D.L., Eds.; U. S. Geological Survey Circular 1306: Reston, VA, USA, 2007; pp. 169–174.

27. Wanless, H.R.; Parkinson, R.W.; Tedesco, L.P. Sea level control on stability of Everglades wetlands. In *Everglades: The Ecosystem and Its Restoration*; St. Lucie Press: Delray Beach, FL, USA, 1994; pp. 199–223.

28. Armentano, T.V.; Doren, R.F.; Platt, W.J.; Mullins, T. Effects of Hurricane Andrew on coastal and interior forests of southern Florida: Overview and synthesis. *J. Coast. Res.* **1995**, *21*, 111–144.

29. Slater, H.H.; Platt, W.J.; Baker, D.B.; Johnson, H.A. Effects of Hurricane Andrew on damage and mortality of trees in subtropical hardwood hammocks of Long Pine Key, Everglades National Park, Florida, USA. *J. Coast. Res.* **1995**, *21*, 197–207.

30. Horvitz, C.C.; Pascarella, J.B.; McMann, S.; Freedman, A.; Hofstetter, R.H. Functional roles of invasive non-indigenous plants in hurricane-affected subtropical hardwood forests. *Ecol. Appl.* **1998**, *8*, 947–974. [CrossRef]

31. Swain, E.D.; Krohn, D.; Langtimm, C.A. Numerical computation of hurricane effects on historic coastal hydrology in southern Florida. *Ecol. Process.* **2015**, *4*, 4. [CrossRef]

32. Hook, D.D.; Buford, M.A.; Williams, T.M. Impact of Hurricane Hugo on the South Carolina coastal plain forest. *J. Coast. Res.* **1991**, *8*, 291–300.

33. Rathcke, B.J.; Landry, C.L. Dispersal and recruitment of white mangrove on San Salvador Island, Bahamas after Hurricane Floyd. In Proceedings of the Ninth Symposium on the Natural History of the Bahamas, San Salvador, Bahamas, 14–18 June 2001.

34. Jiang, J.; DeAngelis, D.L.; Anderson, G.H.; Smith, T.J., III. Analysis and simulation of propagule dispersal and salinity intrusion from storm surge on the movement of a marsh-mangrove ecotone in South Florida. *Estuar. Coasts* **2014**, *37*, 24–35. [CrossRef]

35. López-Hoffman, L.; Ackerly, D.D.; Anten, N.P.R.; Denoyer, J.L.; Martinez-Ramos, M. Gap-dependence in mangrove life-history strategies: A consideration of the entire life cycle and patch dynamics. *J. Ecol.* **2007**, *95*, 1222–1233. [CrossRef]

36. Wanless, H.R.; Brigitte, M.V. *Coastal Landscape and Channel Evolution Affecting Critical Habitats at Cape Sable*; Final Report to ENP; Everglades National Park: Homestead, FL, USA, 2005.

37. Ross, M.S.; O'Brien, J.J.; Ford, R.G.; Zhang, K.; Morkill, A. Disturbance and the rising tide: The challenge of biodiversity management for low island ecosystems. *Front. Ecol. Environ.* **2009**, *9*, 471–478. [CrossRef]

38. Cabral, O.M.; McWilliam, A.; Roberts, J. In-canopy microclimate of Amazonian forest and estimates of transpiration. In *Amazon Deforestation and Climate*; Gash, J., Nobre, C., Roberts, J., Victoria, R., Eds.; Wiley Press: Chichester, UK, 1996; pp. 207–220.

39. White, I.; Falkland, T. Management of freshwater lenses on small Pacific islands. *Hydrogeol. J.* **2010**, *18*, 227–246. [CrossRef]

40. Chui, T.F.; Terry, J.P. Modeling fresh water lens damage and recovery on atolls after storm-wave washover. *Ground Water* **2012**, *50*, 412–420. [CrossRef] [PubMed]

41. Pearlstine, L.G.; Pearlstine, E.V.; Aumen, N.G. A review of the ecological consequences and management implications of climate change for the Everglades. *J. Am. Benthol. Soc.* **2010**, *29*, 1510–1526. [CrossRef]

42. Odum, W.E.; McIvor, C.C.; Smith, T.J., III. *The Ecology of the Mangroves of South Florida: A Community Profile*; Bureau of Land Management Fish and Wildlife Service: Washington, DC, USA, 1982.

43. Meshaka, W.; Loftus, W.F.; Steiner, T. The herpetofauna of Everglades National Park. *Fla. Sci.* **2000**, *63*, 84–103.

44. Zinke, PJ. Forest interception study in the United States. In *Forest Hydrology*; Sopper, W.E., Lull, H.W., Eds.; Pergamon: Oxford, UK, 1967; pp. 137–161.

45. Zhai, L.; Jiang, J.; DeAngelis, D.L.; Sternberg, L.S.L. Prediction of plant vulnerability to salinity increase in a coastal ecosystem by stable isotopic composition of plant stem water: A model study. In review.

46. Swain, E.D.; Wolfert, M.A.; Bales, J.D.; Goodwin, C.R. *Two-dimensional hydrodynamic simulation of surface-water flow and transport to Florida bay through the Southern Inland and Coastal Systems (SICS)*; U.S. Geological Survey Water-Resources Investigations Report 03-4287; U.S. Geological Survey: Reston, VA, USA, 2003.

47. Munns, R. Comparative physiology of salt and water stress. *Plant Cell Environ.* **2002**, *25*, 239–250. [CrossRef] [PubMed]

48. Moreira, M.; Sternberg, L.D.L.; Martinelli, L.; Victoria, R.; Barbosa, E.; Bonates, L.; Nepstad, D. Contribution of transpiration to forest ambient vapour based on isotopic measurements. *Glob. Change Biol.* **1997**, *3*, 439–450. [CrossRef]

49. Harwood, K.; Gillon, J.; Roberts, A.; Griffiths, H. Determinants of isotopic coupling of CO2 and water vapour within a Quercus petraea forest canopy. *Oecologia* **1999**, *119*, 109–119. [CrossRef]

50. Passioura, J.B.; Ball, M.C.; Knight, J.H. Mangroves may salinize the soil and in so doing limit their transpiration rate. *Funct. Ecol.* **1992**, *6*, 476–481. [CrossRef]

51. Hollins, S.E.; Ridd, P.V.; Read, W.W. Measurement of the diffusion coefficient for salt in salt flat and mangrove soils. *Wetl. Ecol. Manag.* **2000**, *8*, 257–262. [CrossRef]

52. Herbert, D.A.; Rastetter, E.B.; Gough, L.; Shaver, G.R. Species diversity across nutrient gradients: an analysis of resource competition in model ecosystems. *Ecosystems* **2004**, *7*, 296–310. [CrossRef]

53. Jiang, J.; DeAngelis, D.; Smith, T.J.I.; Teh, S.Y.; Koh, H.L. Spatial pattern formation of coastal vegetation in response to external gradients and positive feedbacks affecting soil porewater salinity: a model study. *Landsc. Ecol.* **2012**, *27*, 109–119. [CrossRef]

Journal of
*Marine Science
and Engineering*

MDPI

Article

Bias and Efficiency Tradeoffs in the Selection of Storm Suites Used to Estimate Flood Risk

Jordan R. Fischbach [1,2], **David R. Johnson** [2,3,*,†] **and Kenneth Kuhn** [2,3,†]

1   RAND Corporation, 4570 Fifth Ave., Ste. 600, Pittsburgh, PA 15213, USA; jordanf@rand.org
2   Pardee RAND Graduate School, 1776 Main St., Santa Monica, CA 90401, USA; kkuhn@rand.org
3   RAND Corporation, 1776 Main St., Santa Monica, CA 90401, USA
*   Correspondence: djohnson@rand.org; Tel.: +1-765-494-7122; Fax: +1-765-494-7693
†   These authors contributed equally to this work.

Academic Editor: Rick Luettich
Received: 12 July 2015; Accepted: 28 January 2016; Published: 15 February 2016

Abstract: Modern joint probability methods for estimating storm surge or flood statistics are based on statistical aggregation of many hydrodynamic simulations that can be computationally expensive. Flood risk assessments that consider changing future conditions due to sea level rise or other drivers often require each storm to be run under a range of uncertain scenarios. Evaluating different flood risk mitigation measures, such as levees and floodwalls, in these future scenarios can further increase the computational cost. This study uses the Coastal Louisiana Risk Assessment model (CLARA) to examine tradeoffs between the accuracy of estimated flood depth exceedances and the number and type of storms used to produce the estimates. Inclusion of lower-intensity, higher-frequency storms significantly reduces bias relative to storm suites with a similar number of storms but only containing high-intensity, lower-frequency storms, even when estimating exceedances at very low-frequency return periods.

Keywords: flood risk; joint probability methods; storm selection; computational efficiency

---

1. Introduction

Flooding is the most frequently occurring weather-related natural disaster. In the United States, for example, 90 percent of all natural disasters involve flooding [1]. While not all flooding is caused by severe storms, coastal communities are particularly at risk of flooding from storm surge. Sea level rise and, in some areas, land subsidence will compound the problem, making many coastal communities significantly more vulnerable to storm surge hazards in future decades [2–4]. In addition, multiple studies have concluded that the average intensity of Atlantic hurricanes is likely to increase in the future due to projected increases in sea surface temperatures and other factors [5–8].

The Joint Probability Method with Optimal Sampling (JPM-OS) is one approach for estimating the probability distribution function of flooding in a region [9–13]. Storms are characterized using a set of parameters such as their minimum central pressure deficit, forward velocity, and location at landfall. A hydrodynamic model is used to simulate storms whose parameters span the range of plausible values that could occur in nature. A joint probability distribution function (PDF), fit to the historical record of observed storms, is used to estimate the relative likelihood of observing each storm in the suite of simulated, "synthetic" storms. The form of the joint PDF is based on assumptions about conditional relationships between the storm parameters and structural assumptions about how parameters are distributed; for example, the potential intensity of a hurricane is related to its size (radius of maximum windspeed) [14]. The probability masses associated with each synthetic storm are combined with their simulation results to build a cumulative distribution function (CDF) for quantitative metrics of flood risk like surge elevations, flood depths, or direct economic damage.

Long-range planning for flood risk management can involve evaluation of a large number of possible flood control or risk management strategies, such as different combinations of levee and floodwall projects, hazard mitigation, or broader coastal resiliency investments. Benefit-cost analysis entails predicting risk reduction benefits in multiple future time periods over a project's useful lifespan. Uncertainty about climate change, land subsidence, or other deeply uncertain drivers may necessitate that a range of future scenarios should be modeled. All of these factors multiply the number of scenarios, and thus model runs, needed to conduct a thorough and complete comparison of the benefits and costs of different flood risk reduction investments. Computational, budget, and time constraints can thus put downward pressure on the number of storms that can be run per case. We refer to the resulting problem of deciding what storms should be included in the analysis storm suite as the "storm selection" process.

This problem is exacerbated in diverse and complex wetlands ecosystems, such as those present in coastal Louisiana. In these areas, lower-resolution hydrodynamic models may produce biased results, and simplifying assumptions regarding future climate or sea level rise impacts may be invalid [15,16]. A high-resolution hydrodynamic simulation such as the ADvanced CIRCulation (ADCIRC) model is needed to accurately reproduce storm surge behavior from coastal storms [17].

This study has produced estimates of flood depth and damage at different exceedance probabilities associated with storm surge hazard in coastal Louisiana using the Coastal Louisiana Risk Assessment model (CLARA), a quantitative simulation model of storm surge flood risk developed by researchers at the RAND Corporation [18–20]. CLARA was developed to better understand how future coastal changes could lead to increased risk from storm surge flooding to residents and assets in Louisiana and assess the degree to which investments in risk mitigation could reduce this risk.

We have expanded upon the original storm selection approach described in Fischbach *et al.* [18] by examining flood depth and damage exceedances rather than surge elevation exceedances. The storm selection analysis was performed as part of a larger model development effort in support of the State of Louisiana's 2017 Coastal Master Plan. This paper draws largely from Fischbach *et al.* [20], a technical report describing the larger project; interested readers may refer to that document and Fischbach *et al.* [18] for additional details about CLARA.

Flood depth exceedance and expected annual damage estimates were produced using a reference set of 446 storms—the complete corpus of storms developed for use in recent JPM-OS studies of the Gulf coast—and with various subsets of the reference set, ranging from 40 to 304 storms. This paper examines the biases in flood depth and damage exceedances introduced by reducing the storm suite relative to a full implementation of JPM-OS using 446 storms. We also discuss the estimated parametric uncertainty associated with results from different storm suites. The results provide guidance for the types and size of storm suites that can be used to produce a good approximation of the results from the reference case when a full implementation of JPM-OS is computationally infeasible or prohibitively expensive for regions like coastal Louisiana.

2. Methods

The original storm selection problem was to choose a suite of storms to simulate that, when used to fit a surge response surface, produces a probability distribution of surge that closely replicates the true underlying distribution [9]. Prior analyses developed a suite of 304 idealized, synthetic storms that are defined parametrically by their central pressure c_p, radius of maximum windspeeds r_{max}, forward velocity v_f, longitudinal location at landfall x, and angle of incidence at landfall θ_l. For the Louisiana coastline, landfall was defined as the point when the eye of the storm crosses 29.5° N latitude [21,22]. These studies focused on representing storms that produce Category 3 or greater wind speeds on the Saffir-Simpson scale (minimum central pressure of 960 mb or lower); later work increased the number of available storms to 446 by adding intermediate storm tracks and less extreme storms with a minimum central pressure of 975 mb.

The 446 storms were run through a dynamically coupled set of storm surge and wave models: ADCIRC and Simulating WAves Nearshore (SWAN), respectively [17,23,24]. These models were adapted and calibrated to current environmental and landscape conditions by other researchers working with the State of Louisiana [20]. The storms were also run through these models using coastal landscape and sea level conditions in the year 2065 as projected for the "Less Optimistic" scenario used in Louisiana's 2012 Coastal Master Plan [16,25]. An initial screening analysis, described in the Experimental Design section, used results from the current conditions model outputs. The final, detailed analysis focused on the Less Optimistic future scenario, rather than current conditions. The higher level of risk (particularly in areas enclosed by ring levees) associated with the Less Optimistic future allows more information about bias to be drawn from a wider range of return periods, as there are fewer return periods where no flooding occurs.

Using this approach, we estimate flood depth and damage statistics at a large number of spatial grid points. Estimates of flood statistics derived from the full 446-storm suite—referred to as the reference set—are used as a reference standard for comparison. We ran CLARA's modified JPM-OS procedure for a relatively large number of subsets of the reference set to investigate the use of smaller suites of synthetic storms. While we cannot assert that the 446-storm results are themselves unbiased or a true representation of the underlying storm risk, they represent the most complete set currently available for this region and are the best available basis for comparison when evaluating subsets.

In some cases, subsets were formed by eliminating storms from the reference set in ways that we hypothesized would introduce minimal bias. In other cases, subsets were formed by reducing the number of parameters that vary within the storm subset. For example, the storm tracks represented in the reference set are based on analysis of historic storm landfalls, in which a mean angle of incidence θ_l was calculated for each landfall point x. The reference set includes storms that follow the mean historic landfall angles, as well as storms that proceed along tracks 45 degrees to either side of the mean angle; some of the subsets we tested were formed by eliminating these "off-angle" tracks from the reference set. Further details on the subsets of storms considered are provided in the experimental design section of this manuscript.

For each subset of the reference set we proceeded as follows.

1. Fit response surfaces for peak storm surge and peak significant wave heights at each location not enclosed by a levee represented in the CLARA model (the "unenclosed grid points") and at special points ("surge and wave points," or SWP) 200 meters in front of levees and floodwalls along the boundary of enclosed protection systems [20]. Separate response surfaces are fit for surge hydrographs (surge elevations over time) and wave periods at the SWPs. The wave heights and periods are calculated at the time of peak surge in the surge hydrograph; the response surfaces for these quantities are fit using these values, and they are assumed to be constant over time (Estimated wave heights are also limited by the depth of the underlying surge. Holding wave heights and periods constant throughout the hydrograph reduces computational complexity and provides a first-order approximation of wave behavior during the points in the surge hydrograph where surge levels are high (*i.e.*, when overtopping may occur).).

2. Estimate peak surge and significant wave heights, using the response surface fits to obtain the predictions; also predict the surge hydrographs and wave periods at SWPs. In unenclosed areas, the effective flood elevations resulting from each synthetic storm are calculated as the sum of peak surge and the free-wave crest height. At SWPs, the predicted surge and wave characteristics are used as inputs to the flood module (as described in Step 4).The response surface fits are based upon the ADCIRC and SWAN outputs for each storm included in a given subset; these storms are sometimes referred to as the "training set" in the ensuing discussion. In some cases, the response surfaces can be used to predict surge and wave characteristics for a larger number of synthetic storms than those used to fit the response surface. For example, Set D3 includes storms on each track with a central pressure of 960 mb and r_{max} values of 11 and 35.6 nautical miles. The response surface is fit using these storms to estimate the effect of r_{max} on surge and waves. The

predictions also include a storm with an intermediate r_{max} of 21 nautical miles. Predictions are made for all storms in the 446-storm set that can be estimated using a response surface based on a given subset. For example, if a subset only includes storms with $v_f = 11$ knots, the response surface cannot identify the marginal effect of v_f on surge and waves, so synthetic storms with other values for v_f are excluded from the set of synthetic storms used to generate predictions. A general rule is that for any combination of track and angle represented in a training set, the set of predicted synthetic storms run through the rest of the steps listed below includes all storms from the 446-storm corpus corresponding to that track and angle, excluding storms with $v_f = 6$ or 17 knots if the training set does not also include storms with variation in the forward velocity.

3. Partition the JPM-OS parameter space by the synthetic storms used for predictions and estimate the probability mass associated with each synthetic storm. The probabilities assigned to each synthetic storm are derived using maximum likelihood methods on a data set of observed historic storms, under certain structural assumptions about the joint probability distribution of the storm parameters [26].
4. Run the predicted synthetic storms through CLARA's flood module using the predicted SWP hydrographs and wave heights to obtain final still-water flood depths on the interior of enclosed protection systems. The flood module includes analysis of overtopping from surge and waves, system fragility and the consequences of breaches, and routing of water between interior polders, also accounting for pumping capacity and rainfall [20,27]. The probability of system failure is modeled as a function of overtopping rates [28].
5. Combine the flood depths and probability masses associated with each predicted synthetic storm to build the distribution functions of flood depths at each CLARA grid point.
6. Extract the flood depth exceedance values from the cumulative distribution functions at various return periods (A flood depth value with a $(1-1/n)$% annual exceedance probability (AEP)—the chance of occurring or being exceeded in a given year—is referred to as having an n-year return period. For this study, we recorded 5-, 8-, 10-, 13-, 15-, 20-, 25-, 33-, 42-, 50-, 75-, 100-, 125-, 150-, 200-, 250-, 300-, 350-, 400-, 500-, 1000-, and 2000-year exceedances.).
7. Run the flood depth exceedances through CLARA's economic module to estimate economic damage exceedances. Also estimate the expected annual damage (EAD) associated with each storm subset.

Full details on how each of these steps are accomplished (for example, the specification of the response surface model) can be found in Fischbach *et al.* [20]. We have found that deriving flood depth exceedance curves from small numbers of storms can be problematic in enclosed areas behind levee systems, with large, sudden jumps in exceedance values; for example, one might estimate a 12-foot difference in flood depths between the 100-year and 125-year exceedances [29]. If one uses a small number of storms, leaving large gaps in the parameter space, each storm may produce very different results from its "nearest neighbor" storms. This is exacerbated by the nonlinear relationship between surge and wave heights and levee overtopping, and also by the impact of system fragility. As a result, using the response surface to generate predicted synthetic storms with intermediate parameter values not represented in the training set is necessary for "smoothing out" the exceedance curve and making it less sensitive to small changes in the probability weights assigned to each storm. For consistency, when running synthetic storms through the CLARA flood module, we use the response surface predictions for all synthetic storms tested as part of a particular test set (as opposed to using the actual hydrodynamic modeling outputs for storms, when available, and response surface predictions only for storms outside of the training set).

The performance of each storm subset was evaluated by summarizing the bias in predicted flood depths at various return periods, relative to the results generated by the reference set. We also analyzed and compared the estimated standard errors associated with each subset.

3. Experimental Section

We first conducted a screening-level analysis by evaluating a large number of storm suites—sixteen in total, inclusive of the 446-storm reference set. For the initial screening, flood depths were calculated at a total of 18,273 grid points across the coast. These grid points were chosen by selecting a limited sample of 24 watersheds that varied by (a) size, (b) proximity to the coast, (c) degree of economic development, and (d) whether the watershed is protected by some federally-accredited levee or floodwall system. (The full CLARA v2.0 study region for coastal Louisiana consists of 77,643 grid points comprising 117 watersheds (excluding points consisting of open water). Grid point resolution varies with the concentration of population and assets, with a maximum distance of 1 km between neighboring points.) Watersheds protected by a fully-enclosed ring levee system were excluded from the initial screening sample to avoid running CLARA's flood module (step 4 above) as part of the screening exercise. Damage results were also excluded from the screening analysis.

The ten JPM-OS tracks used in the 2012 Coastal Master Plan analysis are sometimes referred to as primary tracks. They were labeled E1 through E5 and W1 through W5 for tracks in the eastern and western halves of the coast, respectively. Secondary storm tracks correspond to paths in between the primary tracks and were denoted by a B at the end of the track name (e.g., track E1B). Tracks also vary by their angle of incidence made with the coastline upon landfall. Tracks following the mean landfall angle are referred to as central-angle tracks; those making landfall at angles 45 degrees less or greater than the mean angle are referred to as off-angle tracks.

The modeled subsets were chosen to be collections of storms with easily interpretable and describable characteristics, to have variation in the total number of storms, and to have variation in the types of storms represented over the subsets. They were also chosen to avoid groups of storms that could cause identifiability or other performance issues in the response surface model. Louisiana's Coastal Protection and Restoration Authority (CPRA) also identified a need for storm sets with fewer than 154 storms—and preferably fewer than 100 storms—that could be used to evaluate a range of individual structural protection projects during 2017 Coastal Master Plan model production using available computing resources. Table 1 describes what types of storms comprise each of the storm suites evaluated in the screening-level analysis. The number of storms listed in the table represents the number of storms in the training set used to fit the response surface in Step 1 of the procedure described in the Methods section; as noted in the description of Step 2, the set of predicted synthetic storms run through the rest of the procedure may be larger than the training set. However, the size of the training set is the relevant number to discuss when comparing storm suites, as it represents the number of storms that must be run through the more computationally intensive ADCIRC and SWAN models.

Table 1. Characteristics of storm suites selected for screening-level analysis.

Set	Storms	Description
S1	446	Reference set
S2	40	2012 Master Plan (MP) storm suite: 10 primary storm tracks, 4 storms per track varying c_p and r_{max}
S3	90	2012 MP storm suite expanded to 9 storms per track that vary c_p and r_{max}
S4	304	LACPR storm suite: all storms in reference set, except those with 975 mb c_p
S5	194	All storms from LACPR storm set with 11-knot v_f
S6	154	All storms from LACPR storm set with 11-knot v_f on primary storm tracks
S7	216	All storms from reference set with 11-knot v_f on primary storm tracks
S8	144	All storms from reference set on primary storm tracks with 900 mb or 930 mb c_p
S9	120	All LACPR storms on primary, central-angle tracks (includes variation in v_f)
S10	264	All storms from reference set with 11-knot v_f
S11	50	2012 MP storm suite expanded to 5 storms per track varying c_p and r_{max}
S12	60	Set 11, plus storms with 975 mb c_p and central values for r_{max}
S13	120	All central-angle, primary-track storms with 11-knot v_f (includes 975 mb storms)
S14	148	Set 13, plus all 960 mb and 975 mb storms on secondary storm tracks
S15	154	Set 3, plus all 960 mb and 975 mb storms on primary, off-angle storm tracks
S16	182	Set 3, plus all 960 mb and 975 mb storms on secondary tracks or primary, off-angle tracks

Notes: c_p—central pressure; r_{max}—radius of maximum windspeed; v_f—forward velocity; LACPR—Louisiana Coastal Protection and Restoration study [22].

The initial findings led to the development of several new test suites, ranging from 60 to 154 storms, for the detailed analysis. The final sets tested, including number of storms and a description of key characteristics, are shown in Table 2 below. Note that the storm set ID numbers for the detailed analysis do not all match up with the IDs used in the screening analysis; although some sets from the screening analysis were included in the detailed analysis, the ID numbers were designated independently. The detailed analysis included estimation of flood depth exceedances at all grid points across the coast, including in areas enclosed by ring levees. Damage exceedances and bias in EAD were also analyzed for these storm suites.

Table 2. Characteristics of storm suites selected for detailed investigation.

Set	Storms	Description
D1	446	Reference set
D2	40	2012 Master Plan (MP) storm suite: 10 primary storm tracks, 4 storms per track varying c_p and r_{max}
D3	60	2012 MP storm suite, expanded to 5 storms per track varying c_p and r_{max}, plus storms with 975 mb central pressure and central values for r_{max}
D4	90	2012 MP storm suite, expanded to 9 storms per track varying c_p and r_{max}
D5	90	2012 MP storm suite, expanded to 7 storms per track (excludes one 930 mb and one 900 mb storm), plus 975 mb storms using extremal (rather than central) r_{max} values
D6	92	Set 3, plus 960 mb and 975 mb storms on off-angle tracks only in E1-E4
D7	92	Set 3, plus 960 mb and 975 mb storms on off-angle tracks only in W3-W4, E1-E2
D8	100	All central-angle, primary-track storms with 11-knot v_f, plus 975 mb storms with central r_{max} values
D9	110	All central-angle, primary-track storms with 11-knot v_f, plus 975 mb storms with extremal r_{max} values
D10	120	All central-angle, primary track storms with 11-knot v_f, including all 975 mb storms
D11	154	Set 4, plus all 960 mb and 975 mb storms on primary, off-angle storm tracks

Notes: c_p—central pressure; r_{max}—radius of maximum windspeed; v_f—forward velocity.

The off-angle tracks for sets D6 and D7 were selected based on their paths passing near to New Orleans, and thus having the potential for significant flooding in urban centers where a large number of grid points and economic assets are densely packed. Our hypothesis was that including off-angle storms affecting the greater New Orleans region would improve accuracy in estimating how likely it is for the city to experience the catastrophic flooding associated with levee failures.

4. Results and Discussion

An example result from the screening analysis is shown in Figure 1. The depicted storm suite (Set S3 from Table 1) is comprised of nine storms from each of the ten primary tracks. Within a track, storms vary by c_p and r_{max}. The central pressure values represented are 900, 930, and 960 mb; storms with 975 mb pressures were not included. All storms have the central value for v_f of 11 knots. The figure illustrates bias in the estimated 100-year flood depths, relative to the reference set. Note that this particular suite tends to strongly overpredict 100-year depths (blue coloring), relative to the 446-storm reference set, in the far eastern portion of the state, east of New Orleans and the Mississippi River, while it tends to underpredict depths (orange coloring) in the watersheds farthest west.

Figure 2 summarizes the findings of the screening-level analysis. It shows the root mean squared error (relative to the reference set), averaged over all grid points in the screening watersheds, for each modeled storm suite; the figure plots this average bias as a function of the number of storms in each suite. Blue points represent storm suites which contain at least one storm with a 975 mb minimum central pressure, the dot shape represent suites which contain off-angle storm tracks. Sets 8 and 9 (as well as the reference set) from the screening analysis also contain storms with non-central values for forward velocity.

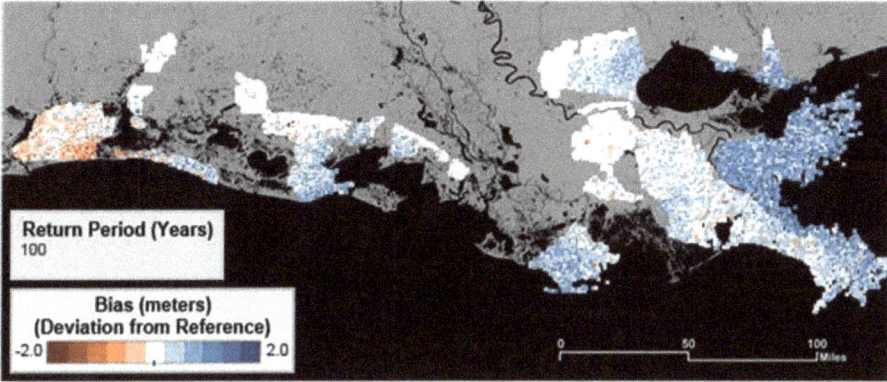

Figure 1. Deviation from reference set in 100-year flood depth exceedances in screening analysis watersheds from a suite consisting of 90 storms (screening set S3).

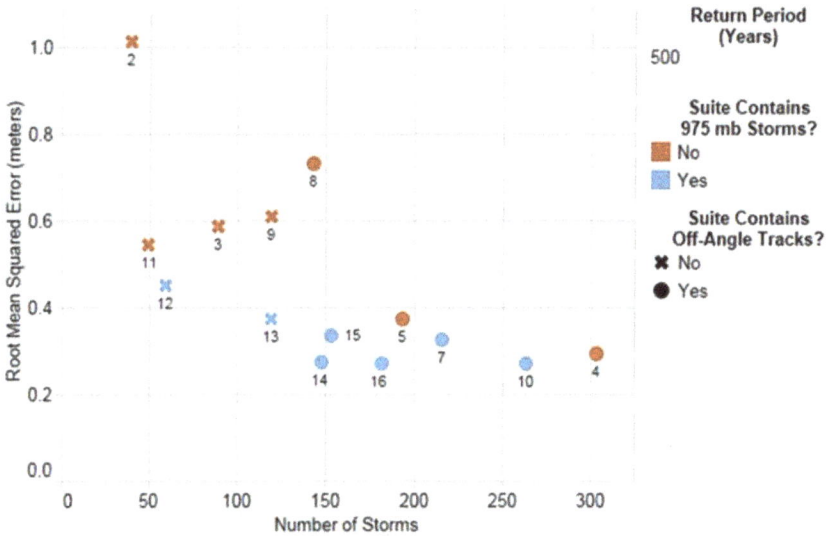

Figure 2. Average coastwide bias and variation by number of storms (screening analysis), 500-year flood depths.

In addition to the dimensions illustrated in the figure, sets S4, S8, and S9 contain storms with non-central values for forward velocity. The screening results suggest that including higher-frequency storms with 975 mb central pressure improved statistical performance. By contrast, secondary or off-angle storm tracks and storms with non-central values for forward velocity did not yield similar improvement and were generally not included in the more detailed final testing and results.

4.1. Flood Depth Bias and Variance Comparisons

Figure 3 summarizes the average coastwide 100-year flood depth bias (root mean squared error, y-axis) and coefficient of variation (point size) for each set in the detailed analysis, plotted against the number of storms in a given storm suite (x-axis). As before, colors indicate whether the set includes

975 mb storms, and shape indicates whether off-angle tracks are included. Bias is estimated relative to the flood depth results from the reference set.

Summary results show that average bias in flood depths at the 100-year interval varies from less than 0.25 m to nearly 1 m, depending on the storm sample. Substantial bias is observed for Set D2, the 2012 Coastal Master Plan suite, compared with the other candidate sets tested. All additional sets with more than 40 storms improve upon the 40-storm results. In fact, increasing the number of storms to 60 (Set D3) leads to substantial improvement, reducing average bias by more than a half a meter. Similar results are observed at other return periods.

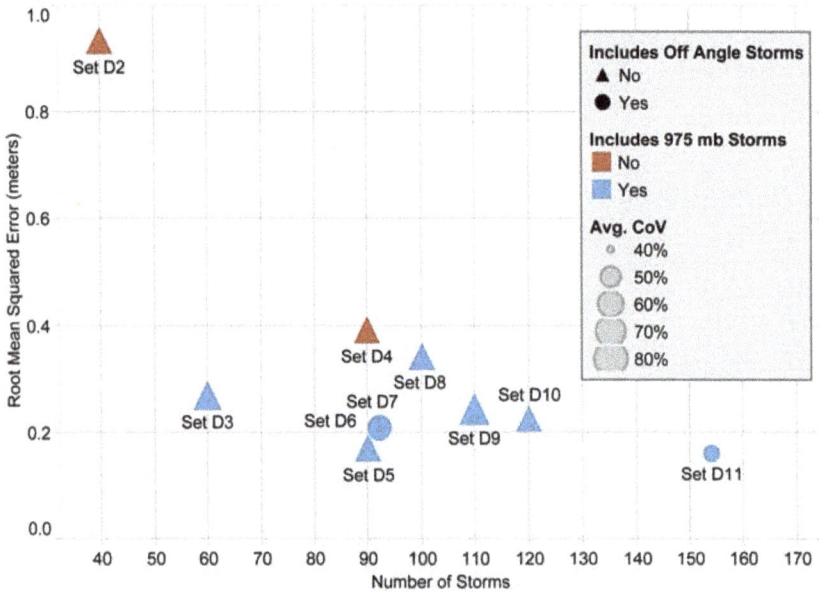

Figure 3. Average coastwide bias and variation by number of storms (detailed analysis), 100-year flood depths.

We focused primarily on suites of storms that include 975 mb storms because the screening analysis results suggested that such suites outperform suites that exclude the less intense storms. The relatively poor performance of sets D2 and D4, the suites with no 975 mb storms, suggests that this is a still a valid conclusion when considering the full coastal study region. Improvement is less apparent when off-angle tracks are included, however, especially for sets in which only certain off-angle tracks were included (e.g., Sets D6 and D7).

Figure 4 illustrates how the average bias, again measured by RMSE, changes for each storm suite at different return periods of the flood depth distribution. The x-axis is the return period, on a logarithmic scale. The thickness of the lines shown in the figure reflects the coefficients of variation, a measure of the uncertainty around derived point estimates; thicker lines denote greater uncertainty.

With few exceptions, the relative order of performance across storm subsets is consistent over a wide range of return periods. Sets D2 and D4 are conspicuous in their poor average performance; neither set contains any storms with central pressures of 975 mb, further emphasizing the importance of including higher-frequency events in the training set. Interestingly, this effect is still particularly apparent in the tail of the distribution beyond the 100-year AEP interval, indicating that exclusion of higher-frequency events from the response surface model also skews the model's predictive accuracy for more extreme synthetic storms.

In general, average bias is smaller at 50-year and more frequent return periods. This is due to the large number of points in which no flooding occurs for these exceedances. The absolute error is more likely to be small when the depth estimates themselves are small. Set D11, with 154 storms, performs well across most of the distribution, and better than any smaller sets. However, several of the sets with fewer than 100 storms are not far behind in terms of average bias across the exceedance distribution.

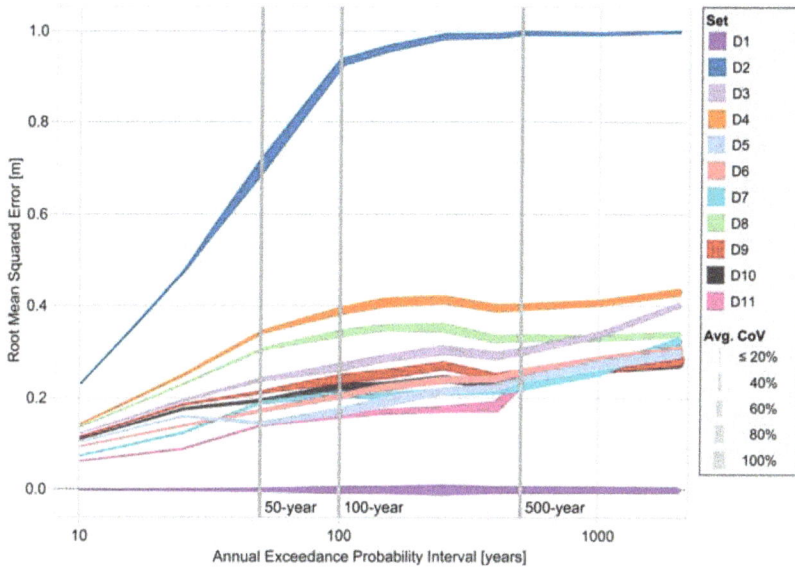

Figure 4. Average bias and variance by exceedance interval for each modeled storm suite.

Figure 5 shows a sequence of maps that display the spatial patterns of bias, relative to the reference set, for three of the storm suites tested. Flood depths at the 100-year return period, as estimated using the reference set, are shown for comparison in Figure 6; when examining the three maps in Figure 5, this gives some indication of the magnitude of the biases relative to the baseline flood depths. When examining the spatial distribution of bias, some interesting patterns emerge. The Set D2 results, representing the 2012 Coastal Master Plan's 40-storm sample, show a substantial overestimate of flood depths across nearly the entire study region. This effect occurs because of the types of storms excluded from Set D2; it only contains extreme central-angle storms on primary tracks, leaving out both 975 mb and off-angle storms. For instance, Set D2 excludes off-angle storms that pass well to the east of New Orleans. These storms would tend to lower the estimated flood depth exceedances in St. Bernard Parish, so as a result Set D2 has positive bias in that region.

Set D3, which includes 60 storms, improves dramatically on the Set D2 results. Some positive bias is still noted in the western portion of the coast and many areas east of the Mississippi River, but the magnitude is notably lower than Set D2. In addition, Set D3 actually underestimates 100-year flood depths compared with the reference in some enclosed areas, including the East Bank of the Greater New Orleans Hurricane Storm Damage Risk Reduction System (HSDRRS) and the Larose to Golden Meadow levee system.

Set D11 yields the lowest overall bias when averaged over all points, and it shows balanced results coastwide when looking at geospatial patterns. Positive bias is still observed in the western parishes, but again with a lower magnitude than Sets D2 or D3. There are some instances where the 100-year flood depth estimates are both positively and negatively biased within the same watershed. When these differences in flood depths are translated into damage, the coastal and parish-level results are more similar to those of the reference set.

The differences between enclosed and unenclosed areas are clearly shown in Figures 7 and 8 (note the different scale on the y-axis between figures). The curves shown in Figure 4 are broken out by showing the same results for points enclosed by (Figure 7) and unenclosed points (Figure 8), respectively.

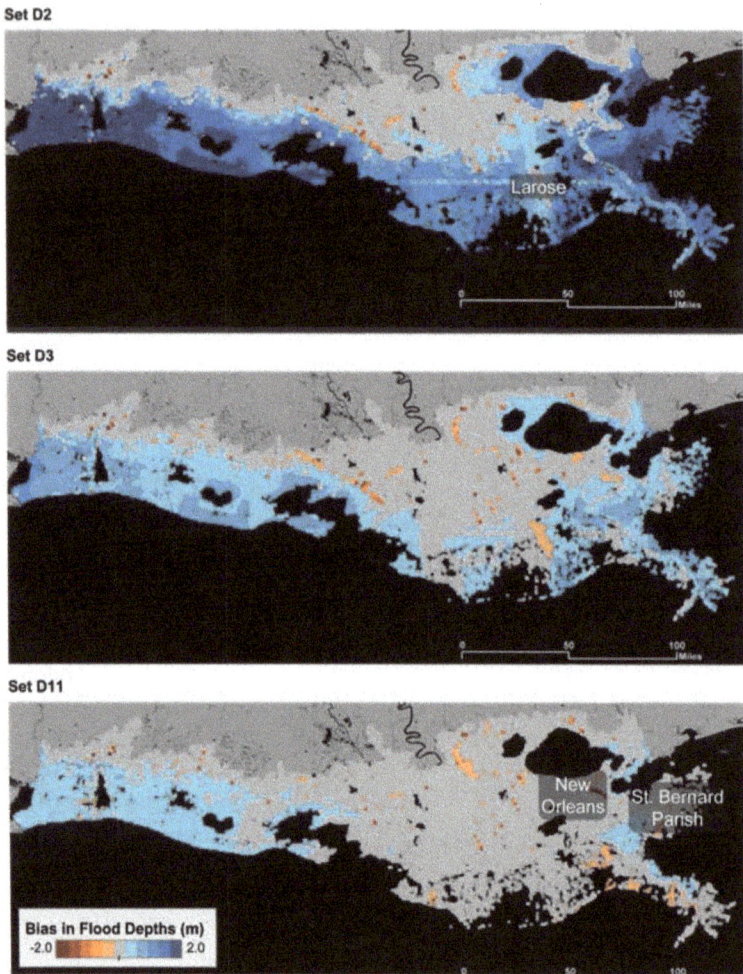

Figure 5. Maps of bias by grid point for detailed analysis sets D2, D3, and D11, 100-year flood depths.

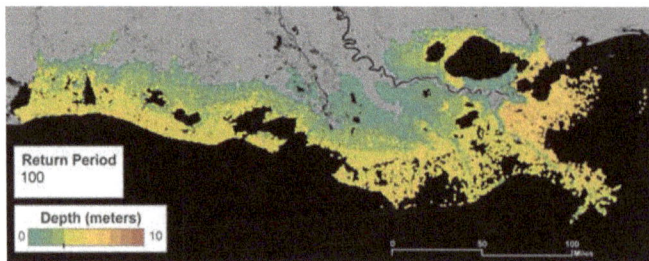

Figure 6. Median 100-year flood depths from reference set D1.

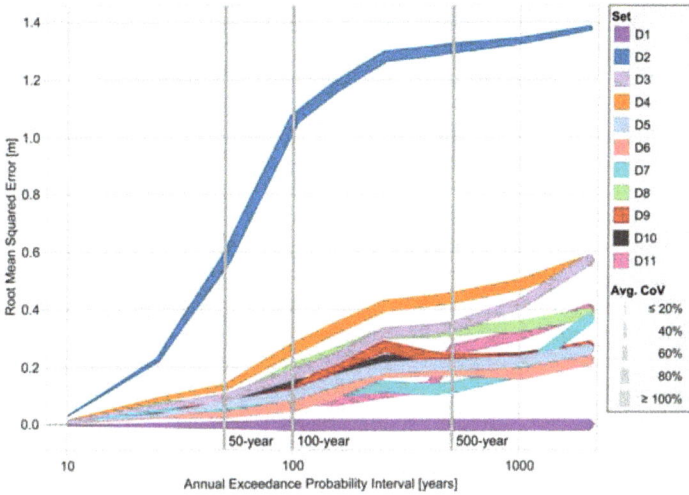

Figure 7. Average bias and variance by exceedance interval (HSDRRS Only).

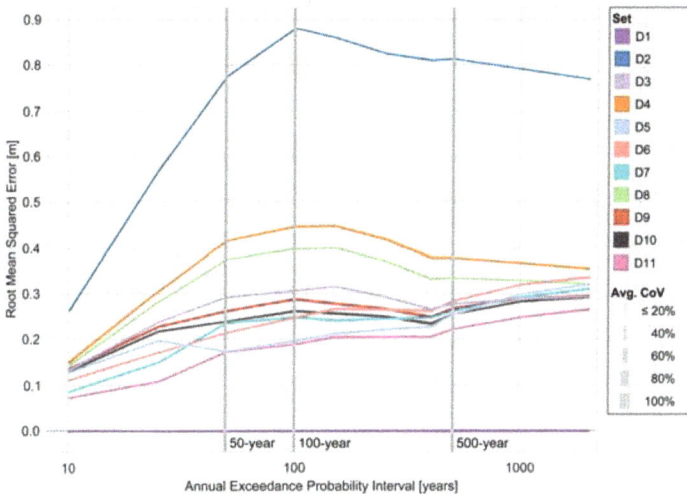

Figure 8. Average bias and variance by exceedance interval (unenclosed points only).

HSDRRS was accredited in 2014 by the United States Federal Emergency Management Agency as protecting the Greater New Orleans area against at least 100-year surge levels [30]. However, the same standard may not be met in 2065 under the Less Optimistic future scenario assumptions in a future without further mitigating actions. The precise level of protection provided by HSDRRS is uncertain [31]. As such, the average coefficient of variation is relatively large, over 100% for many storm suites over large portions of the distribution.

In unenclosed areas, the performance of all suites but Set D2 is relatively good and largely similar. With the exception of Set D2, all storm suites have an average bias of less than half a meter at all exceedance intervals, and the average uncertainty in the predicted values is also considerably smaller than for points within HSDRRS. This reflects that the exceedance curves in unenclosed areas have lower variance and are more stable, in the sense that few points experience sudden, sharp increases in flood depths at any point of the distribution.

The greater variation in performance in enclosed areas illustrates that flood depth estimates are more sensitive to the choice of storm suite when the accuracy and uncertainty in response surface predictions is further compounded by interactions with engineered flood protection systems. Maintaining accuracy with fewer storms is easier for prediction of storm surge elevation exceedances, as in the original storm selection problem faced by JPM-OS, than for flood depth exceedances when accounting for the levee and floodwall failures, overtopping, and interior drainage dynamics.

Figure 9 focuses in on differences in storm suite performance for enclosed points within HSDRRS, on the east and west sides of the Mississippi River, at the 500-year AEP interval. Consistent with what was shown in Figures 7 and 8 results are most uncertain and show the greatest variance in the East and West Bank areas. Except for Set D2, all sets considered yield an average bias of approximately 0.3 meters (1 foot) or less in enclosed locations on the East Bank. Uncertainty is greater on the West Bank. Similarly, Sets D3 through D11 also result in less than 1 meter of average bias in enclosed areas not within HSDRRS, with many sets also yielding less than 0.3 meters of average bias in these areas (results not shown).

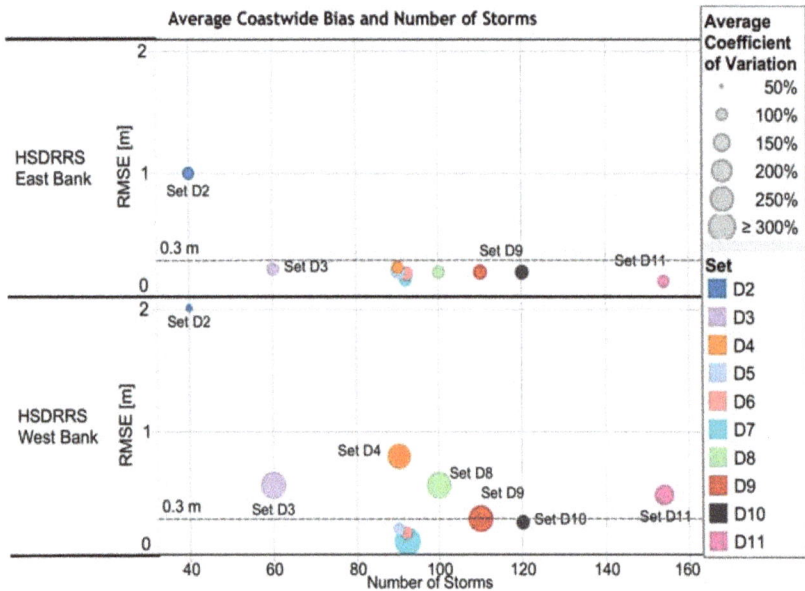

Figure 9. Average bias and variation, 500-year flood depths, east and west banks of the Greater New Orleans Hurricane Storm Damage and Risk Reduction System.

Figure 9 confirms that Set D11 is the best or near-best performer in terms of mean bias across all enclosed areas. For sets with fewer than 100 storms, performance depends on HSDRRS location. Set D3 performs substantially worse, however, on the West Bank, as do many other sets. By contrast, Set D7 was specifically designed to improve performance in the West Bank area of HSDRRS. This suite adds storms to Set D3, including off-angle tracks for the middle of the coast with landfall locations observed to have the greatest effect on this portion of HSDRRS. The results bear this out: Set D7 is the best overall set in terms of bias in the West Bank, even slightly better than the larger Set D11. However, Set D7 produces greater bias in the East Bank HSDRRS, and is among the worst performers when looking at enclosed areas other than the West Bank. It also has a significantly larger coefficient of variation on the West Bank than Sets D5 and D6, which both have nearly identical average RMSE and number of storms to Set D7. This illustrates that while it may be possible to carefully tailor a storm suite to perform well in a particular region, results may be significantly worse in other, even nearby, areas. It can be difficult to predict the overall performance of a storm set, which emphasizes the importance of conducting a storm selection analysis like this on at least one landscape scenario before embarking on a major flood risk assessment encompassing many future scenarios, time periods, and system configurations.

4.2. Damage Bias Comparisons

Next, we sought to understand how the potential bias from different storm suites translates to bias in damage estimates. Somewhat surprisingly, we found that the total bias associated with a suite of storms was typically driven by bias in unenclosed areas, although such regions are more sparsely populated and contain fewer assets than in protected areas like New Orleans. The greater protection in enclosed areas provided by federal levee systems instead meant that there was a lower projected baseline risk in those regions; most storm suites did not predict extensive flooding in enclosed areas except at lower-frequency AEPs, so this resulted in less bias.

Figure 10 below shows a similar spatial breakdown as Figure 9, for example, but instead displays bias in terms of damage (EAD, median results) estimated by the CLARA v2.0 economic damage model. Set D2, which yields very high EAD bias overall, is omitted from the figures below for clarity. Figure 10 confirms the performance noted above, with roughly the same overall ranking of storm suites by EAD bias as with flood depths. Sets D5, D6, D7, and D11 all have a cumulative bias of less than $0.5 billion across all points. D3 shows a total bias over all points of about $1.5 billion; D8 has bias of $2 billion, while D4 has $2.5 billion.

Increasing the number of storms from 60 up to 90–110 does not substantially improve EAD performance for enclosed areas; Sets D4 and D8 consistently produce worse error than Set D3, despite having 90 and 100 storms, respectively. Instead, it is the addition of 975-mb storms that appears to drive improvement. Set D4 has no such storms, while Set D8 contains half as many 975-mb storms than the other suites with 90–110 storms.

Similarly, Figure 11 shows a summary by coastal Louisiana parish (county) of bias in EAD compared to the reference set. Some storm suites, regardless of coastwide performance, produce large bias in specific parishes. For example, Set D4 produces approximately 50% more damage in Jefferson and St. Bernard Parishes than the 446-storm reference set. In turn, this leads to a substantial upwards bias in coastwide EAD because these parishes contain large concentrations of assets.

Sets D5, D6 and D11 are the highest performers when disaggregating results over the parishes; they are the only storm sets with only one parish's bias over 20% and a maximum bias of magnitude 25% or less. The maximum bias for any parish from Set D11 is 22% in Calcasieu Parish, which is in the far west of the state and includes relatively few assets. All storm suites overestimated EAD in Calcasieu, with errors ranging from 14 to 41%, indicating a general difficulty in reproducing estimates of risk so far west with a small number of storms. Each of the other storm suites produces biases as large or larger in several parishes. Few storm sets performed well, with errors of 10% or less, in both Orleans and Jefferson parishes, the two parishes with the largest value of assets at risk. EAD performance in non-HSDRRS parishes, by contrast, is typically more similar across all sets tested in this round of analysis.

The results discussed so far have portrayed the median (50th percentile) outputs over the sampling design, with the coefficients of variation giving some information about the parametric variation across different suites. Figure 12 illustrates the distribution of EAD in another way. The y-axis indicates the bias in coastwide EAD relative to the full 446-storm reference set; the three points for each subset, from bottom to top, represent the bias associated with the 10th, 50th, and 90th percentile values, respectively, of EAD.

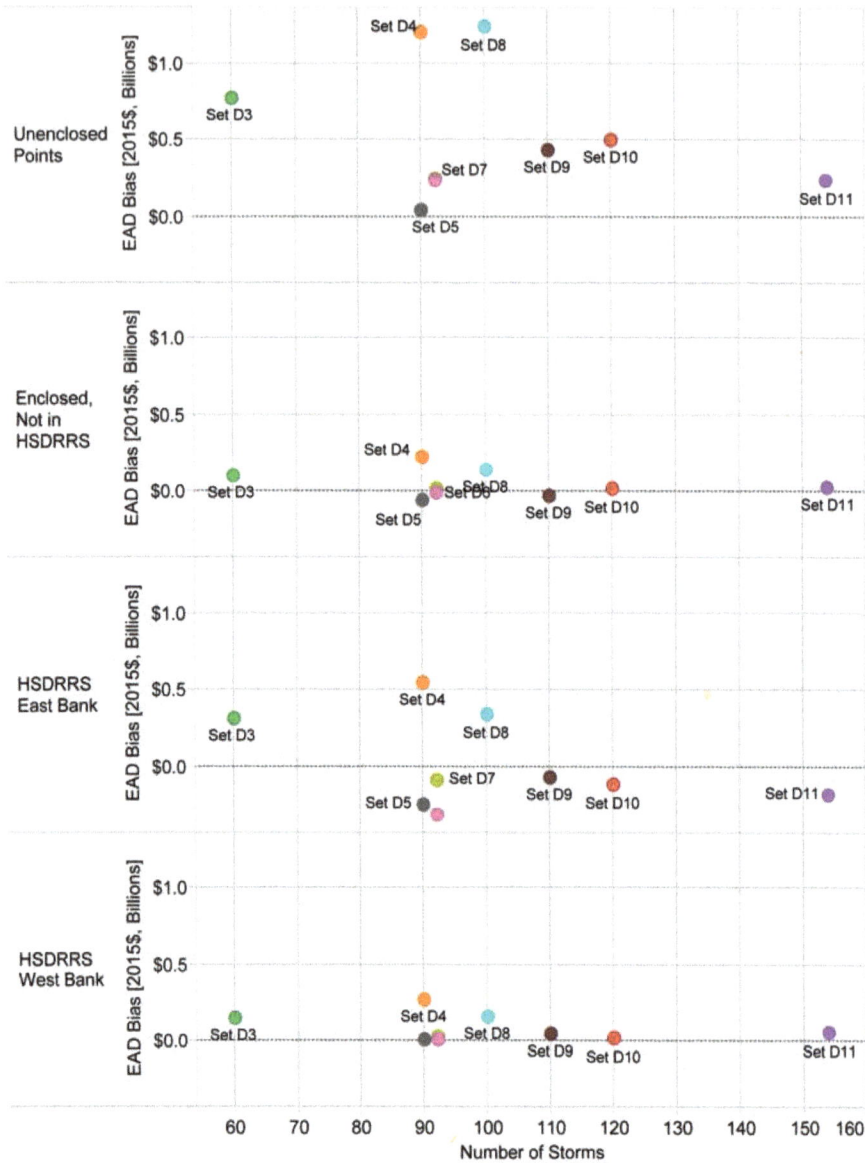

Figure 10. Bias in expected annual damage (median estimates), by location.

The primary takeaway here is that many storm suites end up with similar ranges of uncertainty in terms of EAD bias relative to the reference set, when aggregating damage coastwide. Considering only suites with fewer than 100 storms, Set D5 is the best performer at the median, but includes a wider range of results across the parametric distribution than Set D6, and the 10th percentile estimate is that Set D5 may underestimate EAD more strongly than any other set. Set D3 has the smallest range of variation in bias between the 10th and 90th percentile estimates, but it has a larger median bias than most of the larger sets; Set D6 has the second smallest range of uncertainty. Six sets underestimate EAD at the 10th percentile but overestimate it at the median and 90th percentiles.

Parish Name	D3	D4	D5	D6	D7	D9	D10	D11
Acadia	2%	3%	-2%	0%	8%	-4%	-1%	9%
Ascension	27%	34%	14%	19%	8%	20%	21%	10%
Assumption	2%	3%	0%	3%	0%	1%	1%	-3%
Calcasieu	24%	24%	14%	23%	41%	17%	19%	22%
Cameron	0%	-3%	2%	-2%	2%	-1%	0%	-1%
Iberia	0%	2%	-5%	0%	5%	-3%	-2%	6%
Jefferson	32%	44%	-4%	-17%	-4%	2%	-1%	-5%
Jefferson Davis	9%	13%	1%	9%	18%	8%	8%	15%
Lafourche	5%	11%	0%	2%	2%	1%	4%	2%
Livingston	15%	17%	13%	10%	5%	14%	14%	3%
Orleans	4%	22%	-21%	-15%	0%	-8%	-13%	-7%
Plaquemines	12%	22%	2%	1%	2%	8%	7%	4%
St. Bernard	27%	57%	7%	-8%	0%	15%	9%	7%
St. Charles	5%	8%	2%	3%	0%	5%	5%	1%
St. James	-4%	9%	-12%	-5%	-11%	-12%	-12%	-16%
St. John the Baptist	24%	30%	13%	13%	7%	21%	21%	13%
St. Martin	1%	0%	0%	1%	0%	0%	0%	-1%
St. Mary	0%	3%	-10%	-2%	-3%	-5%	-6%	0%
St. Tammany	12%	14%	9%	7%	7%	10%	10%	8%
Tangipahoa	0%	0%	-1%	-2%	-7%	-1%	1%	-7%
Terrebonne	4%	8%	-4%	2%	-2%	1%	1%	0%
Vermilion	5%	7%	3%	6%	7%	3%	4%	7%

EAD Bias (percent) -50% 50%

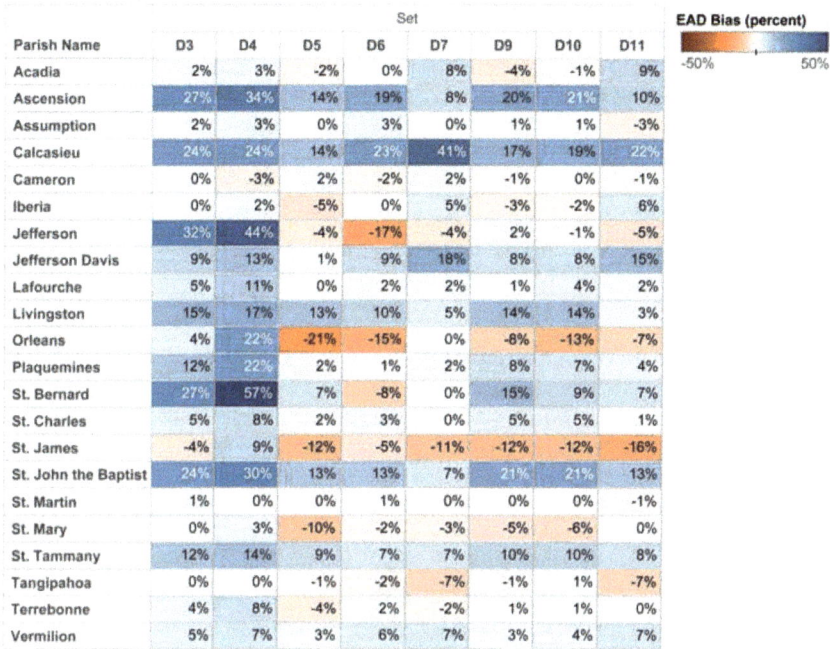

Figure 11. Percentage bias in expected annual damage by parish (median).

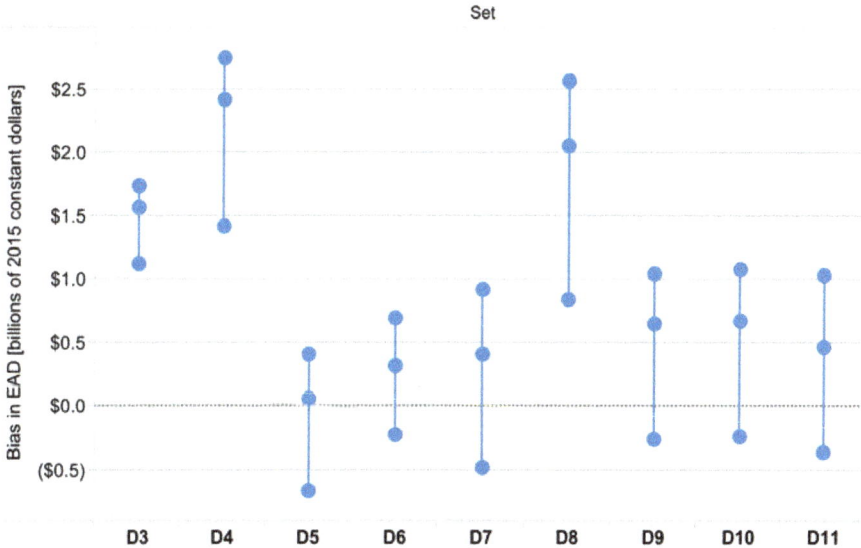

Figure 12. Coastwide bias in terms of expected annual damage (billions of 2010 US dollars).

5. Conclusions

The storm selection analysis shows a tradeoff between the number of storms and the resulting bias when compared with the reference set of 446 storms. Results show that nearly all storm sets tested produce lower bias when compared with the 2012 Coastal Master Plan 40-storm suite (Set D2). Aside from the fact that Set D2 was the smallest storm suite tested, this should not be surprising, given that the 2012 Coastal Master Plan was intended to model damage associated only with Category 3 or greater hurricanes. Substantial improvement is noted when storms with 975 mb central pressure were included, as well as with the addition of off-angle storms in some cases. The inclusion of storms with variation in forward velocity was found to have little impact on performance.

Of the storm suites tested, Set D6 (92 storms) appears to yield the best balance of results. It shows relatively low bias compared with the reference set in terms of both flood depth and damage, no concerning spatial patterns of bias, and reasonable performance in enclosed areas (particularly Greater New Orleans). It also has the second smallest range of uncertainty in damage among the suites analyzed. Set D3 is notable for producing less average error in flood depths than expected given that it contains only 60 storms.

The results of this storm selection analysis suggest some insights for future planning studies using a JPM-OS approach with high-resolution storm surge and wave modeling. First, the number of storms needed to estimate unbiased flood depths or damages can vary greatly depending on the size of the geographic area of focus, landscape type and configuration, presence of hurricane protection structures, and range of exceedance probabilities targeted. In some cases—for instance, in areas without protection structures or with relatively similar landscape characteristics over a broader area—a smaller storm suite may yield acceptably low bias, with results suitable for screening or planning-level decisions.

In some areas, however, large numbers of storms are needed to avoid bias and assess a wide range of plausible storm impacts from storms with different characteristics. Larger storm suites are also needed to support flood protection engineering and design studies, where decision makers need confidence that the widest range of plausible outcomes are simulated and adequate factors of safety have been developed accordingly.

For the 2017 Coastal Master Plan, a hybrid approach will be adopted: a smaller suite of 60 storms (Set D3) will be used for preliminary evaluation and testing of individual risk reduction projects, but coastwide alternatives—including hurricane protection structures, ecosystem restoration, and nonstructural risk reduction (hazard mitigation for buildings)—will be evaluated and compared using a larger set of 92 storms (Set D6). This two-step approach will help to confirm or validate preliminary project-level results and ensure that potential risk and damage reduction benefits from the plan as a whole are relatively unbiased.

Acknowledgments: Acknowledgments: Other RAND researchers and programming staff who have contributed to development of the CLARA model include Ricardo Sanchez, Chuck Stelzner, Rachel Costello, Edmundo Molina-Perez, and James Syme. This work was funded by the Louisiana Coastal Protection and Restoration Authority under the 2017 Coastal Master Plan's Master Services Agreement. Coordination of the process has been managed by The Water Institute of the Gulf. In particular, the authors wish to thank collaborators Hugh Roberts and Zach Cobell at ARCADIS; Mandy Green, Melanie Saucier, Mark Leadon, and Karim Belhadjali at CPRA; Denise Reed at The Water Institute; and Gary Cecchine at RAND Corporation for their intimate involvement in and support of this project.

Author Contributions: Author Contributions: Fischbach is serving as principal investigator for the 2017 Coastal Master Plan's Risk Assessment team. He led the development of the storm selection analysis methodology and results visualizations, with Johnson's assistance. Johnson is the lead technical developer of the CLARA model; he executed the model runs to support the storm selection analysis and developed the storm suites included in the detailed analysis based on the initial screening results. Kuhn assisted in carrying out the analysis and quality assurance process. All authors contributed to the technical report on which this journal article is based; Johnson led its adaptation for journal submission, with assistance from the other authors.

Conflicts of Interest: Conflicts of Interest: The authors declare no conflict of interest.

References

1.   Federal Emergency Management Agency. Fact sheet: Flooding—Our nation's most frequent and costly natural disaster. Federal Emergency Management Agency: Washington, DC, USA, 2010; p. 2.

2. Karl, T.R.; Meehl, G.A.; Peterson, T.C.; Kunkel, K.E.; Gutowski, W.J., Jr.; Easterling, D.R. *Weather and Climate Extremes in a Changing Climate, Regions of Focus: North America, Hawaii, Caribbean, and U.S. Pacific Islands*; NOAA: Silver Spring, MD, USA, 2008.

3. Burkett, V.R.; Davidson, M.A. *Coastal Impacts, Adaptation and Vulnerability: A Technical Input to the 2012 National Climate Assessment*; Burkett, V.R., Davidson, M.A., Eds.; Island Press: Washington, DC, USA, 2012.

4. Neumann, J.E.; Emanuel, K.; Ravela, S.; Ludwig, L.; Kirshen, P.; Bosma, K.; Martinich, J. Joint effects of storm surge and sea-level rise on US coasts: New economic estimates of impacts, adaptation, and benefits of mitigation policy. *Clim. Change* **2014**, 1–13. [CrossRef]

5. Knutson, T.R.; McBride, J.L.; Chan, J.; Emanuel, K.; Holland, G.; Landsea, C.W.; Held, I.M.; Kossin, J.P.; Srivastava, A.K.; Sugi, M. Tropical cyclones and climate change. *Nat. Geosci.* **2010**, *3*, 157–163. [CrossRef]

6. Emanuel, K. Downscaling CMIP5 climate models shows increased tropical cyclone activity over the 21st century. *Proc. Natl. Acad. Sci. USA* **2013**, *100*, 12219–12224. [CrossRef] [PubMed]

7. Villarini, G.; Vecchi, G.A. Twenty-first century projections of north atlantic tropical storms from CMIP5 models. *Nat. Clim. Change* **2012**, 604–607. [CrossRef]

8. Webster, P.J.; Holland, G.J.; Curry, J.A.; Chang, H.R. Changes in tropical cyclone number, duration, and intensity in a warming environment. *Science* **2005**, *309*, 1844–1846. [CrossRef] [PubMed]

9. Resio, D.T.; Irish, J.L.; Cialone, M.A. A surge response function approach to coastal hazard assessment. Part 1: Basic concepts. *Nat. Hazards* **2009**, *51*, 163–182. [CrossRef]

10. Irish, J.L.; Resio, D.T.; Cialone, M.A. A surge response function approach to coastal hazard assessment. Part 2: Quantification of spatial attributes of response functions. *Nat. Hazards* **2009**, *51*, 183–205. [CrossRef]

11. Toro, G.R.; Resio, D.T.; Divoky, D.; Niedoroda, A.W.; Reed, C. Efficient joint-probability methods for hurricane surge frequency analysis. *Ocean Eng.* **2010**, *37*, 125–134. [CrossRef]

12. Ho, F.P.; Myers, V. *Joint Probability Method of Tide Frequency Analysis Applied to Apalachicola Bay and St. George Sound, Florida*; U.S. Department of Commerce: Silver Spring, MD, USA, 1975.

13. Myers, V. *Joint probability method of tide frequency analysis applied to Atlantic City and Long Beach Island, New Jersey*; U.S. Department of Commerce: Silver Spring, MD, USA, 1970.

14. Shen, W. Does the size of hurricane eye matter with its intensity? *Geophys. Res. Lett.* **2006**, *33*, L18813. [CrossRef]

15. Westerink, J.J.; Luettich, R.A.; Feyen, J.C.; Atkinson, J.H.; Dawson, C.; Roberts, H.J.; Powell, M.D.; Dunion, J.P.; Kubatko, E.J.; Pourtaheri, H. A basin- to channel-scale unstructured grid hurricane storm surge model applied to southern Louisiana. *Mon. Weather Rev.* **2008**, *136*, 833–864. [CrossRef]

16. Cobell, Z.; Zhao, H.; Roberts, H.J.; Clark, F.R.; Zou, S. Surge and wave modeling for the Louisiana 2012 Coastal Master Plan. *J. Coast. Res.* **2013**, 88–108. [CrossRef]

17. Luettich, R.A.; Westerink, J.J. *Formulation and Numerical Implementation of the 2d/3d Adcirc Finite Element Model Version 44.Xx*; R. Luettich: Chapel Hill, NC, USA, 2004.

18. Fischbach, J.R.; Johnson, D.R.; Ortiz, D.S.; Bryant, B.P.; Hoover, M.; Ostwald, J. *Coastal Louisiana Risk Assessment Model: Technical Description and 2012 Coastal Master Plan Analysis Results*; RAND Corporation: Santa Monica, CA, USA, 2012.

19. Johnson, D.R.; Fischbach, J.R. Using cost-effective and robust strategies to assess the potential for nonstructural risk reduction in coastal Louisiana. In *Improving Flood Risk Estimates and Mitigation Policies in Coastal Louisiana under Deep Uncertainty*; RAND Corporation: Santa Monica, CA, USA, 2013; pp. 67–101.

20. Fischbach, J.R.; Johnson, D.R.; Kuhn, K.; Pollard, M.; Stelzner, C.; Costello, R.; Molina-Perez, E.; Sanchez, R.; Roberts, H.J.; Cobell, Z. *2017 Coastal Master Plan: Model Improvement Plan, Storm Surge and Risk Assessment Improvements (Subtask 4.9)*; Coastal Protection and Restoration Authority: Baton Rouge, LA, USA, 2015.

21. *Performance Evaluation of the New Orleans and Southeast Louisiana Hurricane Protection System*; US Army Corps of Engineers: New Orleans, LA, USA, 2009.

22. *Louisiana Coastal Protection and Restoration Technical Report*; US Army Corps of Engineers: New Orleans, LA, USA, 2009.

23. Dietrich, J.C.; Tanaka, S.; Westerink, J.J.; Dawson, C.N.; Luettich, R.A., Jr.; Zijlema, M.; Holthuijsen, L.H.; Smith, J.M.; Westerink, L.G.; Westerink, H.J. Performance of the unstructured-mesh, SWAN+ADCIRC model in computing hurricane waves and surge. *J. Sci. Comput.* **2012**, *52*, 468–497. [CrossRef]

24. Booij, N.; Holthuijsen, L.H.; Ris, R.C. The "swan" wave model for shallow water. *Coast. Eng. Proc.* **1996**, *25*. [CrossRef]

25. Peyronnin, N.; Green, M.; Richards, C.P.; Owens, A.; Reed, D.; Chamberlain, J.; Groves, D.G.; Rhinehart, K.; Belhadjali, K. Louisiana's 2012 Coastal Master Plan: Overview of a science-based and publicly-informed decision making process. *J. Coast. Res.* **2013**, *67*, 1–15. [CrossRef]
26. Resio, D.T. *White Paper on Estimating Hurricane Inundation Probabilities*; US Army Corps of Engineers: New Orleans, LA, USA, 2007.
27. Johnson, D.R.; Fischbach, J.R.; Ortiz, D.S. Estimating surge-based flood risk with the Coastal Louisiana Risk Assessment Model. *J. Coast. Res.* **2013**, *67*, 109–126. [CrossRef]
28. *Final Post Authorization Change Report: Morganza to the Gulf of Mexico, Louisiana*; US Army Corps of Engineers: New Orleans, LA, USA, 2013.
29. Johnson, D.R. *Improving Flood Risk Estimates and Mitigation Policies in Coastal Louisiana under Deep Uncertainty*; Pardee RAND Graduate School: Santa Monica, CA, USA, 2013.
30. Holder, K. *FEMA Accredits Hurricane and Storm Damage Risk Reduction System (HSDRRS)*; US Army Corps of Engineers: New Orleans, LA, USA, 2014.
31. Johnson, D.R.; Fischbach, J.R.; Kuhn, K. *Current and Future Flood Risk in Greater New Orleans*; The Data Center: New Orleans, LA, USA, 2015.

MDPI AG

St. Alban-Anlage 66

4052 Basel, Switzerland

Tel. +41 61 683 77 34

Fax +41 61 302 89 18

http://www.mdpi.com

JMSE Editorial Office

E-mail: jmse@mdpi.com

http://www.mdpi.com/journal/JMSE

www.ingramcontent.com/pod-product-compliance
Lightning Source LLC
Chambersburg PA
CBHW051726210326
41597CB00032B/5618